Fundamentals of Renewable Energy

Fundamentals of Renewable Energy

Editors

Dr. Shaji James P.

Dr. Dipak Suresh Khatawkar

SCIENTIFIC PUBLISHERS

Published by
SCIENTIFIC PUBLISHERS (INDIA)
5 A, New Pali Road, P.O. Box 91
Jodhpur 342 001 (INDIA)
E-mail: info@scientificpub.com
Website: http://www.scientificpubonline.com

ISBN: 978-93-92590-01-6 (H/B)
ISBN: 978-93-92590-03-0 (EBook)

Printed in India

Preface

Climate change due to global warming is the most important environmental hazard faced by the world today. Anthropogenic greenhouse gas emission is the major cause for this unprecedented warming of the earth's atmosphere. To a great extent, production of energy from non-renewable sources like fossil fuels for various economic activities is accounted for this alarming situation. The international community is aiming to contain global warming to a fairly safe level, as being advised by the Intergovernmental Panel on Climate Change guided by the global scientific community. Use of energy from renewable sources is one among the most important strategies for climate change mitigation by controlling carbon emission. All countries as well as all responsible citizens should aim at reducing their carbon footprint.

In this scenario, there is a world-wide quest for new and renewable energy sources and their development. Sun is the eternal source of renewable energy for planet earth and efforts for refinement of existing solar energy technologies as well as developing new technologies are attracting more attention. Other sources derived from solar energy like wind, biomass and hydro power are also becoming more significant in the drive to reduce fossil fuel use. Energy from ocean as well as geothermal energy are also considered, even though they have technological and geographical limitations.

Sharing the concern of the global community on the future of our mother earth, India has firmly initiated steps to increase the share of renewable energy significantly. As announced by the Ministry of New and Renewable Energy, India is nearing the target of 175 GW of renewable energy by 2022. The major contribution of 100 GW is expected from solar, followed by 60 GW from wind, 10 GW from small hydro and 5 GW from biomass-based projects.

It is important that the basic knowledge on renewable energy technologies is readily available for all. Apart from engineering curricula, various Universities have taken renewable energy as a compulsory topic to be included in their curricula for various undergraduate and post graduate programmes. But not many books are available which deal the subject from the basics so that a person with basic knowledge of science can understand the theory behind these technologies. This book arises from such a need to provide a simple and comprehensive reading material for students, aspirants of various competitive examinations as well as teachers. It can also serve as a basic guide for entrepreneurs as well. We also expect our book to serve as a beginner's guide to all those who have an interest in renewable energy.

Even though a trifle contribution, we hope that our effort will serve the global community in their battle to save the planet from the catastrophic effects of climate change.

22 April 2022 ***Authors***

A brief profile of Dr. Shaji James P., Former Professor of Farm Power, Machinery and Energy, Kerala Agricultural University

Dr. Shaji James P., Professor of Farm Power, Machinery and Energy was with Kerala Agricultural University for about 30 years and is presently an Adjunct Professor at Jyothi Engineering College, Cheruthuruthy, Kerala. He has developed several technologies and gadgets and his contributions have been recognized by the Government and other professional bodies with several awards including, 'Jawaharlal Nehru Award' of the Indian Council of Agricultural Research, 'Udhaya Award' of the Tamil Nadu Agricultural University, 'Energy Conservation Award' of the Government of Kerala and 'Best Extension Scientist Award' of Kerala Agricultural University. He has authored more than hundred publications including five books. He has also participated in several international programmes in renewable energy.

A brief profile of Dr. Dipak Suresh Khatawkar, Post-Doctoral Researcher Dept. of Farm Machinery and Power Engineering, Kerala Agricultural University

Dr. Dipak Suresh Khatawkar is pursuing his Post-Doc Research in the Department of Farm Machinery and Power Engineering at Kerala Agricultural University with the '*Chief Minister's Nava Kerala Post-Doc Fellowship*' awarded by the Government of Kerala. He is also recipient of CSIR's Senior Research Fellowship for his doctoral research on 'Development of Backpack Electrostatic Induction Sprayer'. His area of research includes Electrification of Agricultural Machinery, Renewable Energy and Robotics applications in Agriculture. He has authored two scientific books in his area and published several research papers in highly reputed peer reviewed journals.

Profile of contributing authors

Dr. Joe Joe L. Bovas, is a post graduate in Farm Power, Machinery and Energy from Kerala Agricultural University and is presently working as an Assistant Professor. Prior to this, he worked as a Lecturer in Farm Power, Machinery and Energy at Hamelmalo Agricultural College, Eritrea. He was the principal investigator for many research and development projects in his specialised area including Renewable Energy. He guided several research projects for graduate and diploma level academic programmes. He has authored five books and about 20 research articles in peer reviewed journals.

Er. Aravind James is a mechanical engineer by profession and is a post graduate in Machine Design from APJ Abdul Kalam Technological University in Kerala, India. He worked as an Assistant Professor in Mechanical Engineering in Focus Institute of Science and Technology, Kerala and as a Design Engineer in Automat Kitchens Company in Kuwait. He has done research in different areas of his discipline including use of Renewable energy for environmental protection. He has also published research papers in Renewable Energy and related areas. Currently he is undergoing a post graduate programme in Industrial Plant Engineering at University of Genoa, Italy.

Er. Athira Prasad is a post graduate in Farm Machinery and Power Engineering from Kerala Agricultural University and has been teaching the course 'Renewable Energy' for B. Tech. programme in Agricultural Engineering at Kerala Agricultural University. She has also developed simple gadgets to reduce drudgery and improve energy efficiency in agricultural operations. She has also conducted research on utilization of solar energy in agriculture and published research papers. Currently she is a research scholar at Kelappaji College of Agricultural Engineering and Technology under Kerala Agricultural University.

Foreword

The global community is much concerned over the catastrophic effects of global warming and renewable energy is the most useful tool to combat climate change. But curtailing carbon emission by switching over to renewable sources of energy is a big challenge to many countries. Even though Edmund Becquerel invented the photo voltaic technology as early as 1839, the technology has found wide spread application only in the recent decades. In this context, development of affordable renewable energy technologies is highly warranted to meet the national targets of emission reduction. Newer technologies incubate from young brains for which the students should be well versed in the basic concepts.

Addressing the requirement of a basic comprehensive book on renewable energy, Dr. Shaji James P., and his team has brought out this very useful book "Fundamentals of Renewable Energy". This can meet the long-felt requirement of such a text book which will be widely utilised across the discipline barriers. I am sure that this publication will serve the purpose of a preferred text book for students of undergraduate and post graduate programmes in engineering and science so that more youngsters will be attracted to the topic. The publication of this book is timely when the nation is striving to reach carbon neutral status and it will give an impetus towards our efforts to become world leaders in the adoption of environment friendly energy technology.

It is not a simple job to bring out a book which could encompass topics like solar energy, wind energy, bioenergy, green hydrogen, ocean energy, geothermal energy, hydro power along with other related topics like fuel cells and energy conservation in a comprehensive manner. The authors have done an exemplary work in dealing with each topic with a systematic and scientific approach. The simple but elegant style of the book make the students interested in renewable energy technology. The good quality sketches, pictures, flow charts and line diagrams along with necessary graphs and tables have been used abundantly to provide clarity.

I congratulate the editor and leading author Dr. Shaji James P., who is well known for his professional contributions in bioenergy development, Dr. Dipak S. Khatawkar, the co-editor and all the co-authors for their contribution for writing the book. I recommend this book to students who want to study Renewable Energy from its basics and teachers who handle this subject in graduate as well as post graduate programmes.

Dr. C.R. Mehta

Director,

ICAR-Central Institute of Agricultural Engineering,

Nabibagh, Bhopal

FUNDAMENTALS OF RENEWABLE ENERGY
Author Index

Chapter 1	Energy, Development and Environment	Dr. P. Shaji James
Chapter 2	The Sun and Solar Radiation	Dr. P. Shaji James
Chapter 3	Solar Thermal Energy Utilization	Dr. Dipak S. Khatawkar
Chapter 4	Solar Photovoltaic Energy Conversion	Dr. Dipak S. Khatawkar
Chapter 5	Fundamentals of Wind Energy	Er. Aravind James
Chapter 6	Wind Turbines	Er. Aravind James
Chapter 7	Wind Resource Analysis	Er. Aravind James
Chapter 8	Energy from Biomass – Bioenergy	Dr. P. Shaji James
Chapter 9	Anaerobic Bioconversion of Biomass for Energy	Dr. P. Shaji James
Chapter 10	High-Rate Anaerobic Bioreactors for Energy Production	Dr. P. Shaji James
Chapter 11	Fuel Alcohol	Dr. P. Shaji James
Chapter 12	Biological Hydrogen Production	Er. Athira Prasad
Chapter 13	Fuel Cells	Er. Athira Prasad
Chapter 14	Biomass Combustion	Dr. P. Shaji James
Chapter 15	Densification of Biomass for Energy Conversion	Er. Athira Prasad
Chapter 16	Thermo-Chemical Gasification and Pyrolysis of Biomass	Er. Athira Prasad
Chapter 17	Biodiesel	Dr. Joejoe L. Bovas
Chapter 18	Energy From Microalgae	Dr. Joejoe L. Bovas
Chapter 19	Hydropower	Dr. Dipak S. Khatawkar
Chapter 20	Ocean Thermal Energy Conversion (OTEC)	Dr. Dipak S. Khatawkar
Chapter 21	Wave Energy	Dr. P. Shaji James
Chapter 22	Tidal Energy	Dr. Dipak S. Khatawkar
Chapter 23	Geothermal Energy	Dr. Dipak S. Khatawkar
Chapter 24	Clean Development Mechanism and Renewable Energy	Dr. P. Shaji James
Chapter 25	Energy Management and Conservation	Dr. P. Shaji James
Chapter 26	Energy Conversion of Municipal Solid Waste	Dr. P. Shaji James

CONTENTS

CHAPTER 1

Energy, Development and Environment

Energy is the most vital resource for any country. The early man used the energy of his muscles and to some extent, the solar energy for his sustenance. The discovery of fire paved a new way in human history. During those days his energy needs were minimal and most of the energy was derived from biological sources. Gradually, as civilization advanced, he utilized more and more energy. Probably, agriculture was the driving force in the early days for man's quest for energy. Around 3000 BC man harnessed animals for deriving mechanical power output for agricultural needs. Later on, he utilized wind to sail in the seas, which led to human inhabitation throughout different continents. All these years the energy sources got easily replenished in nature and nobody was concerned about an 'energy crisis'. Even though there would have been instances of shortage of draught animals and horses or shortened sunshine hours in rainy season and winter, there was no anxiety as we have today.

In 1700 AD, steam engine was discovered and subsequently started using coal, which was also available in plenty in some parts of the world. But when we look back today, the use of coal was the start of a new era in the history of energy development. Towards the second half of the eighteenth-century internal combustion engines were discovered. Rudolf Diesel came out with his new engine in 1900 Paris exhibition. In fact, he used peanut oil as fuel for his engine. But we learn that 'Petroleum' the miraculous fuel for the coming days was discovered in Titusville, Pennsylvania on 27 August 1859 by Col. Edwin Drake. In 1908 the British owned Anglo-Persian oil company discovered oil in Persia (Iran). The availability of this concentrated form of energy in the form of natural gas and oil tied up man on fossil fuels. Now we realize that unlike the biological sources (now we call bioenergy) like fire wood and animal power or the natural physical sources like solar energy and wind, they are not inexhaustible. As these sources are not replenished as fast as they are used, we say that they are not 'renewable'.

The reckless human endeavours to modify the nature for more comfort led to digging out more and more fossil fuels and burned them not caring about their reserves. This extravagant use of energy has grown into such a level that the mineral fuel resources are likely to be exhausted in the near future. Most estimates reveal that, at this level of consumption, petroleum may not last for another half century. If at all available, the cost will be so prohibitive that we can't afford to burn them for utilities such as powering automobiles. It is inevitable to search for other fuel sources and there is a worldwide hunt for alternate energy sources.

Man's continued pursuit for harnessing the environment demanded more and more energy and he even dug very deep into the earth for coal and petroleum. Man spent huge quantities of energy and the widespread use of fossil fuels has brought numerous benefits to modern societies. Man's enhanced mobility, not only in land and water, but also in the air and even in space could be achieved only by these sources. His ability to utilize and exploit resources was strengthened many folds by his invention of electricity produced from different fossil fuel sources.

The man in his extravaganza with energy had little time to bother how the environment will react to this. Of late, he started realizing that *Homo sapiens* – the most powerful species on the earth, the only species capable of even destroying his only home, has done much harm to his habitat. Our planet called 'the mother earth' by the ancient civilizations of the east is put on the verge of catastrophes like global warming, ozone layer depletion and acid rain as a result of a minority's everlasting greed for more comfort. Global warming caused by excessive emission of greenhouse gases to atmosphere is the greatest environmental challenge faced by humans today. Developed societies in many countries as well as rich people in under developed parts of the world following a materialistic life style, shall be regarded as the primary respondents for this calamity endangering the entire flora and fauna of the planet. Still, some of them swear that they want to preserve their 'style' of life. It is high time to ask, at what cost?

The economic crisis due to the Covid 19 pandemic outbreak was instrumental in teaching us a lesson on how the world can survive by curtailing wasteful practices leading to an eco-friendly culture of resource conservation. The words of the great philosopher of peace Mahatma Gandhi, *"The earth has everything for our needs; but not for our greed"* resonates even today.

1.1 Concept of Energy

The simple definition of energy is well known and is indeed the capacity to do work. Let us see this concept from the basic principles. According to basics of Physics and Mechanics, Force (F) is the product of mass (m) and acceleration (a),

$$F = m \cdot a \qquad ...(1.1)$$

Work (W) is defined the product of a force (F) and the distance (S) through which the point of application is moved in the direction of the force.

$$W = F \cdot S \qquad ...(1.2)$$

Power (P) is the rate of doing work and hence the work done per unit time (t). Power is also the product of the force (F) and the velocity (v) with which the point of application moves.

$$P = W/t \qquad ...(1.3)$$

Also,

$$P = F \cdot S/t \qquad ...(1.4)$$

When the point of application of force moves is a straight line, the motion is linear. In some other cases the motion is rotational. In computing power in rotational motion, the angular velocity (ω) is taken into consideration and instead of force (F), torque (T) is taken into account. Torque is the rotational equivalent of linear force. Depending on the situation, it is also referred to as the 'moment', 'moment of force' or rotational force. The concept originated from usage of levers. The power in rotational motion is given by,

$$P_r = T \cdot \omega \quad ...(1.5)$$

Even though in simpler terms 'Energy' is the capacity to do work, it is being used in a much broader sense to indicate the potential for accomplishing useful work. Energy has the same units as work. Energy is available in the nature in different forms such as heat or light energy. Energy transfer occurs in nature. One example is the transfer of thermal energy from a body at higher temperature to one at a lower temperature. Another aspect is the transformation of energy. When certain materials are oxidised, the chemical energy is transformed to thermal energy. These materials are generally called 'fuels' and are often regarded as energy sources. Different systems convert heat into other forms of energy such as mechanical energy as in the case of a 'heat engine'.

The common energy forms other than heat energy and light energy are:

Potential energy, attained by a body by virtue of its higher level from earth's surface i.e., product of mass (m), acceleration due to gravity (g) and the height (h).

Kinetic Energy (K.E.) of matter is the energy possessed by the body of mass (m) by virtue of its velocity (v) i.e., K.E.= $\frac{1}{2} mv^2$

Chemical energy of materials which can be released / converted by a chemical process.

Electric energy of a charge (q) in a potential difference (V) i.e., Product of q and V.

Magnetic energy of a magnetic quantity (M) in a magnetic field with intensity (H) i.e., Product of H and M.

In a hydroelectric power production facility, the potential energy of water is first converted to kinetic energy when it flows through the head race at a certain velocity to reach the turbine where kinetic energy is transferred to it. The turbine is connected to a generator where this mechanical energy is converted to electrical energy. This electrical energy is subsequently converted to light energy when we use an electric light or to heat when we use an electric heater. It is interesting to note that the potential energy of water in the reservoir is converted to a variety of forms and when the electricity is used for pumping water to an elevated point, it finally becomes potential energy of water itself. Similarly in a solar photo-voltaic lighting system, the light energy of sun converted to electricity is finally transformed to light itself.

Units of Energy

The unit of energy in the *Systeme Internationale* (S.I.) is the joule (J) and is defined as the work done when a constant force of 1 newton (N) moves its point of application by 1 metre (m). One joule is also performed when a charge of 1 coulomb (1 C) moves through an electric potential difference of 1 volt (1 V). The units for energy are expressed in varied multiples as we need to quantify energy in various forms derived from diverse sources. It is often required to have very small units to very big units to quantify energy depending on the situation (Appendix).

1.2 Energy and Development

The development and use of energy have paved the way for progress of human civilization till date. Energy use is the indicator for development of a society. Developed countries use much more energy than the developing countries, even though the population of former is less compared to the latter. It has been observed that, the growth in energy use by developing countries was indicative of their economic growth over the corresponding period. In the case of India, the commercial energy demand increased from 200 mtoe (million tonne oil equivalent; 1 toe = 42.6 GJ) to 600 mtoe during the decade 2000 to 2010, and a corresponding GDP growth was also observed during the period. It has been estimated that energy consumption in developing countries is growing at 3 per cent per annum on an average. The commercial energy (coal, oil, gas and electricity) consumption in some countries may outgrow their GDP growth and likely to be severe burden on their economies, as much of the energy needs are to be met from imported coal and oil.

The relationship of energy consumption with development is not often linear in the case of developed as well as developing countries. In the case of India, a 6 per cent increase in GDP is expected to create an increase of 9 per cent on its energy demand. The ratio of energy demand (ED) to GDP is a useful terminology to indicate the energy intensity in national economic development. When ED: GDP ratio is high, it shows the high energy dependency in development and vice versa. It is observed that most of the developed countries maintain ED: GDP ratios less than 1 by improving energy efficiencies in their production systems.

Energy intensity is another term widely used to indicate the development mode of a country and is defined as the energy consumption per unit of GDP. India's energy intensity is presently more than three times that of Japan and is higher than world average.

1.3 Energy and Environment

Large amounts of industrial, agricultural, domestic and urban wastes generated in developed societies have detrimental effects on the local and global environment. The unscientific disposal of wastes as well as uncontrolled use of energy for various economic activities lead to many problems such as global warming, desertification and acid rains. In recent years, air, water and thermal pollutions due to developmental activities including power generation have grown multi-dimensional.

Environmental issues related to energy use received a low priority in developing countries until the discussions leading to Kyoto protocol commenced in the nineties. In the early scenario, pollution issues due to fossil fuel use was of more concern. With the widespread awareness on the role of anthropometric emission of Green House Gases

(GHGs) in causing global warming, the international community became concerned on fossil-fuel based energy production and use. Even though many doubted on the role of energy use in the early discussions, currently there is wide acceptance of the fact that increased level of greenhouse gases in the atmosphere are mostly anthropogenic, resulting from the burning of fossil fuels. The rise in atmospheric temperature is spectacular after the nineteenth century. The polar ice has started melting at an increased pace, rendering the sea level to go higher. This, of course, is not only endangering the humans and his habitat in the coastal areas of the main land, but also foster a multitude of environmental issues resulting in global climate change.

The social issues related to global warming are also much complicated. The rise of sea level for fraction of a metre can cause some of the island nations getting wiped out from the surface of the earth. The cabinet meeting of the island nation, Maldives, was once conducted at the ocean bed in their effort to call international attention on sea level rise caused by global warming. The modern world has miserably failed to settle out the disputes arising from political problems between nations or even issues with different ethnic groups of the same nation resulting in severe unrest by refugees. The problem of environmental refugees, or more specifically, 'energy refugees' could be equally difficult to handle.

1.4 Indian Energy Scenario

India is the second most populous country of the world with about 1.4 billion people (2022) representing 17.6 per cent of the world population. But, the geographic area of India is only 2.4 per cent of the world's land area. The fossil fuel reserves are highly inadequate for the growing Indian population as the coal and oil resources are only 10 and 0.3 per cent, respectively of the global resources. Electricity is the most sought-after energy form for varied applications in domestic, agricultural and industrial sectors. A major portion of the electricity in India is produced from thermal power stations in which fossil fuels are predominantly used as fuel.

Table 1.1 Source wise electric power production in India

Source	Capacity, MW	Share, %
A. Total Thermal	2,33,171	61.5
(i) Coal	2,01,085	53.0
(ii) Lignite	6,620	1.7
(iii) Gas	24,957	6.6
(iv) Diesel	510	0.1
B. Hydro	46,209	12.2
C. Nuclear	6,780	1.8
D. Renewable Energy Sources	91,154	24.5
Total	3,79,130	

Source: Ministry of Power, Govt. of India (2021)

(a) Coal: Coal is the predominant energy source for power production in India, generating approximately more than half of total electricity. Even though energy demand in India keeps on increasing, the power deficit is decreasing over the last decade due to increased installed capacity, especially of renewable power from wind and solar. Still, coal is likely to dominate the thermal power sector in this decade also. The all-India production in 2019-20 was 730 million tonnes (mT) and the imports were about 248 mT according to the Ministry of Coal, Govt. of India (2021).

(b) Petroleum: India's demand for crude oil was 186.55 mT in 2009-12, which increased to 232.86 in 2015-16. The indigenous production marginally increased from 33.69 mT to 36.86 mT over this period. There is a sharp increase of 60 per cent in oil consumption during the decade; 3.29 billion barrels in 2009 to 5.27 billion barrels in 2019. The domestic crude oil production is showing a decreasing trend in the recent years (32.16 mT in 2019-20). India was the second topmost crude oil importer in 2019 and more than 82 per cent of the oil demand of the country is met from imports.

(c) Natural Gas: Natural gas is an important component in India's economy and is being used for fuel as well as non-fuel (manufacture of chemical fertilizers) applications. The indigenous production began in the 1960s mainly from gas fields in Assam and Gujarat. Currently the net imports exceed 45 per cent of the demand. The government is trying to increase the share of natural gas in the energy basket as it is a cleaner fuel compared to oil and coal.

1.5 Classification of Energy Sources

The sources of energy are broadly classified based on different considerations as given below:

- Primary and Secondary energy
- Commercial and Non-commercial energy
- Renewable and Non-Renewable energy

1.5.1 Primary and secondary energy

Primary sources of energy are those which are found in nature. Examples are coal, oil, natural gas and biomass (firewood). These primary sources are often converted to readily usable secondary energy forms such as heat or electricity in utility systems for end use. For example, coal is burned in boilers to produce steam which in turn is used to produce electricity in thermal plants.

1.5.2 Commercial and non-commercial energy

The energy sources that are commercially sold in the market for a price are known as commercial energy. Electricity, coal and petroleum products such as diesel fuel and gasoline (petrol) come under this category. Even fire wood is commercially sold in some urban areas. As commercial energy forms the basis of development, the use of commercial energy is often considered as an indicator for development. The energy sources that are not sold in the market for a price are classified as non-commercial energy. People in rural India collect firewood for domestic use and is not sold. Biogas from domestic biogas plants, solar heat utilized in solar water heaters, dung cakes and agricultural residues as cooking fuel in rural areas are also examples for non-commercial energy.

1.5.3 Conventional and non-conventional energy sources

Few decades before, classification of energy sources as conventional and non-conventional was very common. In those days, sources such as coal and petroleum widely used as primary sources were regarded as 'conventional' and newer sources like solar energy, wind energy and tidal energy were considered 'non-conventional'. Certain sources which are conventional in a locality may be non-conventional in some other area. As the term 'conventional' is arbitrary, this classification has become obsolete. Previously, the central ministry (Government of India) of 'New and Renewable Energy' (MNRE) was named 'Ministry of Non-Conventional Energy Sources' (MNES).

1.5.4 Renewable and non-renewable energy sources

Renewable energy sources are generally regarded as inexhaustible sources like the sun. Specifically renewable energy sources can be defined as those sources which are replenished in the near future. Near future may be regarded as a period of hundred years. Solar energy, wind energy, hydel power, biomass energy, geothermal energy, tidal energy, wave energy and ocean thermal energy are the most common renewable energy sources. Fossil fuels such as coal, oil and gas are finite in nature and hence likely to be depleted if used indiscriminately. They are called 'fossil fuels' as they are derived from dead plant and animal matter trapped under the earth's crust millions of years ago. There is no chance that these fuels will be generated as fast as they are depleted and hence, they are regarded as non-renewable.

1.6 Renewable Energy

Sustainable development is achieved without hindering the environment. The needs of the future generation also have to be met from the environment and man has no authority to finish off all the resources including energy. Agenda 21 of the 1992 Earth Summit in Rio de Janeiro was a comprehensive plan of action to build a global partnership for sustainable development to improve human lives and protect the environment. This philosophy gained momentum in the recent decades and led to formulation of a '2030 Agenda' for sustainable development which was adopted at the UN Sustainable Development Summit held at New York in September 2015. 'Affordable and clean energy' is one among the 17 sustainable development goals and now the world is more interested in 'renewable sources of energy'.

The terminology 'renewable energy' refers to certain forms of energy which are 'renewable' because their sources are constantly replenished and not likely to get exhausted. Renewable energy – like solar and wind energy – is available through-out the world. Renewable energy technologies take this energy and convert it into readily usable forms of energy – most often electricity, but also heat, or mechanical power. These technologies are often described as 'clean' or 'green' because they produce little or no pollutants.

The sun is the ultimate energy provider for the earth and we see that almost all other sources are derived from sun. Even fossil fuels are derived from the sun as they are of biological origin. Sun is an inexhaustible energy source as far as human life on earth is concerned, showering about 1 kW energy per square metre of the earth's surface. Direct use of solar energy has been regarded as one of the most important renewable energy resources. The two energy forms, viz. thermal as well as light can be utilized with available technological options. Thermal energy of solar radiation is tapped by flat plate and concentrating collectors for various purposes like drying materials, heat water, cook food, provide heat in industrial processes, produce steam and even to generate electricity. The sun's rays transmit energy in the form of light energy also, and are used for photovoltaic (PV) conversion into electricity. PV is one among the most environmentally benign technologies for power production, even though there are economic limitations for the use of this technology. The techno-economic bottlenecks are being overcome at a fast pace and this is a thriving energy sector today.

Almost all renewable energy comes directly or indirectly from the sun and they include bioenergy, wind energy, hydel power and wave energy. Heat from the sun causes air temperature differences that, along with the earth's rotation, cause the wind to blow. The wind powers aero-generators that produce electricity or provide mechanical energy for different applications such as water pumping. The most predominant energy source in developing countries viz. energy from biomass is derived from the photosynthetic capture of solar energy by plants. Plants store the energy of sunlight in the chemical bonds of carbohydrate. Reservoirs of hydroelectric projects are filled by rain. As we know the rains result from the condensation of the water vapour evaporated from water bodies by solar heat. Waves are produced by winds caused from differential heating of earth's surface by solar energy. Sun's gravitational pull also has a role in tidal energy even though moon is the major player. Geothermal energy is also a renewable source, even though limited to specific areas.

Photosynthetic energy conversion, in which the solar energy is converted to biomass, is the foremost natural energy conversion phenomenon taking place on earth. Conversion of biomass including organic wastes and crop residues also offer tremendous scope for generation of renewable energy. Different technological alternatives are available for use of energy from biomass. The recycling of waste materials and residues for energy is a basic concept which must be implemented in order to meet the requirements of sustainable development in both developed and developing countries.

CHAPTER 2

The Sun and Solar Radiation

The sun is the eternal energy provider in the solar system which consists of the planets and satellites orbiting around the sun. Sun, the nearest star to earth, is also the heaviest body in the solar system. The sun and the solar system are located in the galaxy called 'Milky Way' as shown in Fig. 2.1. There are billions of galaxies in the universe with different shapes and our galaxy is included in the cluster called 'local group' comprising of two big galaxies and many small galaxies.

Fig. 2.1 The Milky Way (*Courtesy: Exoplanet Exploration – NASA*)

The other big galaxy in the local group is the 'Andromeda nebula' having twice the number of stars compared to Milky Way. Our galaxy is a spiral galaxy with six spiral arms and the sun is situated in one of the two smaller arms, the 'Orion arm'. The galaxy is about 1,00,000 light years wide (one light year ≈ 9460 billion km), whereas the solar system is very small with a width of about 12 light hours only. As said in the Esavasyopanishad, "*Yath kincha jagathyam jagat*" (Everything in the universe is changing / moving), the entire galaxy is rotating in space. The sun makes one lap around the centre of the galaxy in a near circular orbit in about 220 million years with a speed of 230 kilometres per second.

2.1 The Sun

The sun is a sphere of intensely hot gaseous matter (about 75 per cent hydrogen and 25 per cent helium with traces of other elements) which produces its energy by nuclear fusion reaction. Four hydrogen atoms are fused to form one helium atom and the matter converted to energy is radiated in all directions. It has been estimated that about 4×10^6 tonnes of matter is converted to energy every second. The diameter of sun is 1.39×10^6 km and the mass is 1.99×10^{30} kg. The solar interior constitutes the main mass of sun and 90 per cent energy is generated in the inner 0.23 radius. Outer region beyond 0.7 radius is called convective zone with its upper layer called photosphere. The transparent gaseous layer above photosphere is called Chromosphere. The outermost whitish glowing Corona is an aura of plasma which surrounds sun and can be seen during solar eclipse. The temperature of the sun is about 5800° K at the surface. Sun spots are temporarily appearing dark spots on the photosphere caused by regions of reduced surface temperature due to magnetic flux concentration (Fig. 2.2).

The earth is the third planet in the solar system when we count from the centre having a mass of 6×10^{24} kg. The shape of the earth can be described as an 'oblate spheroid'. An oblate spheroid is spherical but slightly flattened. The radius of the earth at equator is 6378.137 km, which is more than the radius at the poles (6356.752 km). The earth rotates about its own axis and completes one rotation in about 24 hours. The earth revolves round the sun in an elliptical path. During the revolution, the earth reaches the nearest position to sun at perihelion around January 4th and reaches the farthest position, aphelion around July 5th. The average distance between the centres of sun and earth is termed as 1 astronomical unit (AU) and is equal to 1.496×10^8 km (Fig. 2.3).

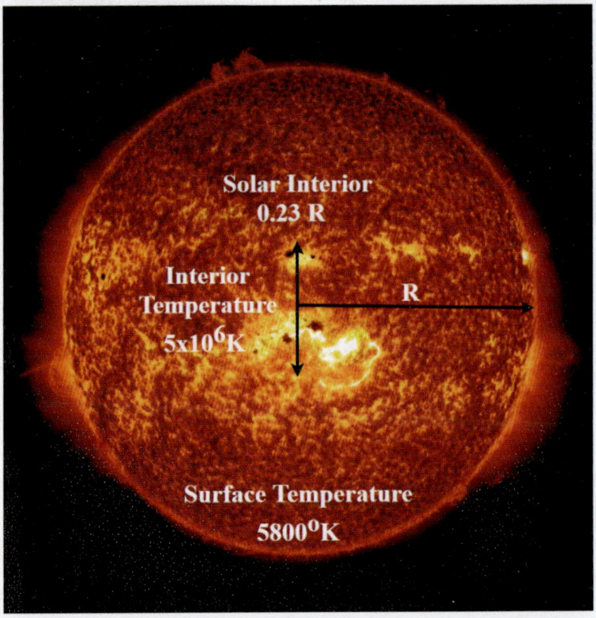

Fig. 2.2 The Sun (*Courtesy: Solar System Exploration – NASA*)

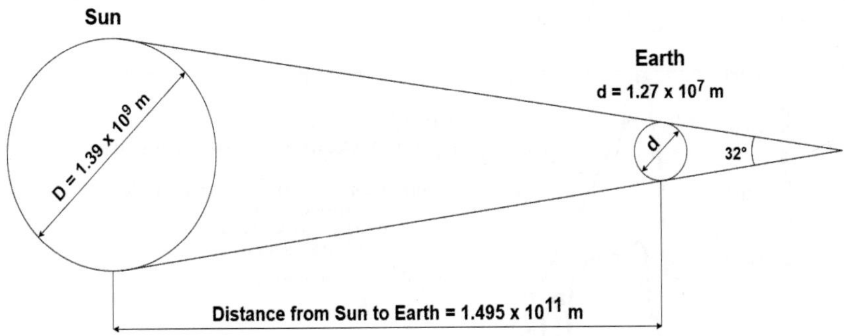

Fig. 2.3 Relative geometrical dimensions of sun and earth

2.2 Solar Radiation

The solar radiation reaching earth can be considered as radiation from a black body at a temperature equal to the surface temperature of sun (5762°K). As sun is far enough from earth, the direct radiations received at the surface are considered parallel. The solar energy transmitted out through electromagnetic radiations ranging from very high frequency cosmic rays to very low frequency radiation. Most of the radiation (99%) is in the wavelength range of 0.2 to 4 μm (1 μm = 10^{-6} m). Broadly, the spectrum can be divided into three regions viz. ultraviolet, visible and infra-red (Fig. 2.4).

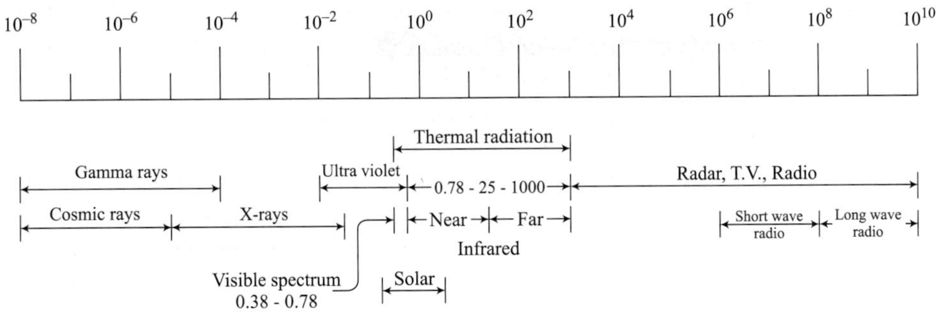

Fig. 2.4 Electromagnetic radiation spectrum

The high frequency-short wave length, ultraviolet (UV) radiation and the low frequency-long wave infrared (IR) radiations are not visible to our naked eyes. IR radiations are regarded as heat waves as they produce heat when incident on surfaces. UV radiations are harmful to living cells as they have high energy to break the carbon-carbon bonds in organic matter. Materials like ordinary plastics get degraded when exposed to UV radiation. Radiation in the wavelength range 0.38 to 0.78 μm is visible light which is responsible for sustaining life on earth. The Photosynthetically Active Radiation, abbreviated as 'PAR' (in the range 0.4 to 0.7 μm), is used for photosynthesis by plants. Out of the total solar energy reaching earth, 47.3 per cent is available in the visible region and 45.7 per cent is in the infrared region. The UV region constitutes only 7 per cent of the spectral distribution (Fig. 2.5).

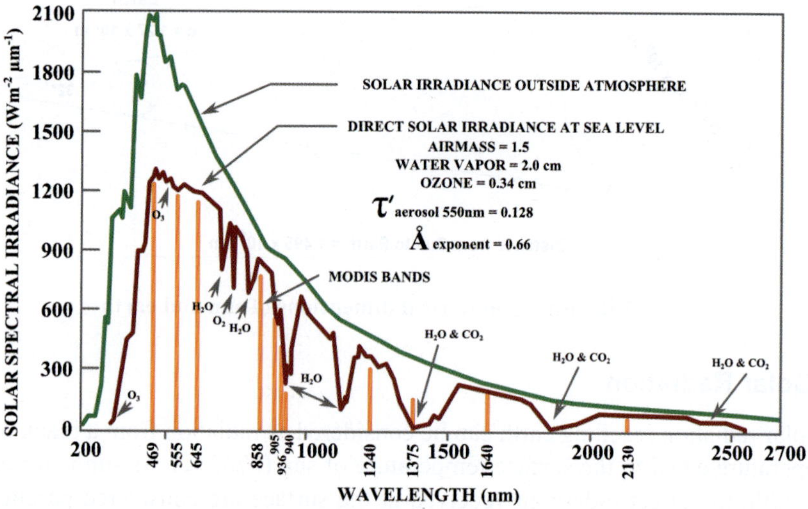

Fig. 2.5 Solar spectral irradiance (*Courtesy: NASA*)

2.3 Solar Constant

The solar energy available at the outer space is called extra-terrestrial radiation and is almost constant. The rate at which energy is incident on the top of earth's atmosphere is called 'Solar Constant'. *The Solar Constant (I_{sc}) is defined as the total energy received from the sun per unit time on a unit area, perpendicular to the rays of the sun, just outside the earth's atmosphere, at the mean distance of the earth from the sun* The standard value of I_{sc} fixed by the National Aeronautics and Space Administration (NASA) was 1353 Wm^{-2} until 1970. At present the revised value is 1367 Wm^{-2}. The solar constant is inclusive of all radiations. It may be noted that it is not a physical constant like speed of light (C). It is the average value of actual measurements by scientists.

Due to the change in relative position of earth with respect to the sun, the value of solar constant changes throughout the year. The value of solar constant (I'_{sc}) on any day, 'n' can be obtained using the following relationship:

$$I'_{SC} = I_{SC}\,[1 + 0.033 \cos (360n/365)] \quad ...(2.1)$$

Where,

'n' is the day of the year (reckoning January 1st as 1)

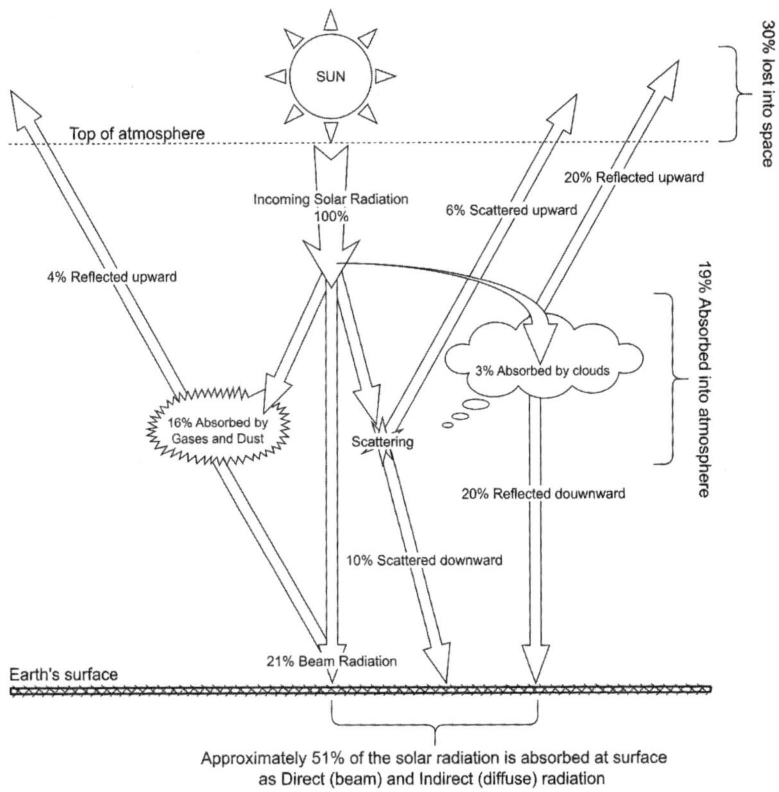

Fig. 2.6 Attenuation of solar radiation

2.4 Variation of Solar Radiation on Earth's Surface

The solar energy availability on any point on the surface of earth is determined by the solar radiation incident on that point at the specified time. This is dependent on the sun's position with respect to that point. As the radiation passes through the atmosphere, loss of energy occurs due to 'attenuation of radiation' (Fig. 2.6). This occurs due to scattering as well as absorption by gas molecules and aerosols. Reflection and scattering by clouds and dust particles also reduce the quantity of energy received at the surface. The solar radiation that has not been subjected to alteration and reaches the surface directly without any deflection is called '*Beam radiation*' or direct radiation. As a considerable quantity of solar radiation is deflected and scattered, indirect radiations also reach the surface. These radiations whose direction has been changed due to reflection and scattering are called '*Diffuse radiation*'. The total solar radiation received on a point on the surface of earth is the sum of beam and diffuse radiations and is termed '*Global radiation*'. Usual terminology to refer to this total available radiation on unit area of a surface is '*insolation*'.

The insolation on a particular point in a given time is dependent also on the position of that point with respect to sun. If we consider a location on the earth's surface, the position of sun is determined by the 'altitude' of sun. The sun's altitude is changing with respect to the time, day, season and latitude. When the altitude is large, the radiation has to pass through a greater distance through the atmosphere and hence, the loss of energy will be more. The path length of radiation passing through atmosphere is called '*Air mass*'.

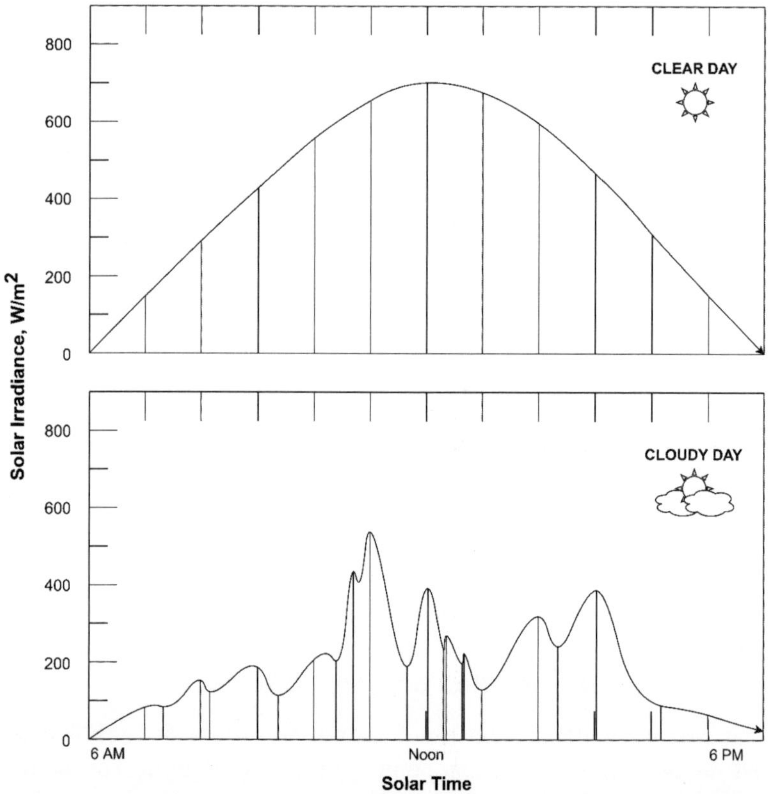

Fig. 2.7 Variation of solar radiation due to change in weather conditions

It can also be observed that different types of particles in the atmosphere behave differently with the incident rays. Some particles scatter the radiation in different directions. Some other particles absorb part of the radiation and emit it as long wave radiation leading to heating of atmosphere. This phenomenon when exceeds the normal limits produce global warming. Water vapour, dust, CO_2, methane, ozone and some other gases absorb up to 16 per cent of the radiation.

Variation in solar radiation occurs at a location on hourly basis due to change in the position of sun with respect to the reference point. Daily variation is due to change in weather as well as the change in orientation of earth. Larger variations occur on seasonal basis due to the change in relative positions of sun with respect to earth. It is also obvious that the availability of solar energy decreases from a location at equator to that in the poles. Thus, energy availability varies greatly with time and location (Fig. 2.7). Maximum solar energy receivable on a clear day on a surface at sea level in a location on equator can be about 1.0 kWm^{-2}. Hence the solar radiation is considered to be a dilute form of energy. All these aspects are regarded as limitations of solar energy use. Fortunately, India receives adequate solar energy in most of its geographical regions.

2.5 Solar Radiation Geometry

The relative position of a location on the surface of Earth changes due to the diurnal and seasonal changes in its orientation and position with respect to sun. As we know that the earth rotates about its own axis as well as revolve around the sun in an elliptical orbit, both in anti-clockwise directions (if observed from a point above the North Pole). The horizontal plane containing earth's orbit is called the 'ecliptic'. The earth reaches the farthest point from sun called 'aphelion' in July whereas it reaches the nearest point in the orbit in January. Due to gradual changes in the eccentricity of the earth's orbit, and the mismatch between calendar year and the duration of one revolution of earth, the dates of aphelion and perihelion are not fixed and they can vary up to two days from one year to another. The aphelion is likely to be between 4th and 7th July and perihelion falling on 3rd to 4th January in the years from 2022 to 2026.

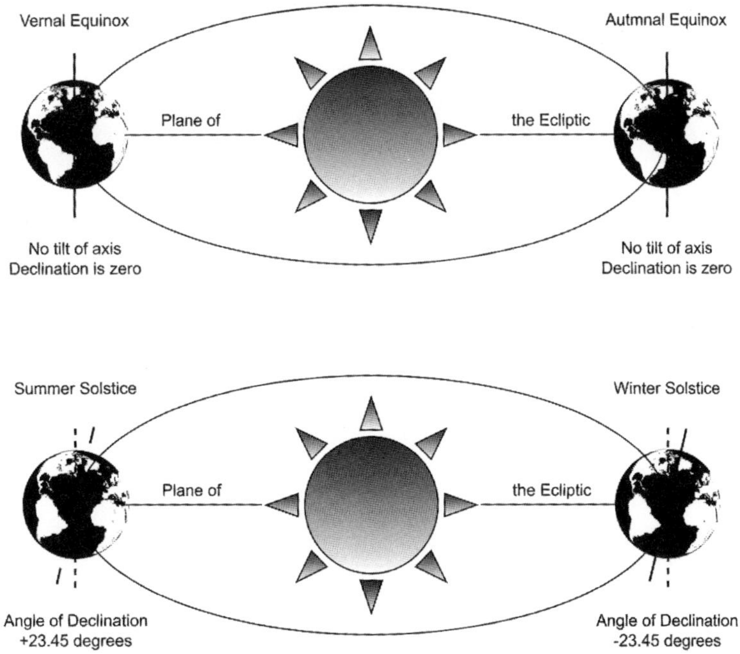

Fig. 2.8 Orientation of Earth's axis to the plane of ecliptic

The earth's axis of rotation is inclined to the ecliptic plane and it varies continuously during the revolution of earth in its orbit. The angular value of the inclination of the axis of Earth to the normal to the ecliptic is called '*declination*'. From Fig. 2.8, it can be seen that on equinox days the axis of the earth is straight. The declination keeps on changing and is positive or negative with respect to the position of sun in the sky as seen from the northern hemisphere. When the declination is positive, the sun appears to be in the north for an observer in the northern hemisphere and sun is seen in the south when declination is negative. During the two equinox days, the declination is zero, whereas it acquires the maximum values during the two solstice days. The northern hemisphere is leaning towards the sun to the maximum on summer solstice day (between June 20 to 22) and the declination is +23.45°, whereas it leans away maximum from sun on the winter solstice day (around December 21 to 22) with the declination -23.45°. After the vernal equinox (between March 20 to 21) the sun appears to proceed northward and Indians use the Sanskrit word '*Uttharayana*' (meaning: northward travel) to describe this phenomenon. Once the earth passes the autumnal equinox position between September 22 to 23, the southward travel (*Dakshinayana*) of sun starts. These dates are also not fixed as in the case of aphelion and perihelion, primarily due to the fact that earth is not taking exactly 365 days for completing one lap around sun (Fig. 2.9).

Apart from declination, there are also other phenomena like 'Precession' due to wobbling of earth's axis. Due to precession, the constellation in which sun appears in a particular day of the year varies in a period of 2160 years, and the precession cycle is completed in about 26,000 years.

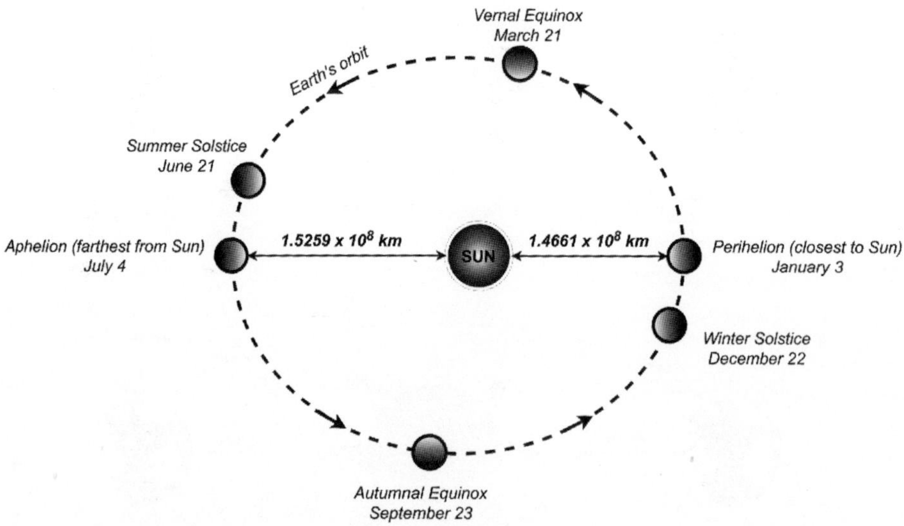

Fig. 2.9 Elliptical orbit and the seasonal positions of Earth around Sun

2.6 Basic Sun – Earth Angles and Solar Time

The effect of positional variations on the availability of solar energy can be understood from the basic solar radiation geometry governed by Sun - Earth angles. The position of a location on the earth's surface is known by the coordinates, viz. latitude and longitude.

2.6.1 Latitude (Φ)

Latitude is defined as the angular distance of a point on the surface of the earth from the equator. It is the angle between the line joining the centre of the earth and the point on the earth's surface to its projection on the earth's equatorial plane. If the point is on the north of equator, it is considered as north latitude and taken as positive. If the point is in the south of equator, it has south latitude which is taken as negative (Fig. 2.10).

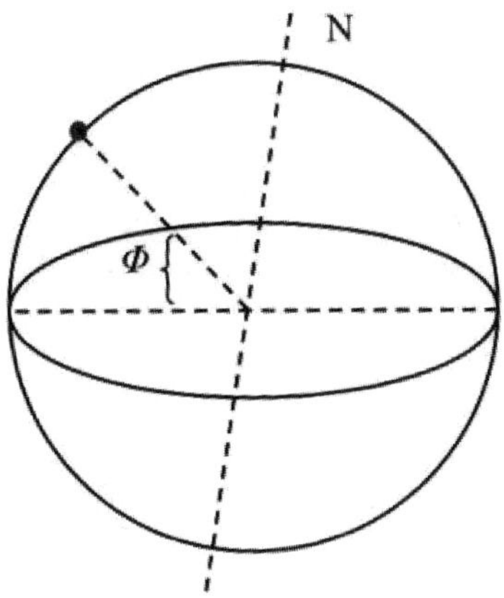

Fig. 2.10 Latitude

2.6.2 Longitude

Longitudes are semi - great circles passing through the poles. Longitude angle is the angular distance of a location on the surface of earth measured eastward or westward from the 'Prime meridian'.

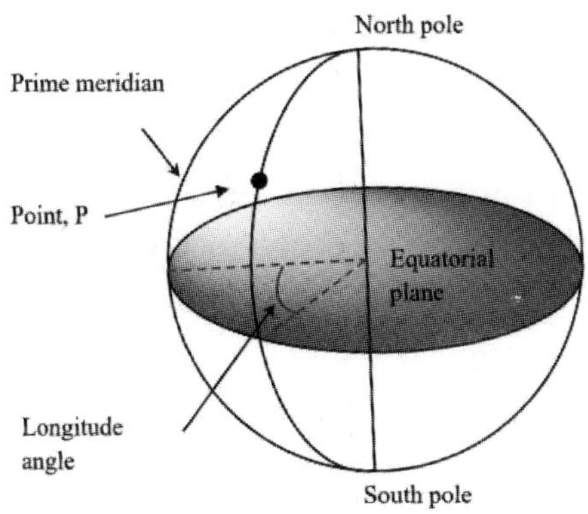

Fig. 2.11 Longitude

The zero-meridian passing through Greenwich is the prime meridian. In other words, the horizontal angle made by the meridian passing through the point 'P' with the prime meridian on the equatorial plane is the longitude of a point. When the point is east of the zero meridian the longitude is referred as 'East' and longitudes to the west are denoted as 'West' (Fig. 2.11).

2.6.3 Solar time

The earth makes a complete 360° rotation on its imaginary axis once in 24 hours. So, earth moves through 15° in one hour. As the earth covers one degree (each meridian) in (24×60)/360 minutes, the rotational speed is 4 minutes per degree.

The actual position of sun with respect to a point at a specific time of the day is required for solar energy calculations for which 'Solar time' is used. Time as shown by a sun dial or measured by the apparent diurnal motion of the sun in the sky, from the eastern horizon to the western horizon, is the solar time. It is different from the standard time of countries which is dependent on standard meridian of the country. Some big countries have different time zones depending on the corresponding standard meridian of that zone. India has only one time zone and Indian Standard Time (IST) corresponds to 82° 30' East longitude. The Solar time can be calculated from the standard time by applying corrections.

As every degree difference is equivalent to 4 minutes, the Local Civil Time (LCT) can be obtained by applying the following correction:

$$\text{LCT} = \text{IST} \pm 4\,(\text{Standard Longitude-Longitude of the location}) \qquad ...(2.2)$$

The deviation of actual solar time from the LCT occurs seasonally due to the variation in rotational as well as angular velocities of the earth. The correction factor to obtain local solar time (LST) from local civil time (LCT) is called 'equation of time', which is given by:

$$\text{Equation of Time} = 0.258 \cos X - 7.416 \sin X - 3.648 \cos 2X - 9.228 \sin 2X \text{ minutes} \qquad ...(2.3)$$

Where,

$$X = \frac{360(N-1)}{365.242} \text{ degrees} \qquad ...(2.4)$$

And, N = day number of the year

Equation of time correction is available from standard tables or charts as well.

$$\text{LST} = \text{LCT} + \text{Equation of Time} \qquad ...(2.5)$$

Example 2.1: Determine the solar time at a location at 75° E longitude at 11.30 am (IST) on May 22.

Solution:

Local Civil Time = $11^{h}\,30^{min} - 4(82°30' - 75°) = 11^{h}\,30^{min} - 30^{min} = 11^{h}$

Equation of time correction for May 22nd (from standard tables) = $3^{min}\,18^{s}$

Local Solar Time = $11^{h} + 3^{min}18^{s} = 11^{h}\,3^{min}\,18^{s}$

2.6.4 Solar declination (δ)

Declination on any day is the angular measure of the deviation in the direction of sun rays north or south of the earth's equator. The declination can also be described as the angle between the line joining the centres of Sun and Earth and the projection of this line on the earth's equatorial plane (Fig. 2.12).

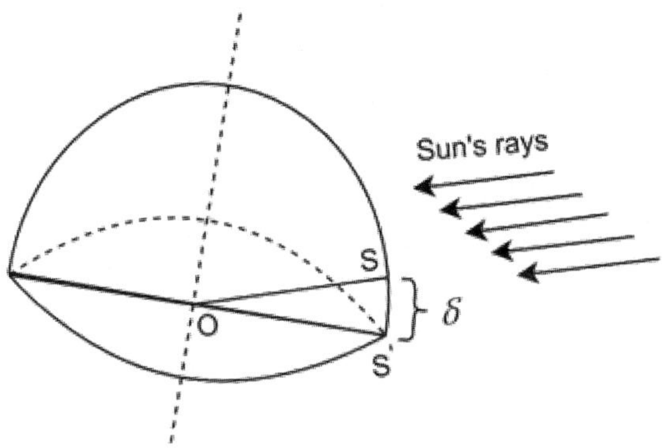

Fig. 2.12 Declination

It varies from –23.45° on winter solstice in December to +23.45° on summer solstice occurring in June. It is 0 on the two equinox days falling in March and September.

A reasonably accurate value of 'δ' on any day can be calculated using the Cooper equation. According to Cooper (1969),

$$\text{Declination} = 23.45 \sin\left[\frac{360}{365}(284+n)\right] \quad ...(2.6)$$

Where, 'n' is the day of the year.

Example 2.2: Determine the declination on August 1.

Value of 'n' for August 1st = 31+28+31+30+31+30+31+1 = 213

$$\delta = 23.45 \text{ Sin } [360\times(284 + 213)/365] = 17.91°$$

2.6.5 Hour angle (ω)

Hour angle is the angular measure of time, reckoning one hour equivalent to 15°. It is the angle through which the earth must turn to bring the meridian of a point directly in line with the incident sun's rays. So, the hour angle is 0° at solar noon. It is the horizontal angle measured on the earth's equatorial plane between the projection of the line OP and the projection of a line connecting the centre of the earth to the centre of sun. It is measured positively westward from the observer (Fig. 2.13).

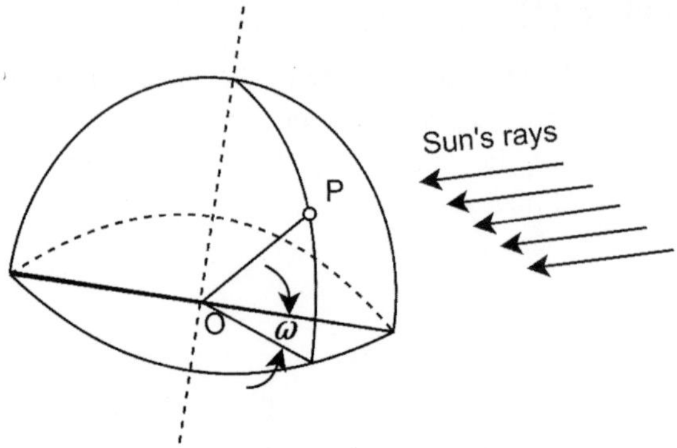

Fig. 2.13 Hour angle

Where,

$$\omega = 15(t - 12) \text{ degrees} \quad ...(2.7)$$

and 't' is the solar time.

The hour angle at sun rise on a particular day is the angular measure of time to solar noon. Sunrise hour angle (ω_s) on a particular day is important in solar energy calculation as this is an indication of the theoretical day length. In fact, the day length in angular measure will be twice the sun rise hour angle. It can be calculated in hours by dividing by 15.

Thus, the Day length, N on a particular day,

$$\text{Day length} = \frac{2 \times \omega_s}{15} \text{ hours} \quad ...(2.8)$$

The sun rise hour angle is a function of latitude (Φ) of the location and declination (δ) on that day and is given by the relation:

$$\cos\omega_s = -\tan\Phi \cdot \tan\delta \quad ...(2.9)$$

We can calculate the theoretical day length on a particular day if we know the latitude of the location and the declination on that day.

$$\text{Theoretical Day length} = \frac{2}{15}\cos^{-1}(-\tan\Phi \cdot \tan\delta) \text{ hours} \quad ...(2.10)$$

Example 2.3: Determine the day length at Ramakkalmedu (Longitude 77° 14' E, Latitude 9° 49'N) in Idukki District of Kerala on August 1st.

Solution: Declination on August 1st (calculated using Cooper equation) = 17.91°

Sunrise hour angle, $\omega_s = \cos^{-1}[-\tan\Phi \tan\delta] = \cos^{-1}[-\tan(9.817)\cdot\tan(17.91)]$

$= 86.794233°$

Day length, $N = (2\times\omega_s)/15 = 11.573$ hours

2.7 Derived Solar Angles

Derived angles are relevant in solar energy calculation as these angles are useful to find the solar flux on a surface of interest. For an observer watching the movement of sun from the eastern horizon to the western horizon, the path of the sun on the equatorial plane of the earth is a semi-circular arc (Fig. 2.14).

2.7.1 Altitude angle (solar altitude)

It is the vertical angle *(α)* between the projection of the sun's rays on the horizontal plane and the direction of sun's rays.

2.7.2 Zenith angle (θ_z)

It is the angle between the direction of sun's rays and the vertical at that point on the surface. It is also the complementary angle of sun's altitude angle.

$$\theta_z = \frac{\pi}{2} - \alpha \quad ...(2.11)$$

2.7.3 Solar azimuth angle (γ_s)

It is the horizontal angle measured from north to the projection of the sun's rays on the horizontal plane, measured west wise or east wise (positive when west wise).

2.7.4 Surface azimuth angle (γ)

It is the angle measured on the horizontal plane from due south to the horizontal projection of the normal to the reference surface. It is the angle of deviation of the normal to the surface from the imaginary North- South line on the surface reckoning zero at the south. The angle will be positive when measured eastward and negative when measured westward.

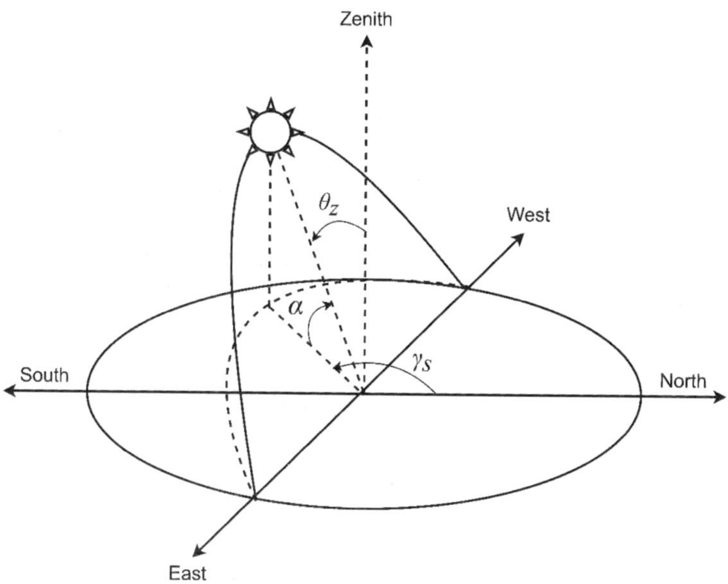

Fig. 2.14 Derived angles

2.7.5 Incident angle (θ)

It is the angle measured between the direction of Sun's rays and the normal to the horizontal plane. If an incident radiation of intensity 'I' is incident on a surface at an incident angle 'θ', then the equivalent solar flux on the surface is $I \cdot \cos\theta$.

2.8 Estimation of Solar Radiation

The extra-terrestrial radiation incident on a surface kept normal to the sun's rays when Earth is at the mean sun-earth distance (1 AU) is the solar constant, I_{sc}. The actual value of solar constant on a particular day, I'_{sc} varies with the day of the year and is given by the following relation:

$$I'_{sc} = I_{sc}\left[1 + 0.033\left\{\cos\left(\frac{360n}{365}\right)\right\}\right] \qquad ...(2.12)$$

Where, 'n' is the day of the year

But as the surface is not kept normal to the sun's rays, the zenith angle has to be considered to calculate the instantaneous value of radiation falling on a horizontal surface. From spherical geometry and by integrating, the instantaneous radiation for different time scales (hour or day) can be estimated.

2.8.1 Estimation of monthly average daily global radiation

The value of daily global radiation averaged over a month is convenient to be used to assess the availability of solar energy. *Page's equation* can be used to estimate the monthly average daily global radiation (H_g).

$$H_g = H_o\,[a + b\,(n/N)]\ \text{kJ}\cdot\text{m}^{-2}\ \text{per day} \qquad ...(2.13)$$

Where,

H_o is the monthly average of the daily extra-terrestrial radiation (outside the atmosphere) which would fall on a horizontal surface, at a particular locality, whereas 'a' and 'b' are constants.

Also, 'n' is the monthly average of actual sunshine hours per day at the location and 'N' is the monthly average of maximum possible day length hours per day at the location estimated from the value of ω_s using the relationship, $N = (2/15)\omega_s$

In the Page's equation, H_o is the monthly mean of the values for each day of the month. The global radiation on a particular day is given by:

$$H_o = \frac{24}{\pi} I_{sc}\left\{1 + 0.033\left[\cos\left(\frac{360n}{365}\right)\right]\right\} \qquad ...(2.14)$$

$$\cdot(\omega_s \sin\Phi \cdot \sin\delta + \cos\Phi \cdot \cos\delta \cdot \sin\omega_s)$$

It has been seen that on a particular day of every month (dates close to the middle of the month) H_o is equal to the average value for the month. Hence, the value of H_o on that day can be estimated from the above equation and considered as the average value for that month (Table 2.1).

Table 2.1: Days on which the daily global radiation is equal to monthly average

Month	Date	Month	Date	Month	Date
January	17	May	15	September	15
February	16	June	11	October	15
March	16	July	17	November	14
April	15	August	16	December	10

Example 2.4: Estimate the monthly average daily global radiation on a horizontal surface at Thiruvananthapuram (Latitude 8°31' N, Longitude 76°56' E) during the month of March (Average Sunshine hours per day for the month is 10.5, value of constants, a = 0.37 and b = 0.39)

Solution:

$$H_0 = H \text{ on March 16, value of n for March 16} = 75$$

$$\delta = 23.45 \sin[360\ (284+75)/365] = -2.42°$$

The hour angle corresponding to sunrise, ω_s can be calculated as:

$$\cos \omega_s = -\tan \Phi \tan \delta$$

$$\omega_s = \cos^{-1} [-\tan(76.93) \cdot \tan(-2.42)]$$

$$= 89.02° = 1.554 \text{ radians} \qquad (1° = \pi/180 \text{ rad})$$

Theoretical maximum day length, $N = (2/15)\times\omega_s = (2/15)\times 89.02° = 11.87$ hours

$$H_0 = \frac{24}{\pi}\times 1.367\times 3600\left\{1+0.033\cos\frac{360}{365}\times 75\right\}$$

$$\times\left\{\omega_s \sin(8.52)\sin(-2.42)+\cos(8.52)\cos(-2.42)\omega_s\right\}$$

The value of ω_s in the first term is in radians.

$$H_0 = 34206 \text{ kJ}\cdot\text{m}^{-2} \text{ per day}$$

$$H_g = 34206\,[0.28 + 0.48(9.5/11.87)] = 22718 \text{ kJ}\cdot\text{m}^{-2} \text{ per day}$$

2.9 Measurement of Solar Radiation

The intensity and duration of solar radiation can be directly measured using various instruments. Certain instruments measure beam radiation only and they are called 'Pyrheliometers'. Instruments which can measure global radiation are called 'Pyranometers'. They can also be modified to measure only the diffuse radiations by shading the aperture. Sun shine recorders are used to measure actual hours of bright sun shine in a day.

2.9.1 Pyrheliometers

In order to measure beam radiation, the instrument should allow the passage of direct radiation only into the aperture of the instrument. The receiving surface should be normal to the direction of sun's rays. The instrument essentially comprises of a hollow tube with two axis tracking mechanism. The blackened absorber plate is at the bottom of the tube. Abbot and Angstrom pyrheliometers are the earlier versions. The 'Normal Incidence Pyrheliometer' is a widely used instrument which has a thermopile mounted at the bottom of a brass tube (Fig. 2.15 and Fig. 2.16).

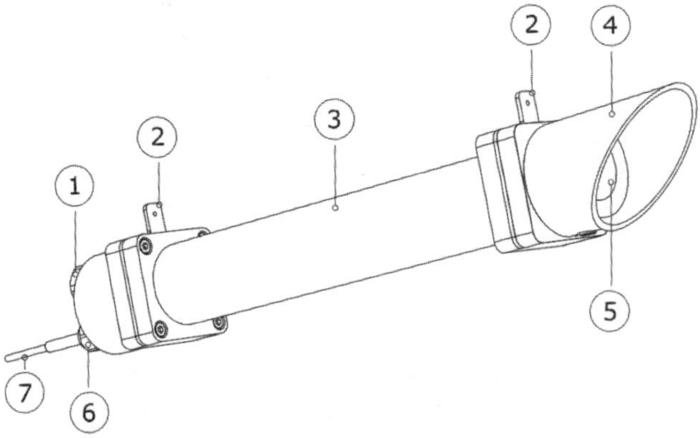

1. Humidity indicator
2. Sights
3. Aperture tube
4. Protection cap
5. Window assembly with heater
6. Cable adapter
7. Power and Data acquisition cable

Fig. 2.15 Constructional details of a typical pyrheliometer

(Author: Hukseflux Thermal Sensors, Description: Drawing of a typical pyrheliometer, showing essential parts, https://commons.wikimedia.org/wiki/File:DR01_pyrheliometer_linedrawing.pdf, CC BY-A 4.0 International, via Wikimedia Commons)

Fig. 2.16 Pyrheliometer

(Author: NREL-U.S. Dept. of Energy, Description: NPC, SERRL, Pyrheliometer Comparisons, https://www.flickr.com/photos/nrel/6214808068/, CC BY-NC-ND 2.0, via Flickr)

2.9.2 Pyranometers

Pyranometers measure the rate of global radiation inclusive of direct and diffuse radiations. The instruments measure radiation over a hemispherical field. Most instruments work on the principle of temperature difference occurring on black and white surfaces measured using a thermopile. 'Eppley black and white pyranometer' is a common instrument in which alternate black and white discs are arranged in a circular fashion. If a shading is provided for preventing beam radiation, they can measure diffuse radiation only. The new generation advanced pyranometers (Fig. 2.17 and Fig. 2.18) working with PV cells are useful to measure global solar radiation with better accuracy.

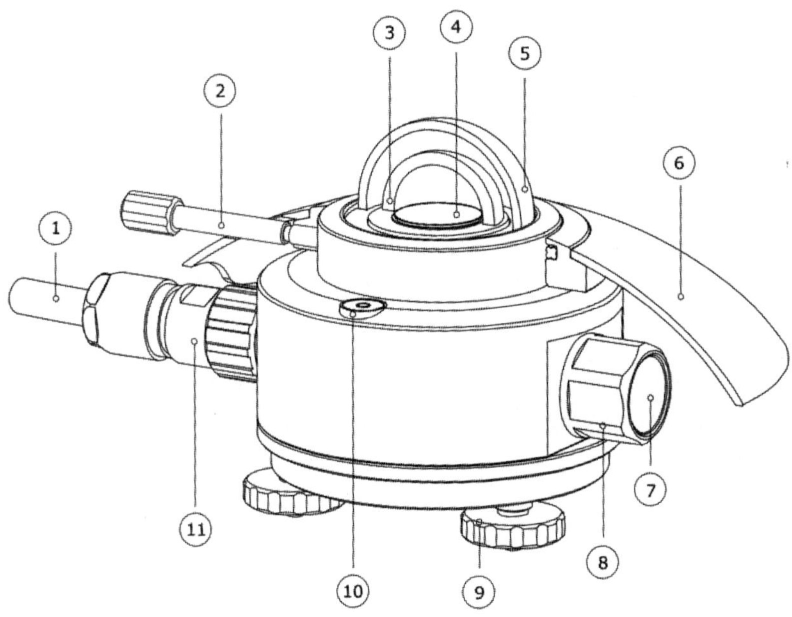

1. Data acquisition cable
2. Thumb-screw for fixing sunscreen
3. Inner glass dome
4. Thermal sensor with black coating
5. Outer glass dome
6. Sunscreen
7. Humidity indicator
8. Desiccant holder
9. Levelling screw leg
10. Bubble level
11. Cable adapter

Fig. 2.17 Constructional details of a typical pyranometer

(Author: Hukseflux Thermal Sensors, Description: Drawing of a typical pyranometer, showing essential parts, https://commons.wikimedia.org/wiki/File:SR20_pyranometer_linedrawing.pdf, CC BY-A 4.0 International, via Wikimedia Commons)

Fig. 2.18 Pyranometer by Hukseflux Thermal Sensors, Netherlands

(Author: Hukseflux Thermal Sensors, Description: Typical pyranometer, for measurement of global solar radiation, https://commons.wikimedia.org/wiki/File:SR20_pyranometer_1.jpg, *CC BY-A 4.0 International, via Wikimedia Commons)*

2.9.3 Sunshine recorder

Sunshine recorders measure hours of bright sunshine by making a burning mark on a recording media placed at the focus of an optical device. Originally, in 1853, John Francis Campbell invented such a device to measure the sunshine hours using a glass sphere set into a wooden bowl with the sun burning a trace on the bowl. Campbell's design was further refined in 1879 by Sir George Gabriel Stokes, using a recorder comprising of metal housing with a card holder set at focal point of the glass sphere, burning a hole through the card. The standard instrument (also known as Stokes's sphere) has a glass sphere of 10 cm diameter and the cards used for recording is graduated in a time scale. The cards used for different seasons are configured differently (Fig. 2.19).

Fig. 2.19 Sunshine recorder (Campbell – Stokes type)

(Author: Daderot, https://commons.wikimedia.org/wiki/File:Unidentified_object_-_Museum_of_Science_and_Industry_(Chicago)_-_DSC06431.JPG, CC0 1.0 Universal Public Domain Dedication via Wikimedia Commons)

CHAPTER 3

Solar Thermal Energy Utilization

A solar collector is an apparatus which utilise the energy of solar radiations to heat a fluid (typically air or water). The incident solar radiations are engrossed by an absorber plate to raise its temperature. The absorbed thermal energy is then transferred to a circulating fluid, which may then be used for suitable applications. The principle of greenhouse effect is often utilized for low temperature applications whereas concentration of radiation is required to get high temperatures.

When the aperture area to the solar radiation is approximately equal to the absorber surface, the arrangement is called *non-concentrating type* of solar collector, commonly known as *flat plate collector*. An efficient solar collector traps the maximum fraction of incident solar radiation and transforms it to heat energy for use with minimum losses.

Essentially, a solar thermal collector is a heat exchanger which absorbs the solar radiations (both beam and diffuse) and convert it into useful heat energy by transmitting to a working fluid for utilization. Though profound geometric differences exist in construction, their ultimate aim remains alike – conversion of solar radiation into useful thermal energy.

The major types of solar collectors for thermal energy utilization are:

(i) Flat plate collectors
(ii) Evacuated tube collector
(iii) Concentrating collectors

The solar thermal energy collectors are utilized mainly in the following applications:

(i) Solar dryers
(ii) Solar chimney
(iii) Solar architecture (space heating and cooling)
(iv) Solar cookers
(v) Solar still
(vi) Solar pond
(vii) Solar cooling
(viii) Solar thermal power production

3.1 Flat Plate Collectors

The elementary working principle of a flat plate collector (FPC) is based on heat transfer, greenhouse effect and fluid dynamics. Assume a flat metal plate with its top surface coated black (Fig. 3.1), exposed to solar radiation. A fraction of the incident radiation is absorbed by the metal plate and the remnant is reflected back into the environment. Due to absorption, the temperature of the metal plate rises. Since the metal plate absorbs solar radiation, it is usually called as the *absorber plate*. Any hot body having a temperature higher than the surroundings tends to lose heat until it reaches a thermal equilibrium with the surroundings. Thermal losses/leakage can take place through any three means of the heat transfer processes:

By *conduction,* if it has physical contact with a cooler body

By *convection,* to the air surrounding the absorber plate

By *radiation,* to the surroundings

The convective heat loss is augmented by air movement over the absorber surface, whereas the heat is radiated at longer wavelengths in the infrared region.

The objective of the technology is to attain high temperatures in the plate, so that it can be utilized for different applications. To attain this objective, primarily, the plate should be capable of absorbing as much radiation as possible. To increase the energy input, the plate should be properly oriented to the incoming solar radiation and it should not reflect any radiation. In addition, the heat losses through conduction, convection and radiation should be arrested. For the heat energy to be utilized, it has to be transferred to a working fluid (*thermal vector fluid*) or media which can be circulated easily. An arrangement of components to achieve the above objectives to a practical level is done in a flat plate collector.

The input energy to absorber plate can be improved by using an appropriate material having high absorptivity as well as low reflectivity. A selective coating for the absorber plate is a better technical option so as to have maximum energy collection. The proper tilt of the collector may be decided considering the location and other related parameters. In northern hemisphere, it is advised to tilt the absorbing surface towards south at an angle equal to the latitude of the location.

To control the heat loss through convection, a transparent cover is provided. The cover should have maximum transmissivity and minimum reflectivity, so as to transmit the maximum radiation on to the absorber plate. Glass is transparent to the visible spectrum of solar radiation and is opaque to infrared wavelengths. This property of glass is the cause for the *greenhouse effect.* While allowing most of the solar radiation to pass through, it does not allow the long-wave radiation from the absorber plate to be radiated out. At the same time, the cover should have minimum absorptivity for such radiation. If a single glass-sheet cover is given, convective circulation of air occurring in between the cover and the absorber plate would heat up the cover. Consequently, heat from the cover will be taken away by the exterior moving air in contact with it and also gets radiated out. Hence, convection losses can be minimized by providing two cover plates or having a vacuum between the absorber plate and the transparent cover. As there is some reflectivity and absorptivity for glass (or other materials used for transparent cover), more than two covers may not be advantageous. It is also possible that heat loss by conduction occur through the sides and bottom of the collector. Hence, proper insulation is provided at the bottom as well as on the four sides.

Heat transfer to the working fluid should be efficient and must be as quick as possible. For this, proper bonding of the conduits (pipes) with the absorber plate is necessary. This can be achieved by welding or clamping the tubes at the bottom or top surface of metal plate. In the case of a water heater, the absorber plate transfers a major part of its energy to water in the tube and therefore, the water gets heated. The two ends of tube are coupled to headers connected to a thermally insulated water storage tank. In a simple *thermo-siphon* type water heater, as the heated water becomes lighter, it rises to the storage tank which in turn brings-in fresh cold water to the tube. This freshly entered water receives heat from the absorber plate and rises, and the progression continues.

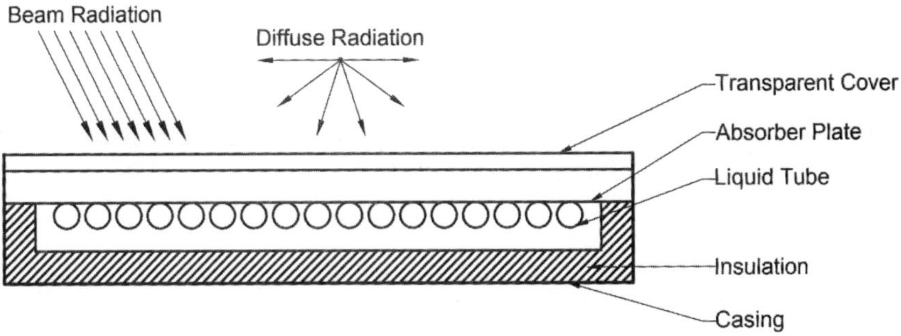

Fig. 3.1 Liquid flat plate collector

A solar air heater is also a similar flat plate collector. There is no need of tubes as in the case of water and wider passages are provided for air movement. The air passage can be simply a parallel plate duct as shown in Fig. 3.2.

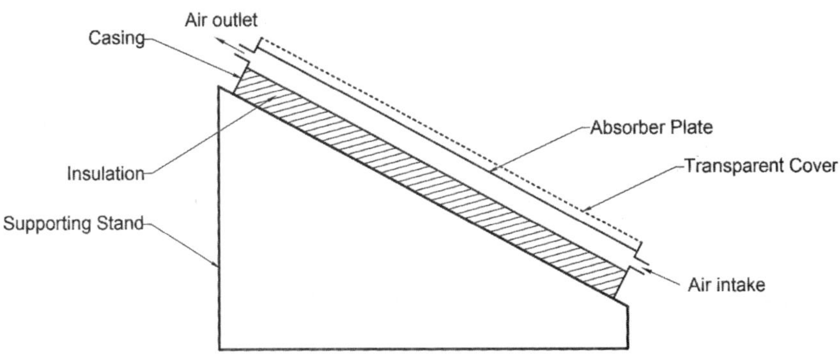

Fig. 3.2 Solar air heater

FPCs are the most important and common type of solar collectors, owing to their simplistic design without any moving parts and reliable working with minimal maintenance. These are always mounted in a fixed-tilt position facing south if installed in northern hemisphere, augmenting the energy absorption. Flat plate collectors could be installed on roof, as a roof itself, or be stand-alone. These are most commonly used for domestic water and air heating purposes, where temperature requirements are in the range of 40 to 100°C.

3.1.1 Energy balance criteria in FPCs

The collector receives energy as the solar radiation enters through the glazing and strikes the absorber plate. Glazing material with superior transmittance (τ) allows more radiation to reach the absorber plate. Only a fraction of this radiation equal to the absorptivity (α) of the absorber plate is absorbed and converted into actual thermal energy. The instantaneous thermal energy (Q_r) acquired by the absorber is determined by,

$$Q_r = A_c \cdot I_T \cdot (\tau\alpha)_{\text{eff}} \qquad ...(3.1)$$

Where, A_c is the collector aperture area, I_T is the solar irradiance, and $(\tau\alpha)_{\text{eff}}$ is the effective optical fraction of the energy absorbed.

As the absorber plate becomes hotter than its surroundings, it begins to lose the heat at a rate of $\frac{T_p - T_a}{R_L}$, where R_L is the resistance to heat loss from the absorber plate (at temperature T_p) to the surroundings at the ambient temperature (T_a). Thus, the net heat flow inside the absorber is given by,

$$Q_{\text{net}} = (\tau\alpha)_{\text{eff}} \cdot I_T \cdot A_c - \frac{T_p - T_a}{R_L} \qquad ...(3.2)$$

$$Q_{\text{net}} = \eta_{\text{col}} \cdot I_T \cdot A_c \qquad ...(3.3)$$

$$\therefore \quad \eta_{\text{col}} = \frac{Q_{\text{net}}}{I_T \cdot A_c} \qquad ...(3.4)$$

Where, η_{col} is the collector efficiency (<1)

The fraction of net thermal energy captured (Q_{net}) at the absorber is transferred to the working fluid (usually water) at temperature T_f and referred as the *transfer efficiency*. Hence, the useful thermal energy output from such collector is given by,

$$Q_u = \eta_{\text{tr}} \cdot Q_{\text{net}} \qquad ...(3.5)$$

In the case of static mass (m) being heated in time (t):

$$Q_u = m \cdot C_p \cdot \frac{dT_f}{dt} \qquad ...(3.6)$$

In terms of dynamic heating, mass flow rate ($\dot{m}$) being heated in unit time:

$$Q_u = \dot{m} \cdot C_p \cdot (T_2 - T_1) \qquad ...(3.7)$$

Where,

C_p is the heat capacity of the fluid at constant pressure,

T_1 is the initial temperature of the fluid at which it enters the collector and T_2 is the temperature at which it leaves the collector.

3.1.2 Application of FPC principle for solar drying

Drying is the most common method for preservation of food materials, whether raw or partly processed. By removal of moisture, we can prevent the microbial growth in otherwise perishable materials. Traditional method of drying of food materials, grain and other agricultural produce are by open sun drying. Sun drying refers to unprotected drying of materials often spread in a thin layer on a surface. Even though this is the cheapest and simple method for moisture removal from products, this has got a number of disadvantages. Contamination of food material from external dirt, dust or birds and rodents is one of the major shortcomings of open sun drying practice. Unexpected rains can damage the entire product and there is no uniformity in drying as some of the particles will be exposed to solar radiation more. The direct exposure to radiation, especially the UV component, can deteriorate the nutritional value and colour of the product.

By the use of a specially made FPC such as a solar air heater, most of the above disadvantages of open sun drying can be overcome. The method incorporating an arrangement for moisture removal of agricultural/fishery/dairy products and partly processed or value-added food products by the direct use of solar thermal energy is called 'solar drying' and the equipment is referred as 'solar dryer' (Fig. 3.3).

The amount of moisture removed from a product depends upon the temperature as well as the air flow rate. At different temperatures, the capacity of air to hold moisture varies. The information on these aspects can be obtained from a 'psychrometric chart'. Air at higher temperatures can hold more moisture and thereby remove it from the drying chamber, if there is sufficient air flow. So, the crux of drying lies in matching the air flow with optimum temperature for a particular product.

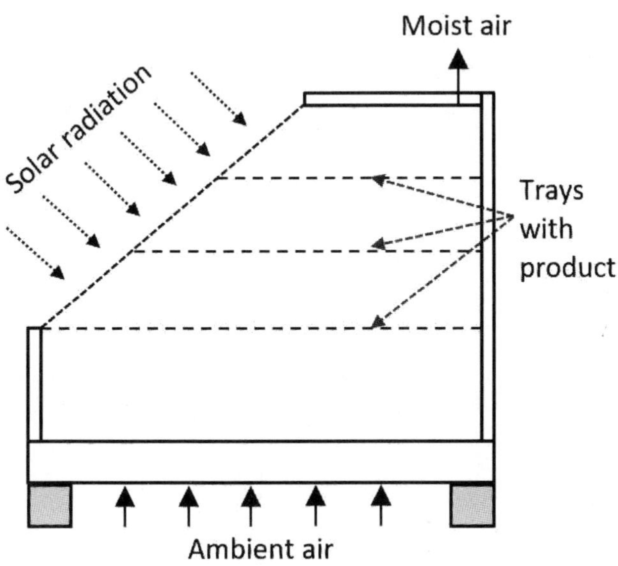

Fig. 3.3 Direct heating type solar batch dryer

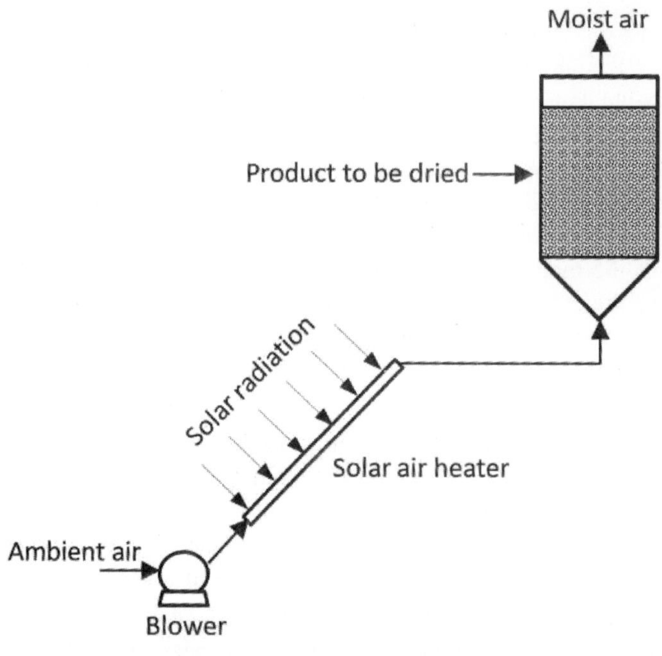

Fig. 3.4 Forced circulation indirect heating type solar batch dryer

(Source: Janjai, 2012)

Fig. 3.5 Solar greenhouse dryer

Solar dryers are generally classified as natural convection dryers and forced circulation dryers. In natural convection type dryers, the heated air moves out of the drying chamber carrying away the moisture due to natural convection. Sometimes, natural convection may not be sufficient to properly drive out the moisture in the material. In such cases, a blower is incorporated in the design to force the moisture laden air out of the drying chamber. They are also referred as 'cabinet dryer' and 'batch' dryer. In cabinet dryer we have a number of trays. In batch dryer, the product is loaded as a batch. Another classification is directly heated and indirectly heated solar dryers. In direct heated dryers, the solar radiation directly strikes the material through a transparent cover. In indirect heating, a solar air heater is integrated with the drying chamber so that the hot air produced by the air heater is circulated through the drying chamber (Fig. 3.4). Another popular concept is the solar greenhouse dryer as shown in Fig. 3.5, in which the high temperature attained in a greenhouse is utilized for drying various products. Instead of glass, UV stabilized polythene sheets are used for these low-cost dryers. A combination of many of these concepts have been made use of in some dryer designs. In forced convection dryers, a blower is an integral part of the drying system.

3.2 Evacuated Tube Collectors (ETCs)

The major factor hindering the overall collection efficiency of solar heat collectors is convective thermal leak through glazing. This can be considerably reduced by evacuating the air gap between the glazing. This problem was resolved significantly as William L. R. Emmet invented a vacuum tube solar thermal collector (US Patent 980505) in 1911.

Evacuated tube collector (ETC) is made of two concentric glass tubes sealed at one end. The space in between is evacuated to a medium to high vacuum. A metal spacer is placed between the tubes to support the tubes and to act as a source for *getter*, typically of barium. Getter is a reactive material placed inside a vacuum system, for effecting and maintaining the vacuum. As the trace gas molecules strike the getter, they react with it chemically or by absorption. Thus, the getter eliminates minor traces of gas from the evacuated space and maintains the vacuum. The advantage in creating and being able to maintain a vacuum is that convection losses between glazing and absorber can be eliminated.

The outer surface of inner tube (radiation absorbing surface) is coated with selective material to limit the reflective and emissive losses. The internal space of the inner tube is provided with the flow paths, through which part of the heat from the absorber surface is transferred to the working fluid. Using non-directional evacuated tube photo-thermal collectors, temperatures between 100 and 200°C could be achieved with selective coating. Fig. 3.6 and Fig. 3.7 illustrate the different possible internal configurations of an evacuated-tube solar thermal collector.

ETCs do not necessarily require a tracking arrangement and can absorb both direct and diffuse radiations. Generally, these collectors are installed in tilted arrays with tube-to-tube clearance equal to the diameter of the external tube. It is desirable to place a reflecting surface (*e.g.,* flat fin, V-channel, tubular or involute reflector, etc.) behind the array to usefully collect some of the solar energy, which may otherwise be lost, thus providing a small amount of concentration. Glass being a brittle and delicate material, is susceptible to splintering and vacuum leak caused by thermal expansion.

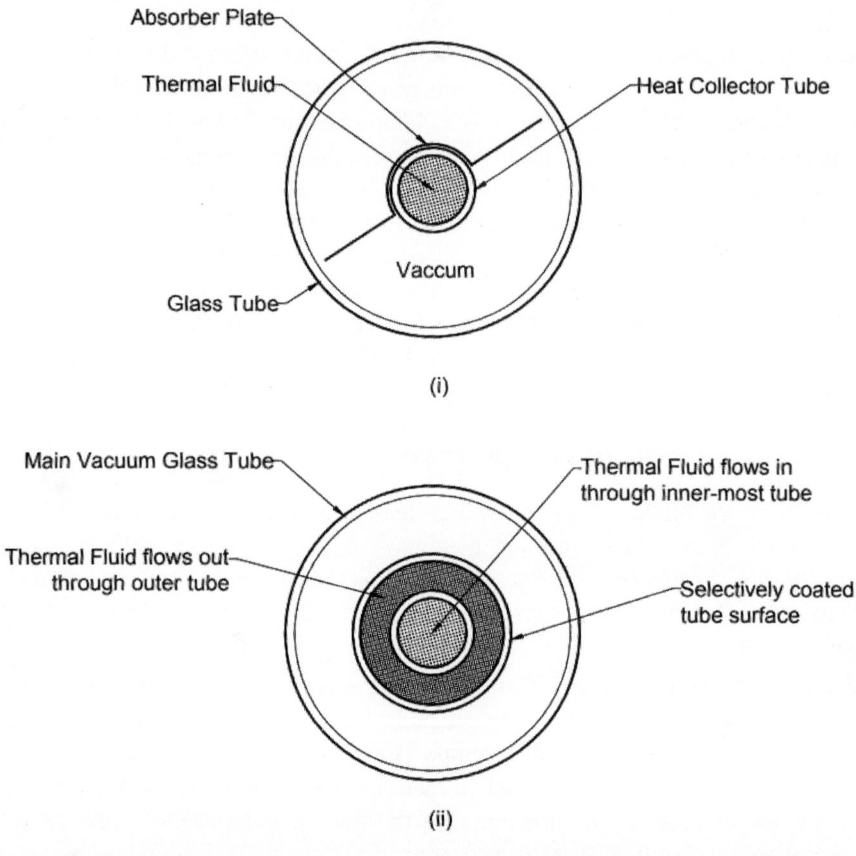

Fig. 3.6 Evacuated-tube solar thermal collector configurations (cross-section)

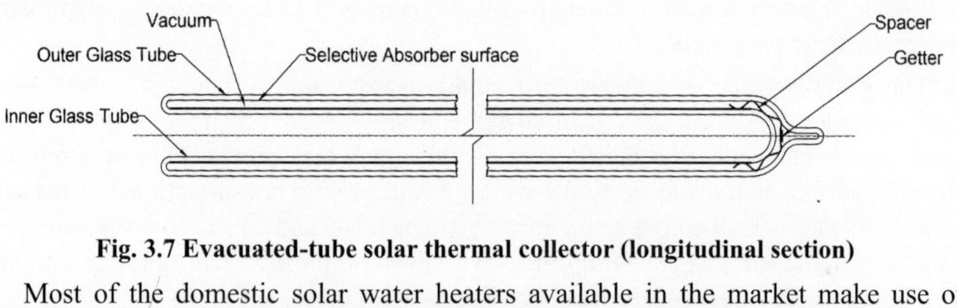

Fig. 3.7 Evacuated-tube solar thermal collector (longitudinal section)

Most of the domestic solar water heaters available in the market make use of evacuated tube collectors as they are low cost. This has also helped the solar water heating technology to become more popular in India and elsewhere.

3.3 Concentrating Solar Collectors (CSCs)

In FPCs, natural solar energy flux (*irradiance*) is not sufficient to generate the thermal gradient to an extent where the working fluid temperatures more than 100°C is a prerequisite. In order to produce electric power using solar thermal energy high temperature levels are to be achieved, for which some kind of concentration or focusing of the incident solar energy flux must be done. Concentration of solar radiation can be achieved using appropriate optical devices like compound reflectors or lenses. Therefore, the quotient of energy flux received at the absorber to that captured by the collector aperture would be greater than one.

Concentrating type solar photo-thermal collectors can be of non-imaging or of the focusing type. Also, based on their construction geometry, they either can be line concentrators (2D) or point concentrators (3D).

The important parameters defining the performance a concentrating solar collector are,

(a) Concentration ratio (C); and

(b) Acceptance angle (θ).

The concentration ratio (C) is expressed as either the ratio of collector aperture area to the absorber area, or as the ratio of the radiation flux density at the receiver to that at the aperture. These two definitions are not equivalent; however, the latter is scientifically superior.

The acceptance angle (θ) is the angle through which the system can be misaimed without affecting the power at the receiver much. In a simpler manner, it is the maximum angle at which the sun's rays may fall over the reflector surface to reach the receiver/absorber surface. There is a theoretical relationship between the concentration and the acceptance angle for the ideal case:

$$C_{\text{ideal}} = (\sin\theta)^{-1} \quad \text{for a 2-dimensional concentrator} \quad ...(3.8)$$

$$C_{\text{ideal}} = (\sin\theta)^{-2} \quad \text{for a 3-dimensional concentrator} \quad ...(3.9)$$

However, it is advisable to have a continuous or discrete tracking system to optimize radiation flux throughout the sunshine hours.

As mentioned earlier, CSCs have major applications in the area of mechano-electric power generation. Therefore, the upper limit of the effectiveness of thermo-mechanical power conversion is the Carnot efficiency (*according to the second law of thermodynamics*):

$$\eta_c = 1 - \frac{T_L}{T_H} \quad ...(3.10)$$

Where, T_L is the temperature at the reservoir and T_H is the temperature of the solar receiver.

To achieve better conversion efficiency, the temperature at the receiver must be as high as possible. In practice, T_L cannot be lower than the ambient temperature, considered to be 23°C or 300K on an average. Therefore, from equation 3.10, T_H should at least be 327°C (or 600K) to maintain the efficiency of 50 per cent minimum.

3.3.1 Concentrating collector designs for power production

It is conceivable that, to run a solar power plant efficiently, concentration of solar radiation becomes the mandate. The most common CSC configurations to energize the stand-alone solar power systems are:

(a) Parabolic trough concentrator;
(b) Parabolic dish (or bowl) concentrator;
(c) Compound parabolic concentrator (CPC)
(d) Heliostats with central receiver
(e) Fresnel lens concentrator

(a) *Parabolic trough concentrator:* Parabolic trough or *linear parabolic concentrator (LPC)* consists of a linear concentrator with a reflective parabolic profile with a receiver tube placed at the optical focus line. The reflecting surface (the trough) trails the sun by single-axis tracking mechanism fixed on a support structure (Fig. 3.8).

The reflecting panel is normally composed of a common glass mirror with suitable thickness. In working, solar flux is focused on the receiver tube and heat absorbed is then transferred to a thermal vector fluid inside. Generally, synthetic oils are employed as thermal vector fluids (instead of water) due to their high temperature stability and lesser volatility.

In a solar thermo-electric power plant, LPCs are connected in series with multiple parallel rows of few hundred metres length called solar field (Fig. 3.8 and Fig. 3.9). The synthetic oil pumped in the warmed up solar field reaches a temperature of about 390°C, which is then fed to the power unit. The part of heat content is transferred to the boiler producing high pressure steam, which in turn drives the turbines generating electricity. After thermal dissipation, the oil (at 290°C) is then recirculated through the solar field to continue the heat transfer cycle in loop. Using LPC solar field, a maximum working temperature of 600°C is achievable, depending on the thermo-physical properties of the vector fluid used.

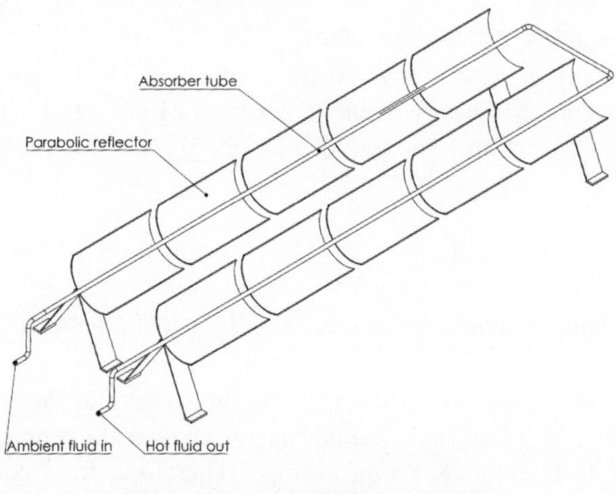

Fig. 3.8 General configuration of LPC

Fig. 3.9 Solar Energy Generating Systems (SEGS) power plants III-VII at Kramer Junction – Mojave Desert, California

(Author: A.R. Akradecki; https://commons.wikimedia.org/wiki/File:Solarplant-050406-04.jpg , CC BY-SA 3.0 Imported via Wikimedia Commons, under GNU Free Documentation License)

(b) *Parabolic dish (or bowl) concentrator:* In this type of CSC configuration, parabolic dish/bowl shaped reflector set to track the sun by turning about two orthogonal axes i.e., *polar* and *declination* axes to harness optimum beam radiation flux. The reflector dish concentrates solar radiation flux to a receiver which is mounted at the focal point (Fig. 3.10 and Fig. 3.11). Resultant high temperature (>650°C) thermal energy absorbed at the receiver is taken up by the vector fluid (helium or sodium vapor) for further power conversion using a suitable heat engine.

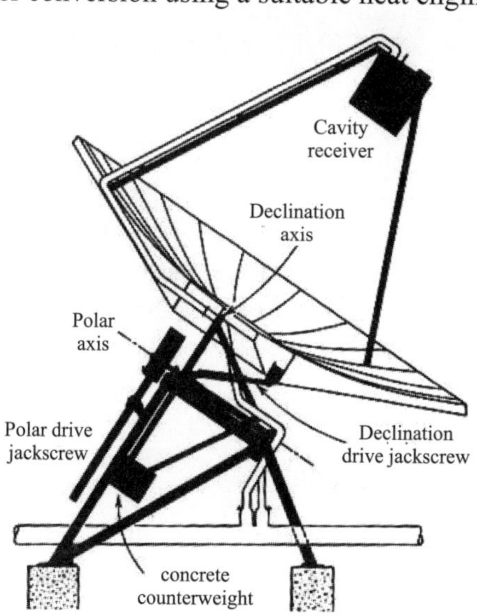

Fig. 3.10 Schematics of Parabolic dish concentrator (PDC) unit

(Reprinted from Solar Energy, Vol. 152, Joe Coventry and Charles Andraka, 2017, Dish systems for CSP, 140-170, Copyright 2021, with permission from Elsevier – License No. 5165761352709)

Fig. 3.11 Parabolic dish solar thermal concentrator

(Photograph courtesy: Schlaich Bergermann Partner, Stuttgart-Germany)

Generally, a solar electric power plant with parabolic dish diameter of 15 m, outputs about 25 to 30 kWe. Employing such multiple collectors as solar field, power plants of higher capacities can be established. Apart from the urban and industrial applications, this technology has immense potential as a decentralized power plant in electrifying distant rural community zones.

In the parabolic dish concentrators, the thermal vector fluid at temperature even higher than 1000°C is practically achievable. With such highly concentrated thermal energy, thermo-chemical dissociation of water can be executed to produce hydrogen. Its environment friendly working principle became the key factor of interest in Europe, since the hydrogen economy has become a mandate in EU's sustainable energy policy. Utilization of solar energy for sustainable production and the uniqueness of hydrogen as a clean fuel would transcend the stressed fossil fuel economy in the near future.

(c) *Compound parabolic concentrator* (CPC): In contrast with the other CSCs those establish a focal line or point, the CPC collector involves two parabolic reflectors that funnel down the solar radiation from an aperture to the receiver. The right and left reflector halves are the exact mirror set (hence called *compound*) with the edges of the receiver being the foci of the opposite parabola as shown in Fig. 3.12.

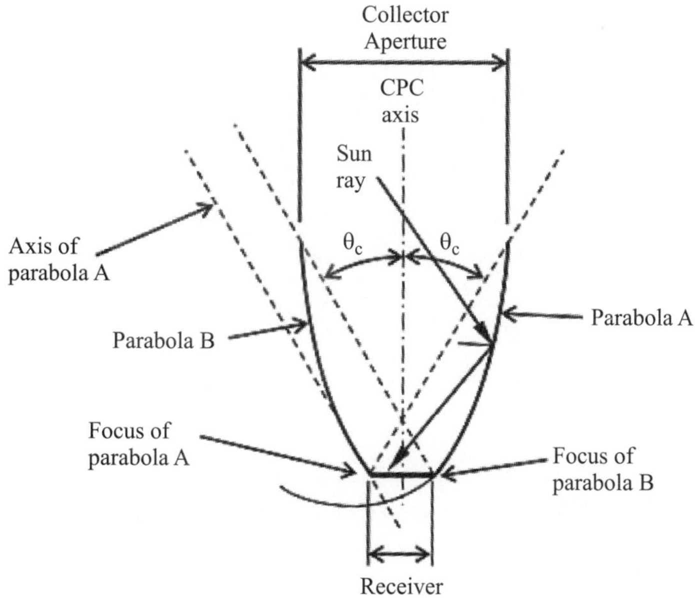

Fig. 3.12 Compound parabolic concentrator (CPC)

(*Reprinted from Advances in Solar Heating and Cooling, S. A. Kalogirou, 2016, 4 – Non-tracking solar collection technologies for solar heating and cooling systems, 63-80, Copyright 2021, with permission from Elsevier – License No. 5165830878000, CC BY 4.0*)

It is evident that any solar ray, beam or diffuse, incident on the CPC's aperture within the acceptance angle would always reach the absorber; while the remnants would bounce back to and fro within the compound space and re-emerge through the aperture. CPCs are also termed as *non-imaging* concentrators, since they do not create a focal point or line or clearly defined images of the solar disk on the absorber surface.

In practice, CPCs can be designed both for either low-concentration with large acceptance angles or high-concentration with small acceptance angles. CPCs with low-concentration ratios ($C \approx 2$) and with east–west tilt can be operated as static devices throughout the year or at most with seasonal tunings only. Distinct from other concentrators, CPCs are capable of collecting both the beam and a large portion of the diffuse radiation. Moreover, they do not involve highly sophisticated optical surfaces and therefore can tolerate dust and degradation. A characteristic module consisting of CPCs in array is shown in Fig. 3.13.

A heat receiver or absorber tube is welded at the lowermost part of CPC trough, and the entire module may be enclosed in a glass cover. Absorber component could also be of various configurations, such as FPC or cylindrical or U-tubes or integrated evacuated tubes collectors. CPCs have considerable potential for water heating close to the boiling point leading to low-pressure steam applications.

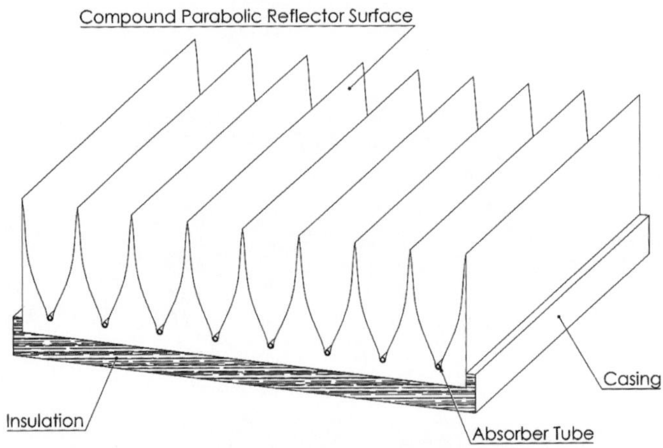

Fig. 3.13 A common CPC module configuration

CPCs offer a geometric concentration ratio (CR_g) in the range of 1.5 up to 10 times the solar radiation without tracking the Sun. The geometric concentration ratio of a CPC is a function of the acceptance angle (θ_{accept}), and is given by:

$$CR_g = \frac{1}{\sin\left(\frac{1}{2}\theta_{accept}\right)} \quad ...(3.11)$$

A linear equation was put forward by U.S. National Renewable Energy Laboratory (NREL) to determine the instantaneous efficiency of an east–west orientated CPC. The following equation is valid for CPCs with a concentration ratio of 5 to 6 and an acceptance angle of about 19°.

$$\eta_{CPC} = 0.73 - 0.64\left[\frac{T_r - T_a}{I_a}\right] \quad ...(3.12)$$

Where,

T_r = Average temperature of receiver (K)

T_a = Ambient temperature (K), and

I_a = Global solar irradiance entering the collector aperture, Wm^{-2}

(d) *Heliostats with central receiver:* This particular CSC configuration, also known as *central tower system* receives the radiant flux from the large solar field of flat reflecting panels called *heliostats* (Fig. 3.14). Heliostats track the sun by rotating on two axes and concentrate the sunlight towards a sole receiver located at the centre of the solar field. The receiver is installed at the apex of a high tower, inside of which a thermal vector fluid flows to absorb the solar heat.

Similar to the *parabolic dish concentrator*, the thermal energy harnessed by this process can be utilized in diverse aspects and profoundly for electric power generation. Central receivers are distinguished by large power levels (1 to 500

MW) in which the thermal vector fluid can reach temperatures in the range of 540 to 840°C. As discussed earlier, such high temperatures assist in achieving high conversion efficiencies (*Carnot efficiency*) in the electricity production via thermo-mechanical power conversion using *steam-turbine* generators. Fig. 3.15 shows the Crescent Dunes Solar Energy Project installed near Tonopah (USA) with a capacity of 110-MW.

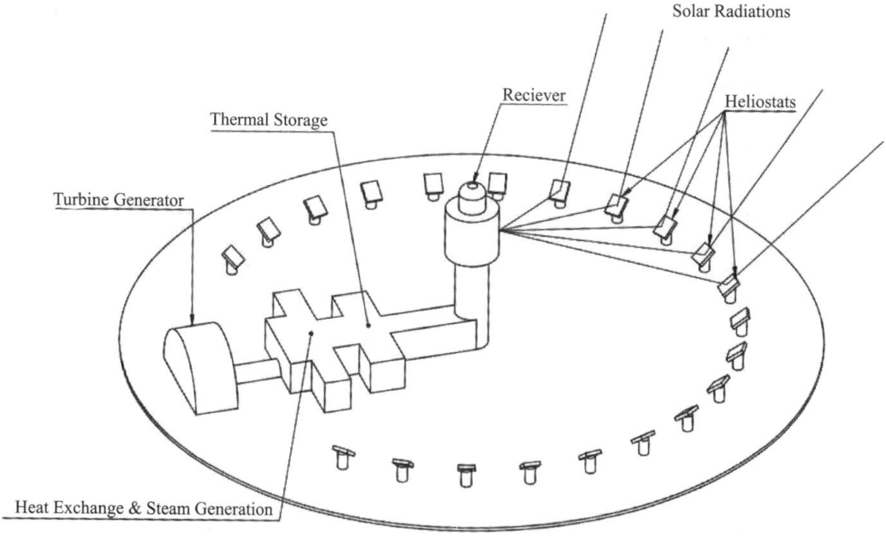

Fig. 3.14 Solar electric power plant – Heliostats with central receiver configuration

Fig. 3.15 Crescent Dunes Solar Energy Project

(Author: Amble, https://en.wikipedia.org/wiki/Crescent_Dunes_Solar_Energy_Project#/media/File:Crescent_Dunes_ Solar_December_2014.JPG , CC BY-SA 4.0 via Wikipedia)

Since the central receiver reaches such extreme temperatures, vector fluid must have excellent thermo-physical stability in order to transport the energy safely to the power plant. Over a long period, varieties of thermal vector fluids (e.g., water, air, melted salts) have been investigated for stability and conversion efficiency. However, at present, the molten-state salt mixture [60% Sodium nitrate ($NaNO_3$) + 40% Potassium nitrate (KNO_3)] technique stands out as the best option available. It has advantages of reaching a higher processing temperature (up to 565°C) and provides more flexibility with a thermal energy storage tank. It should be noted that, beyond the temperature of 565°C the nitrates start to transform into nitrites, triggering severe corrosion problems. But still, the upper limit of the working temperature is significantly higher than that allowed by synthetic oil, i.e., 390°C.

(e) *Fresnel lens concentrator:* Fresnel lenses have been used in solar thermal energy systems as well. By dividing the lens into a set of concentric annular portions, as shown in Fig. 3.16 and Fig. 3.17, these solar collectors require less material and space than a traditional spherical lens. Despite the fact that such canted facets are brought to the plane, there are gaps between them. The volume is drastically reduced while maintaining optical qualities that are nearly identical to those of a regular lens. The better the optical approximation achieved, when the number of facets is large. There should be more than 1,000 sections per centimetre in a high-quality linear Fresnel lens. Flatness results in significant material savings and lowering of production costs.

The sharpness of facets can limit the efficiency of Fresnel lenses. Any ray that strikes the rear side of a facet, as well as the tip or valley of a facet, is not aimed at the receiver. The Fresnel collector, or any other lens system, requires at least one single-axis tracking device to keep the incident light rays normal to the lens aperture in order to keep the refracted rays focused on a receiver that is fixed with regard to the lens.

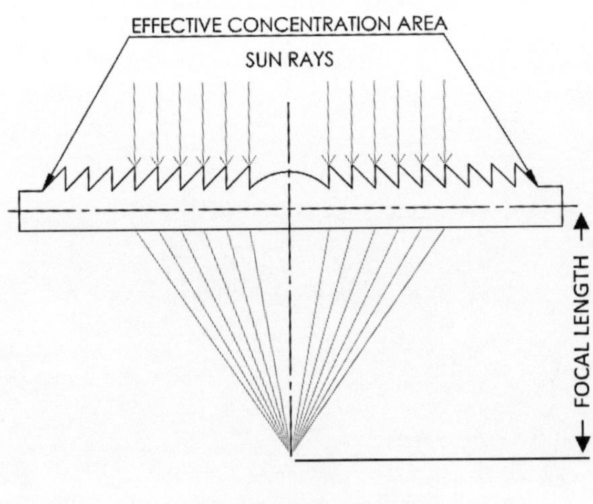

Fig. 3.16 Fresnel lens

A holographic plate, like a Fresnel lens, can be formed to produce a similar optical effect. The holographic technique has the advantage of being able to diffract and scatter light at the same time, creating a rain-bow of focused light with each region of the spectrum directed to a collector designed for that colour range. When employing photovoltaic cells, this offers a substantial advantage.

Fig. 3.17 Fresnel lens at Centro Nacional de Investigaciones Metalúrgicas (CENIM), Spain

(Reprinted from Renewable and Sustainable Energy Reviews, W.T. Xie et al., 2016, Fig. 8 - Fresnel lens installation at CENIM, Concentrated solar energy applications using Fresnel lenses: A review, 15(06):2588-2606, with permission from Elsevier – License No. 5200710255182, https://www.sciencedirect.com/science/article/abs/pii/S1364032111001341?via%3Dihub)

3.4 Solar Chimney

The Solar Chimney Power Plant (SCPP) is a member of indirect solar energy conversion technologies. It utilizes the natural phenomenon involving solar heating of the earth's surface and thereby the adjacent atmospheric air heating. This expanding hot air (of low density) develops an upward buoyancy force causing air currents due to mass density gradient within the Earth's atmosphere. This upward buoyancy force associated with the planet's revolution is so enormous that can initiate intense *tropical cyclogenesis*. Scientifically, such an energy-intense phenomenon can be superimposed with a control mechanism for sustainable power generation.

SCPPs generate electric power by virtue of such hot air buoyant streams. They consist of a hollow tower with the base having a wide spread greenhouse, usually circular in shape that shields a remarkable ground surface as shown in Fig. 3.18. Practically, the air heated by the greenhouse effect, surges inside the chimney due to two physical phenomena, namely:

The air rises by floating (based on the phenomenon that low density hot air tends to rise high)

The air rises due to the pressure gradient between the base and the top of the tower

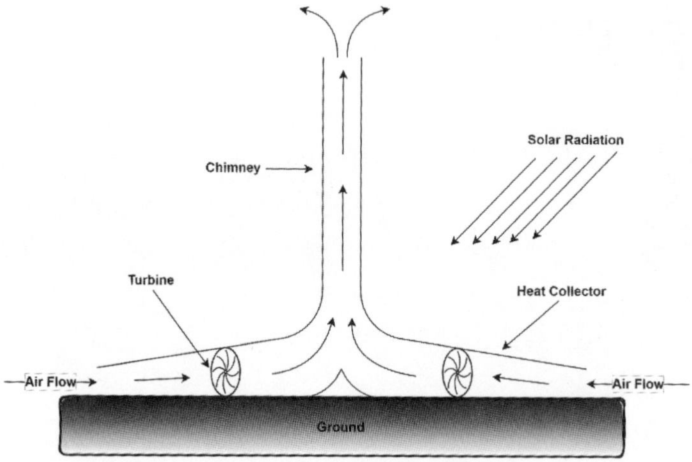

Fig. 3.18 Solar chimney power plant (SCPP) configuration

As it rises inside the chimney, the hot air flow continues to accelerate and spins a stack of turbines installed at the inner base of the chimney to generate electricity. The turbo-generator transforms the kinetic energy and the air potential into electrical energy, by mechanical means (Fig. 3.19).

The solar heat collector (greenhouse) in this case can have plastic or glass glazing. The pilot plant studies have suggested that glass is a better glazing material, owing to its excellent weather resistance. The performance of SCPPs is greatly influenced by the chimney height; higher the chimney – better is the performance. Provided with sufficient chimney height (100 m and more), the SCPP could work 24 hours, 7 days, generating a maximum of its energy output capacity during the summer, when energy consumption is at peak. Due to the natural pressure gradient and ground heat release during the night, SCPPs can function continuously.

Fig. 3.19 Pilot SCPP (50-kWe) with 194m tower height, Manzanares, Spain

(*Photograph courtesy: Schlaich Bergermann Partner, Stuttgart-Germany*)

3.5 Solar Architecture (Space Heating and Cooling)

Proper architecture of buildings is a key aspect in reducing overall energy use. The following considerations must be made in order to effectively use solar thermal energy for room heating and cooling:

(a) *Exposure control:* The orientation of the building must be determined by the insolation characteristics in the area. To offer ambient heating, equator-ward facing windows with overhangs to screen the sun in the summer can be used extensively. Heat loss can be reduced or eliminated by reducing or eliminating pole-facing windows. In the summer, deciduous trees can give shade while permitting insolation in the winter.

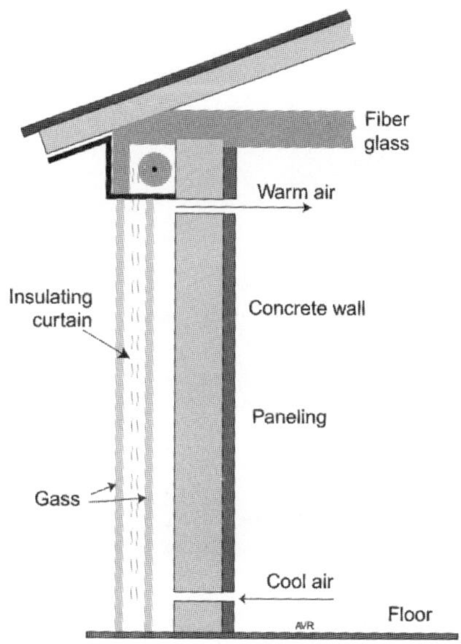

Fig. 3.20 Trombe heat collection wall (passive system)

(Reprinted from Fundamentals of Renewable Energy Processes (3rd Ed.), Aldo da Rosa, 2013, Figure 12.5 Heat storing wall (Concept Construction, Ltd.), Chapter 12 - Solar Radiation, pp. 485-532, with permission from Elsevier – License No. 5201170292504 https://www.sciencedirect.com/science/article/pii/B9780123972194000126)

(b) *Heat Storage:* In the winter, structures that are exposed to the sun can store heat. Also, in summer, the heat saved could be used to pump cold air by establishing natural convection currents. Roof ponds can be used for both heating and cooling purposes. Fig. 3.20 is a detailed rendering of the well-known *'Trombe'* wall, which serves as a heat collector. It comprises of a 25 cm thick concrete wall with a 5 cm air space in front of which a glass pane has been laid. During the summer, the warmed air is vented outdoors, creating a circulation draught that draws cooler air in from the house's pole-ward side.

(a) *Circulation:* Natural circulation set up in a building by convection currents adjusted via vents is referred to as a passive system, where no mechanical components such as a pump is employed to control heat transmission. On the other hand, when an external solar collection and heat exchange system is employed with the forced air circulation system coupled to an auxiliary heating element, the arrangement is referred as an *active* system (Fig. 3.21).

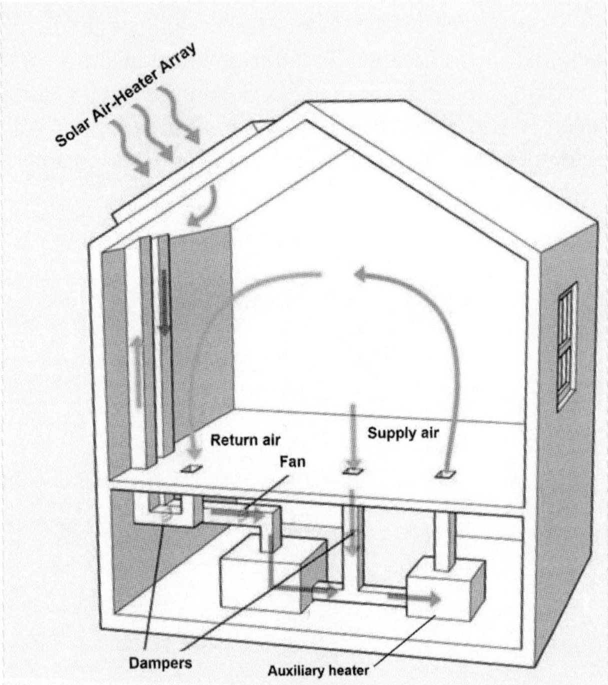

Fig. 3.21 Active Solar space heating system

From a health standpoint, circulation is critical; attempts to save energy by sealing dwellings may result in an increase in the quantity of radon and carbon dioxide that emerges from the earth in some locations but is generally released through leaks. Other unwanted gases build up, one of which is water vapour, which causes excessive wetness. Furthermore, toxic substances must be ventilated. This is especially true of formaldehyde, which is commonly found in paints, varnishes and carpets.

Air-to-air heat exchangers or recuperators are used to reduce heat losses associated with air renewal. The outflowing air pre-heats the incoming fresh air in this way and approximately 70 per cent of the heat may be recovered using this method.

(d) *Insulation:* Thermal power (P), transferred by a material is proportional to the thermal conductivity (λ), the area (A), and the temperature gradient dT/dx:

$$P = \frac{\lambda A dT}{dx} \quad ...(3.13)$$

In the S.I., the unit of heat conductivity is $WK^{-1}m^{-1}$. Under constant temperature gradient, $dT/dx = \Delta T/\Delta x$ where ΔT is the temperature difference across a material Δx units thick. We have,

$$P = \frac{\lambda A dT}{\Delta x} = \frac{A\Delta T}{R} \qquad \text{...(3.14)}$$

Where,

$$R = \frac{A\Delta T}{P} \qquad \text{...(3.15)}$$

R has units of $m^2 KW^{-1}$. Insulating materials are rated by their *R*-values. Fiberglass insulation, 8 cm thick, for instance, is rated *R*-11 (in the American system, multiply by 0.178 to convert into S.I. system).

Example 3.1: Consider a house with interior temperature of 20°C and an attic temperature of 0°C. The ceiling has an area of 100 m^2 and is insulated with *R*-11 material. What will be heat loss through the attic?

Solution:

In the S.I., *R*-11 corresponds to $11 \times 0.178 = 2\ m^2KW^{-1}$. Thus, the heat loss under the assumed 20 K temperature difference would be:

$$P = \frac{100 \times 20}{2} = 1000 \text{ watts}$$

As a result, if the only heat losses were through the ceiling, it would only take a small amount of energy to keep a house warm. Of course, there are significant losses via walls and, particularly, through windows, which could be reduced by appropriate thermal insulation.

3.6 Solar Cookers

Solar cooker is a device that reflects, absorbs the solar radiation and retains the heat inside a confined insulated space to cook the food materials. Depending upon the reflector and absorber characteristics and with various optical arrangements, solar cookers are broadly classified into following kinds:

(i) Box-type solar cooker
(ii) Panel-type solar cooker
(iii) Parabolic dish solar cooker
(iv) Vacuum tube solar cooker

3.6.1 Box-type solar cooker

The exterior box with thermal insulation, interior cooking box or tray, double glass aperture top, mirror with angular adjuster and heat absorbent cooking vessels are the main components of a box-type solar cooker (Fig. 3.22). In most cases, the outside box is made of G.I. or aluminium sheet, or fibre reinforced plastic. The interior absorber box or tray, on the other hand, is made of black-painted aluminium sheet (or coated with selective material) to absorb solar radiation and transfer heat to the cooking containers. The inner box's aperture is glazed with a double glass cover (approximately 20 mm apart in a single frame). The air space between the glass panes acts as an insulator, preventing heat loss by convection. To prevent conductive heat loss, the area between the outer box and the inner absorber (sides and bottom) is filled with insulating material such as glass wool or polyurethane foam. Furthermore, the top cover is hinged with an adjustable plane mirror to reflect solar energy onto the absorbing region.

The cooking containers are usually made of aluminium and coated black on the outside to absorb direct solar energy. Pulses, rice, vegetables, meat, fish and other foods can all be cooked in the solar cooker. Cooking time is determined by the type of food and the amount of sunlight available. On a clear sunny day, the inside space can reach temperatures of 80 to 140°C, with cooking times ranging from 45 to 120 minutes for the majority of food items.

Fig. 3.22 Box-type solar cooker
(Courtesy: VeSat Solar Energy Systems, Coimbatore)

3.6.2 Panel-type solar cooker

It is a simplified and inexpensive version of solar cooker, in which an enclosed heat absorbent cooking vessel is shielded by compound reflector panels (relatively larger reflector area than box-type). The compound multi-reflector panel exhibits combined effects of both parabolic and box type solar cookers. The cooking vessel is generally contained in a special low thermal expansion borosilicate glass shell to reduce heat losses and prevent heat losses by wind effect (Fig. 3.23). This type of solar cookers can reach cooking temperatures in the range of 95 to 125°C.

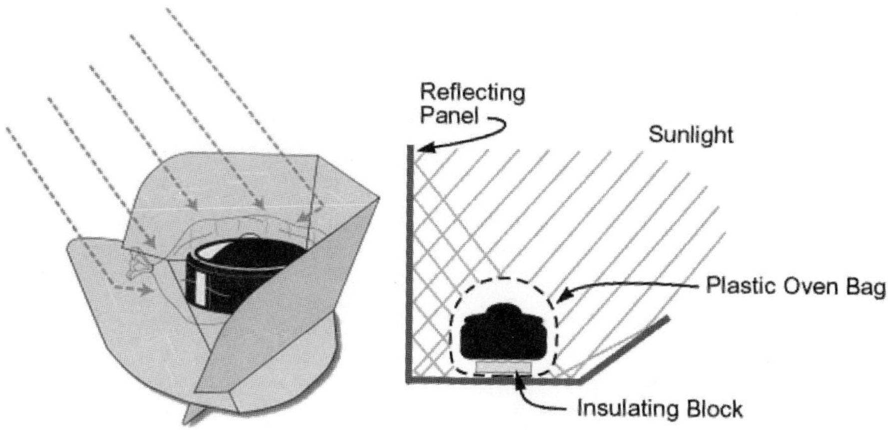

Fig. 3.23 Panel-type solar cooker
(Author: Tom Sponheim, https://solarcooking.fandom.com/wiki/Introduction_to_solar_cooking, CC BY SA 3.0 via Fandom, Inc., under GNU Free Documentation License)

3.6.3 Parabolic dish solar cooker

It utilizes a parabolic concentrator (usually, made of aluminium reflector sheets mounted on a steel structure) focusing the solar radiations towards a single point at which a cooking vessel (heat receiver) is mounted (Fig. 3.24). These type of concentrating solar cookers are also known as *solar oven* as it is possible to reach the same temperature as a conventional oven (about 200°C) at shorter cooking times.

Fig. 3.24 Parabolic solar cooker
(Author: Iweargreen, https://solarcooking.fandom.com/wiki/Introduction_to_solar_cooking, CC BY SA 3.0 via Fandom, Inc., under GNU Free Documentation License)

3.6.4 Vacuum tube solar cooker

The vacuum tube solar cooker takes the advantages of evacuated tube solar collectors as the cooking shaft is backed by the parabolic trough concentrator. The specialized large diameter evacuated collector tube is mounted along the linear foci of the concentrating trough. A semi-cylindrical heat absorber tray loaded with the food items is inserted into the evacuated tube exposed to solar radiation (Fig. 3.25). As a combined effect of concentrated solar flux and minimal heat loss through receiver, vacuum tube solar cooker heats up quickly to higher temperatures (150 to 220°C), thereby speeding up the cooking process.

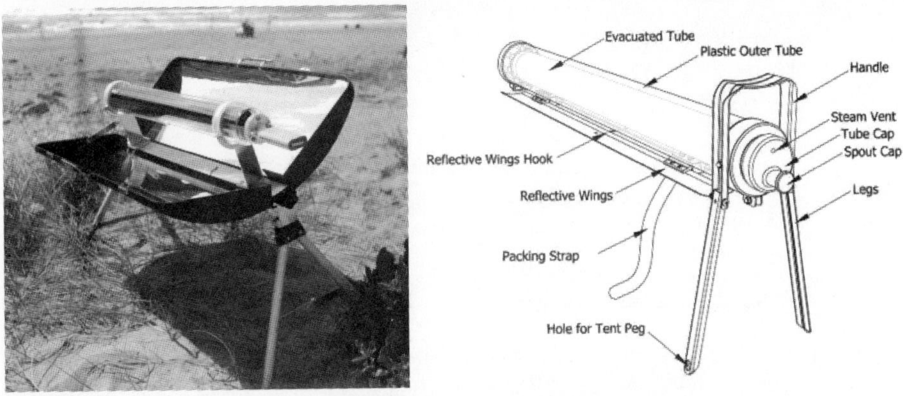

Fig. 3.25 Portable vacuum tube solar cooker
(Author: Paul Hedrick, https://solarcooking.fandom.com/wiki/SLiCK_SM70, CC BY SA 3.0 via Fandom, Inc., under GNU Free Documentation License)

3.7 Solar Still

Distillation is a method of purifying certain components of a solution using variations in their volatilities. Distillation is carried out by evaporating the solvent in a specific region of the device and collecting the condensate to produce pure solvent when the solutes have considerably lower volatilities than the solvent. Solar distillation occurs when radiant heat is used in the process, and such a device is known as a solar still. Water desalination has been the most extensive use for solar stills.

They have an advantage over traditional systems in the distillation of simple fluids like saline or brackish water because they do not require a fuel nor do they have moving parts. Most crucially, these systems may be constructed in isolated locations to meet the freshwater demands of smaller villages that are not connected to the power grid.

There are several types of solar stills, each with its own distinct geometry and structural features. The operating principle is identical for most types but there are distinguishing characteristics for each type. The three distinct elements which distinguish each design are the solar absorber, evaporator and condenser. In the cabinet type solar still shown in Fig. 3.26, the absorber basin also serves as an evaporator and is glazed with inclined glass or clear polycarbonate sheet to provide a greenhouse effect.

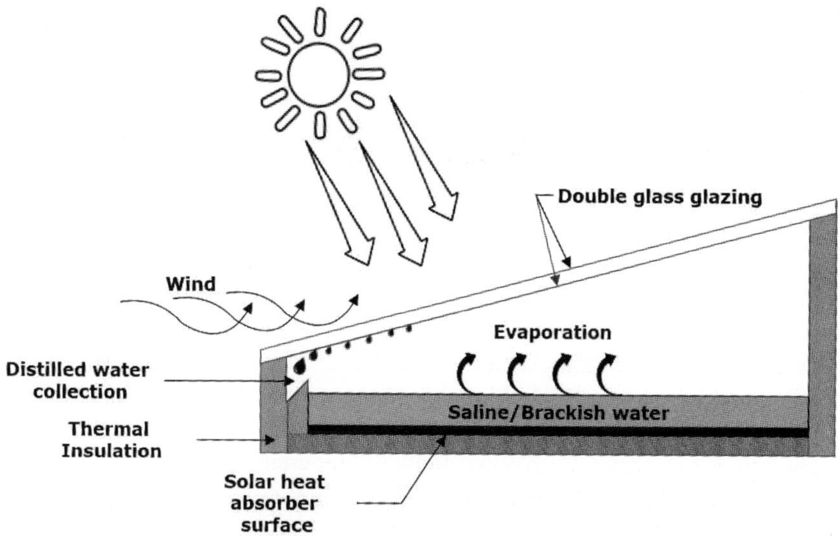

Fig. 3.26 Basic operation of a solar still.

Because it is in close interaction with the atmosphere, the temperature of the glazing is always lower than the temperature of the water vapour. The condensed water droplets flow off into the collection drain at the bottom edge due to the tilted glazing cover. Impurities such as salts and heavy metals are precipitated and harmful microorganisms are destroyed. The maximum temperature within the basin type solar still is always below 80°C, preventing the solution from boiling. It is the most typical passive solar still, with a single basin that runs only on solar heat.

The most important factor influencing distilled water production is the amount of solar energy falling on the still. The ratio of amount of energy used in vaporising water in the still (Q_e, $J\,m^{-2}\,day^{-1}$) over the latent heat of vaporisation of water (L, $J\,kg^{-1}$) gives the daily distilled-water output (M_e, $kg\,m^{-2}\,day^{-1}$). The quantity of energy used in vaporising water in the still (Q_t, $J\,m^{-2}\,day^{-1}$) is a measure of the solar still efficiency (η). Thus,

$$M_e = \frac{Q_e}{L} \quad \text{...(3.11)}$$

and

$$\eta = \frac{Q_e}{Q_t} \quad \text{..(3.12)}$$

Single-basin solar stills have a typical efficiency of around 60 per cent. Distilled water production in a solar still is determined by the amount of solar energy available and the ambient temperature. For example, in sunny places such as Australia, South Asia and the Middle East, output rates for a square metre can range from about 6 litres per day in the winter to over 15 litres per day in the summer.

3.8 Solar Pond

Water heated by the sun can be utilised to create energy using the turbine operation principle used for OTEC. The solar ponds can produce quite high temperatures. The deeper layers of water in a solar pond with a blackened bottom surface warm up and rise to the surface due to their lower density. As a result, internal mixing occurs, neutralising any temperature differential formation. The heat loss through surface evaporation limits the temperature rise in this situation.

Salt gradient solar ponds can offset the temperature loss due to convection currents within pond water. If a vertical salinity gradient is established in the pond, the deeper layers will contain more salt and become more denser (Fig. 3.27). The salinity gradient can overcome the quick convective mixing as the water in bottom layers have a high salt concentration and are heavier. This allows for the prevention of convection, resulting in higher temperatures of up to 80°C in the deeper layer. A heat engine working on OTEC principle has a Carnot efficiency of about 20 per cent when working against a 20°C cold sink, with an attainable practical efficiency of about 10 per cent.

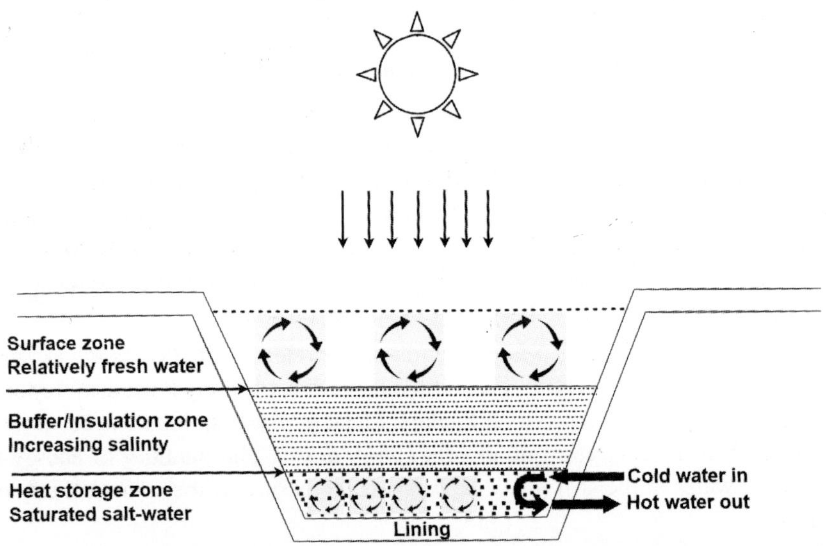

Fig. 3.27 Schematics of solar pond for power generation

3.9 Solar Refrigeration

In a conventional refrigeration system, a vapour compression system is utilised in which the working fluid is compressed with a hermetically sealed compressor using electrical energy. The solar refrigeration system is different from them as a vapour absorption system is used, in which the working fluid (refrigerant) is dissolved in an absorber. There are two types of absorption cooling systems commonly employed for industrial as well as domestic applications. One type makes use of Lithium Bromide (LiBr)-Water combination and the second one is Ammonia-Water combination. The

LiBr-H_2O system is considered more suitable and low-cost due to lower operating temperature.

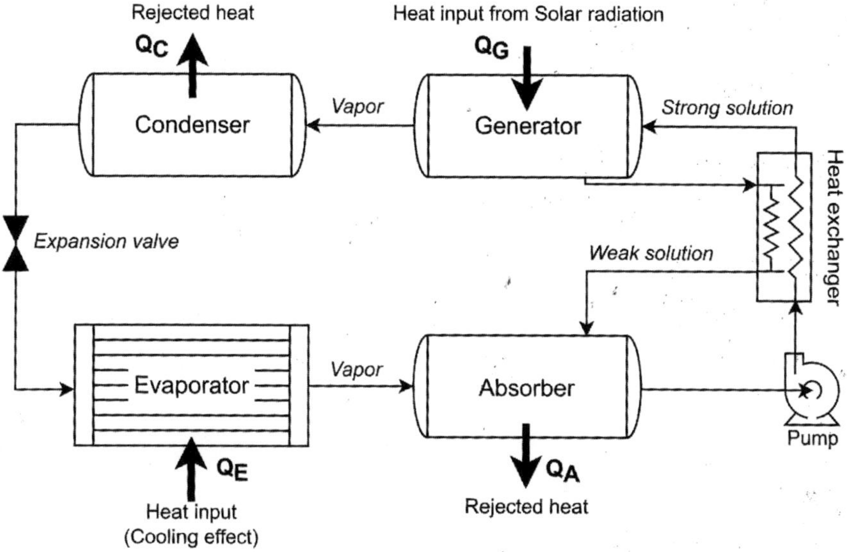

Fig. 3.28 Flow diagram explaining vapour absorption refrigeration system

The mixture or solution of absorbent and refrigerant is contained in the 'absorber' unit, for example, LiBr and water. This strong solution rich in refrigerant is pumped to the generator, where it passes through a heat exchanger. The temperature of the solution increases as it absorbs heat. The generator heats the absorbent-refrigerant mixture using the external solar energy (from a flat-plate collector). The solar heat will boil the solution converting water to vapour which flows to the condenser. The solution retained will be a weak solution (LiBr with some water) and is returned to the absorber through the heat exchanger (Fig. 3.28).

There is a condenser where the water vapor coming from the generator gets condensed and the condensate flow to an evaporator through an expansion valve. In the evaporator the refrigerant evaporates at a low pressure and absorbs the heat from the space which is to be cooled. The vaporized refrigerant then passes to the absorber, where it joins with the weak solution of LiBr, thereby making it a rich mixture for the cycle to repeat.

CHAPTER 4

Solar Photovoltaic Energy Conversion

The term '*photovoltaic*' is a combination of the Greek word '*phōtos*' – representing light, and the universal unit of electromotive force – '*volt*'. Technically, the word *photovoltaic* denotes the phenomenon effecting direct conversion of light into electricity. In this chapter, the word 'light' profoundly indicates '*sunlight*', or more scientifically as '*solar radiation*'.

In 1839, a young French physicist Alexander Edmond Becquerel, observed the photovoltaic effect for the first time. He discovered that, on exposure to the solar radiation, an electric voltage was induced between two metal electrodes immersed in an electrically conductive solution. It took decades to acquire a better understanding of the phenomenon. The progress in its development gained momentum in late 1870s, when scientists found the semiconductor properties in selenium. They observed the presence of light could change the direction of the current flowing through selenium connected to a battery. This led to the invention of the first solid-state selenium solar cell, by an American scientist Charles Fritts in 1883. But the practical utility was still in question, as Fritts estimated that the solar cell could convert not more than 1 per cent of the sunlight into electricity.

It was not until 1904, when Albert Einstein published a paper on the photoelectric effect, that the general scientific community started looking at photovoltaic concept as a revolutionary field of research. The first silicon based solar cell was built in 1954, which converted around 6 per cent of the incident solar radiation into electricity, demonstrating it as a feasible source of energy for the first time. Today, modern photovoltaic cells have reached the conversion efficiencies (laboratory experimental) ranging between 12.6 to 47.1 per cent, depending on their semiconductor junction, optical and structural characteristics.

In this chapter, basics of solar photovoltaics, construction, working principle, types of solar cells and their performance characteristics are discussed in detail.

4.1 Photovoltaic Effect

In terms of experimental physics, when two dissimilar materials in close contact produce an electric potential difference when struck by light or other radiant energy, they are demonstrating the photovoltaic effect. Electrons in crystalline silicon or germanium semiconductors are usually not free to move from one atom to the next within the crystal. The energy required to free certain electrons from their atomic bonds is provided by incident solar radiation. These charged free electrons can more easily pass the junction (the boundary between two different semiconductor material regions) between two dissimilar crystals in one direction than they can in the other. As a result, one side of the junction becomes negatively charged and has a negative electric potential in relation to the other. In semiconductor physics, such a junction is referred as a '*p–n junction*' (*p*-positive, *n*-negative). This is similar to one of a battery's two electrodes having a negative potential in relation to the other due to an active electrolytic chemical process. The electron flow in a solar cell, on the other hand, is solely due to solar radiation striking the cell (Fig. 4.1). As long as light is incident on the two materials, the photovoltaic effect can continue to excite the electrons within the cell. This current can be used to detect the intensity of incident light or to generate electricity for residential and industrial usage.

For ease of comprehension, the photovoltaic (PV) effect in a solar cell can be demonstrated with an analogy of a youngster on a slide. Both the electron and the infant are initially in their 'ground states'. The electron is elevated to its excited state by absorbing energy from solar radiation, just as a youngster climbs up the slide using muscular energy (Fig. 4.2). In both circumstances, there is now energy that may be expended in the excited state.

There is no stimulus for the excited free electrons to go in a certain direction in the absence of junction-forming materials; they would gradually fall back to the ground state. Under solar radiation, however, when two distinct semiconductor materials come into contact, an internal electric field is formed along the contact plane. This is the so-called *built-in electric field*, which drives excited free electrons into an external electrical load, allowing them to dissipate their excess energy. A simple resistor and perhaps other electrical device can be used as the external load. Similarly, the slide is where the youngster expends its surplus (potential) energy.

When the surplus energy is dissipated, both the electron and the youngster are returned to the ground state, where they can resume the process. The electron's mobility, like that of a youngster, is only in one direction. In a nutshell, the photovoltaic effect generates a constant direct current (DC) that always runs in one direction only.

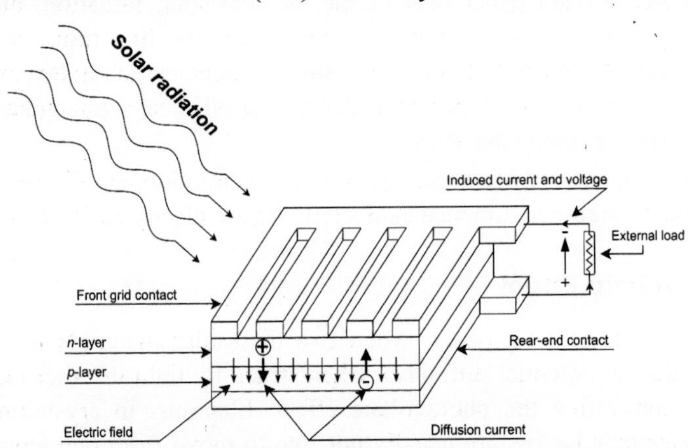

Fig. 4.1 General schematics of a solar cell

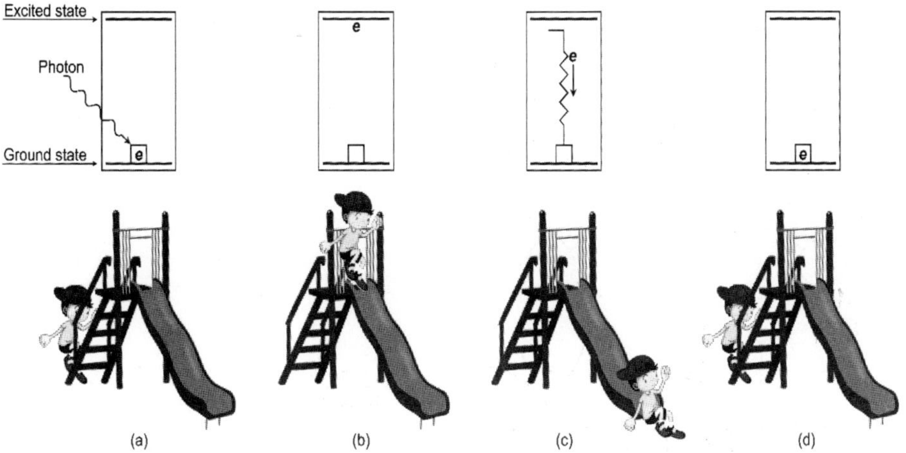

Fig. 4.2 Representation of electron movement in a solar cell with 'child and slide' analogy

4.2 Physics of Solar Cell – Construction and Working Principle

A solar cell is the fundamental component of any solar PV system. As discussed earlier, photovoltaic effect is the conversion of light directly into electricity through the special arrangement of semiconductor materials.

The establishment of the built-in electric field is the factor which prevents the electrons from crossing over the boundary directly (short circuiting). Electrons in the atom are bound to the nucleus by electromagnetic forces. When energised it may wander around, but will return to its original position. If we consider the case of Silicon (Si), it has got four electrons in its outermost orbit. Silicon can attain an 'Octet' configuration in the outermost orbit if they share one electron each from four neighbouring Si atoms. Thus, Si forms a crystal lattice, which is quite orderly and stable. If you expose the Si semiconductor to light, electrons may wander within the crystalline structure, but no electricity flow can occur in the external circuit.

If we add some impurity to this orderly Si crystalline structure (called *doping*), some imbalance will occur within the structure. Donor elements like Antimony (Sb) and Phosphorus (P) atoms have five electrons in its outer most orbit. The neighbouring four Si atoms can share four electrons only in the outermost orbit of Sb atom as shown in Fig. 4.3. Or, one electron is free which is not shared by Si atoms. This creates an apparent negative charge to the Si semiconductor material doped with donor atom. Hence, this is called N-type semiconductor.

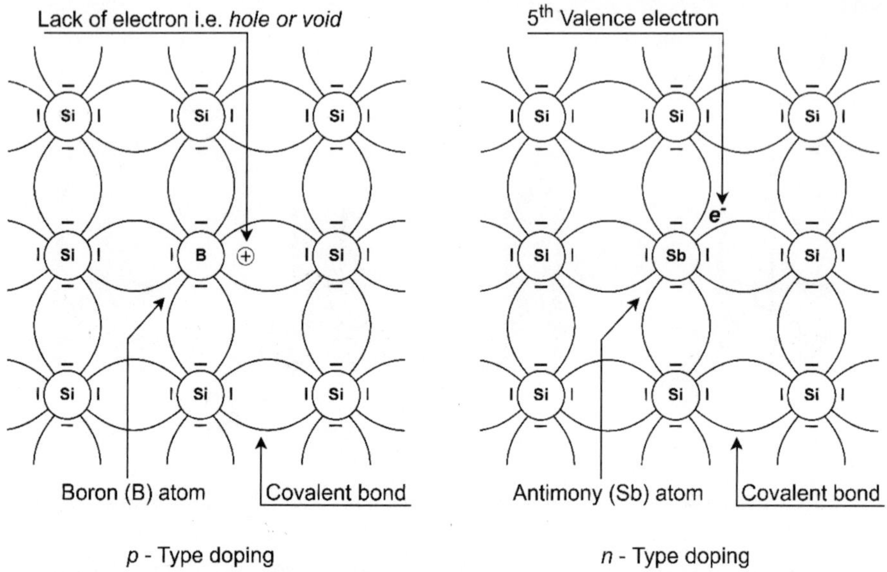

Fig. 4.3 Doping of Silicon to create N-type and P-type semiconductors

As Si and Sb are both neutral atoms, there is no actual negative charge in the material. On the other hand, if we dope Si semiconductor with Boron (having only three electrons in the outermost orbit), an empty space occurs in the crystal lattice. Only three Si atoms in the vicinity can share electrons and as one Si atom do not get electron from boron atom to share. This empty space is called a 'hole' which behaves as if it is positively charged. This kind of a semiconductor is hence called P-type. If we join N-type and P-type semiconductors, we get a P-N junction which dramatically changes the scenario with respect to electron flow when exposed to light. When the N-type semiconductor is exposed to light, a number of electrons cross over the junction to the P-side. The region from where the electrons have moved is called the depletion region. Same thing occurs at the P-side also and holes cross over to the N-side.

But, atoms in both N-side and P-side were actually neutral and this transit of electrons and holes has created an imbalance in the charge. The depletion region in N-side is actually positively charged now as there is a shortage of electrons there. Similarly, the depletion region at P-side has a negative charge. This forms a barrier and no more electrons from N-side can cross over the depletion region in P-side, as there is repulsion by the excess negative charge created there. Similar hindrance is created for holes from P-side to cross over the depletion region at N-side due to the existing positive charge in the depletion region (Fig. 4.4).

In short, when sufficient amounts of solar radiation (*photons*) fall upon the N-side of the P–N junction, the PV effect causes excited free electrons to emerge and these electrons continue moving through the cell to the outer surface under the influence of the p-n junction's internal electric field. Hence this side becomes negative side of the PV cell. When these electrons are collected by a metallic grid contact, an electric current would flow, provided the grid is connected to the metal contact on the P-side (*positive side*) of the cell through an external load.

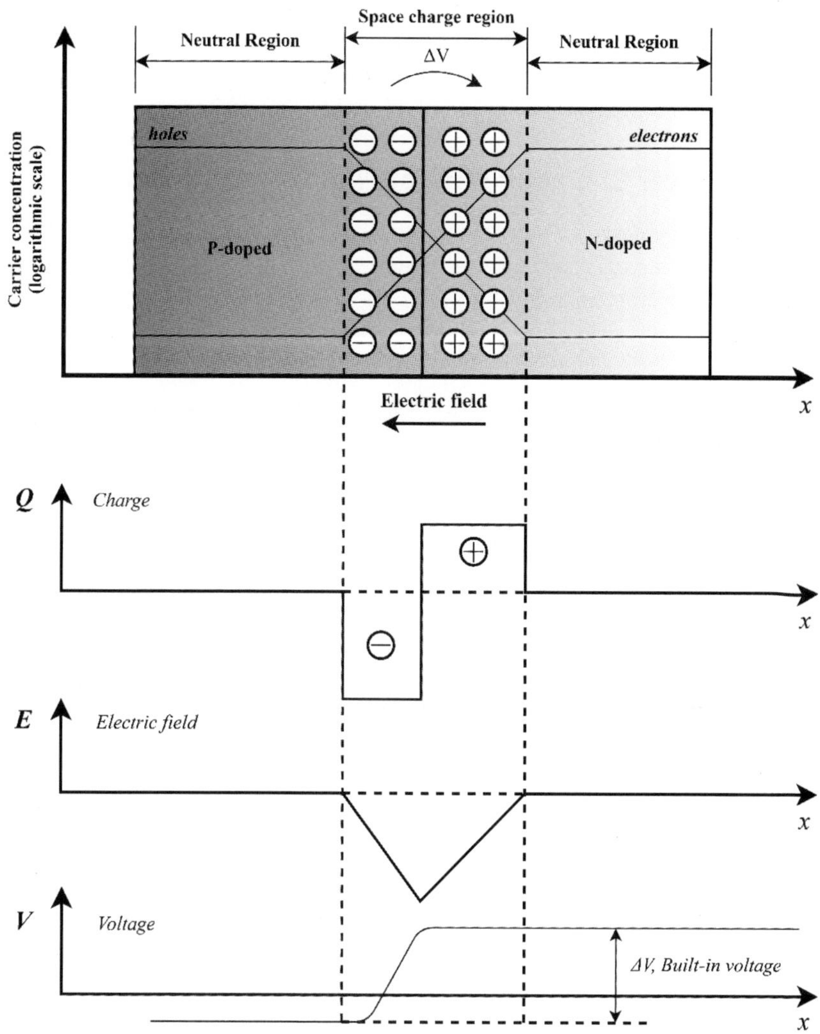

Fig. 4.4 The P-N junction with space charge depletion region

If we contemplate the phenomenon that occurs when a PV cell is exposed to solar radiation, this radiation can be considered as a flux of photons, which have energy $E=h\nu$, where h is Planck's constant (6.6207×10^{-34} $m^2\,kg\,s^{-1}$) and ν is photon frequency. If the photon energy is greater than the energy band gap of the semiconductor, then, the electron from the valence band will jump into the conduction band.

This creates a gap in the valence band i.e., the space without electron, referred as a *hole*. Thus, under the action of photons, generation of *hole–electron* pairs occur, known as the internal photovoltaic effect. Free moving excited electrons are separated from the internal electric field of the p-n junction. This charge separation is characterized by the barrier potential (or also known as *built-in potential*) U_0 of the semiconductor used, which is usually within 200 to 700 millivolts (mV). Once the induced electric field exceeds the barrier potential, the electrons will be directed to the *n*–zone, while the holes towards the *p*–zone. Thus, when connected with the external load circuit between the positive and negative sides of the solar cell, an electric current tends to flow. This is known as the direct conversion of light into electricity through the photovoltaic effect.

4.3 I-V and Power Characteristics of Photovoltaic Cells

The current (*I*) and voltage (*V*) characteristics of a photovoltaic cell, module, or array are shown in *I-V* characteristic curves, which provide a complete description of its solar energy conversion capacity and efficiency. Knowing *I-V* parameters of solar cell (especially P_{max}) is crucial for evaluating its output performance and photovoltaic efficiency.

Fundamentally, *I-V* characteristic curve is a graphical depiction of a solar cell or module's functioning that illustrates the relationship between current and voltage under solar irradiance and temperature circumstances.

These characteristic curves provide the data needed to design a solar PV system so that it can function as near as possible to its ideal peak power point, as illustrated in Fig. 4.5.

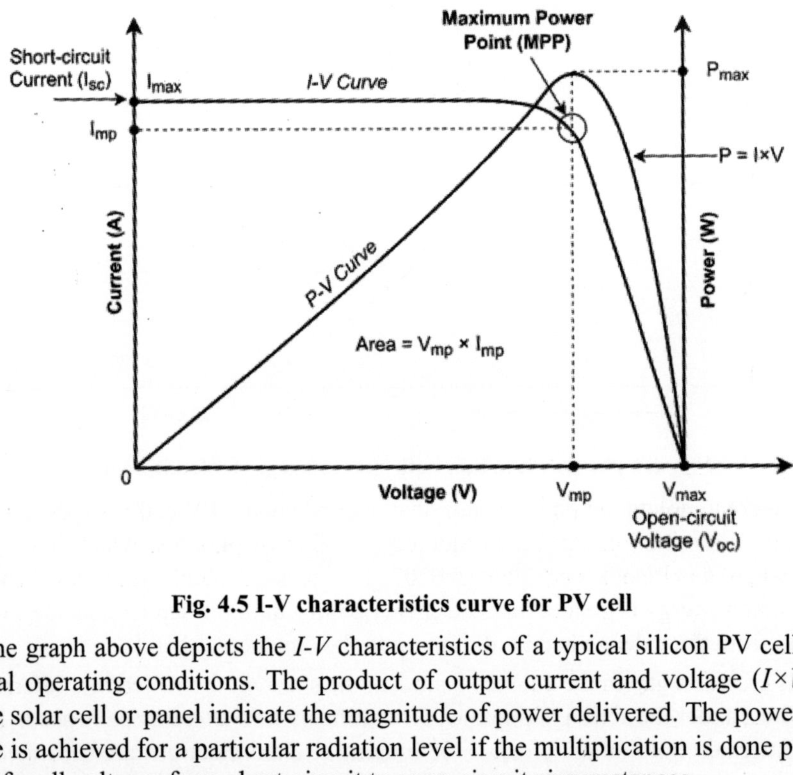

Fig. 4.5 I-V characteristics curve for PV cell

The graph above depicts the *I-V* characteristics of a typical silicon PV cell under normal operating conditions. The product of output current and voltage ($I \times V$) of a single solar cell or panel indicate the magnitude of power delivered. The power curve above is achieved for a particular radiation level if the multiplication is done point by point for all voltages from short-circuit to open-circuit circumstances.

The current shall be at its lowest (zero) and the voltage across a solar cell shall be at its greatest when the solar cell is in open-circuit state, i.e., not coupled to any external load. This is called as the solar cell open circuit voltage (V_{oc}). When the solar cell is short circuited, that is, the positive (+) and negative (-) leads are directly connected, the voltage across the cell drops to zero, but the current flow-out reaches its peak value (Fig. 4.5), which is known as the solar cell short circuit current (I_{sc}).

A solar cell's I-V characteristics curve thus spans from zero output current at open circuit with maximum voltage (V_{oc}) to the short circuit current (I_{sc}) at zero output volts (V_{sc}). In other terms, during open circuit and short circuit, a cell's peak voltage and current are correspondingly achievable. Obviously, neither of these configurations generates any power, but there must be a coordinate point where the solar cell yields the maximum power.

Practically, there is a specific combination of output current and voltage of a solar cell resulting in its maximum power output, i.e., at I_{mp} and V_{mp}. Such combination points of output current and voltage represented with circular markers at the top right rectangular areas shown in Fig. 4.5 and Fig. 4.6 are referred as the '*maximum power point*' or MPP.

Typically, MPP of a solar cell occurs close to the curvature in the *I-V* characteristics curve. The equivalent values of V_{mp} and I_{mp} can be projected from the open circuit voltage and the short circuit current. For most of the photovoltaic cells, $V_{mp} \cong (0.8–0.90)V_{oc}$ and $I_{mp} \cong (0.85–0.95)I_{sc}$. In practice, output voltage and current of a solar cell are influenced by the cell temperature; therefore, the actual power output (*IV*) fluctuates with variations in ambient temperature.

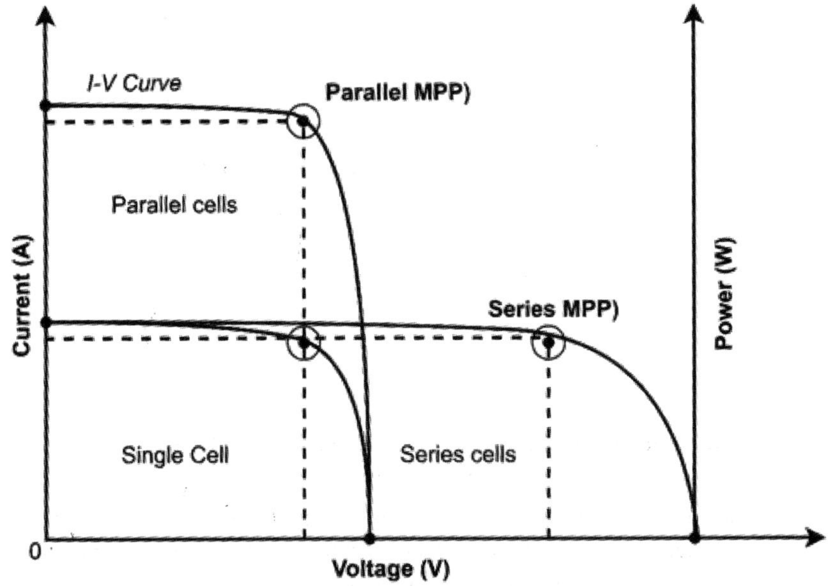

Fig. 4.6 I-V characteristics curves for series and parallel connected modules vs. single PV cell

To expand the voltage or current capacity of a solar array, photovoltaic modules or panels can be wired in series or parallel configurations, or both. When the panels are joined in a series configuration, the output voltage of the combination gets added up, whereas when they are connected in parallel, the current capacity is added up. The electrical power generated in watts by the various cell combinations would always remain the result of voltage times current. However, because the solar panels are linked, the system's MPP will always be in the top right-hand corner.

4.3.1 Prediction of characteristic properties of PV cell

As we know, the charge carrier pairs are created whenever light is absorbed at the *p-n* junction of a solar cell, resulting in a current flow. Since, the semiconductor device is not at thermal equilibrium, electric power can be provided to an external load. As illustrated in Fig. 4.7, a typical solar cell comprises a *p-n* junction with a diode characteristic. This characteristic property could be predicted using ordinary solid-state physics.

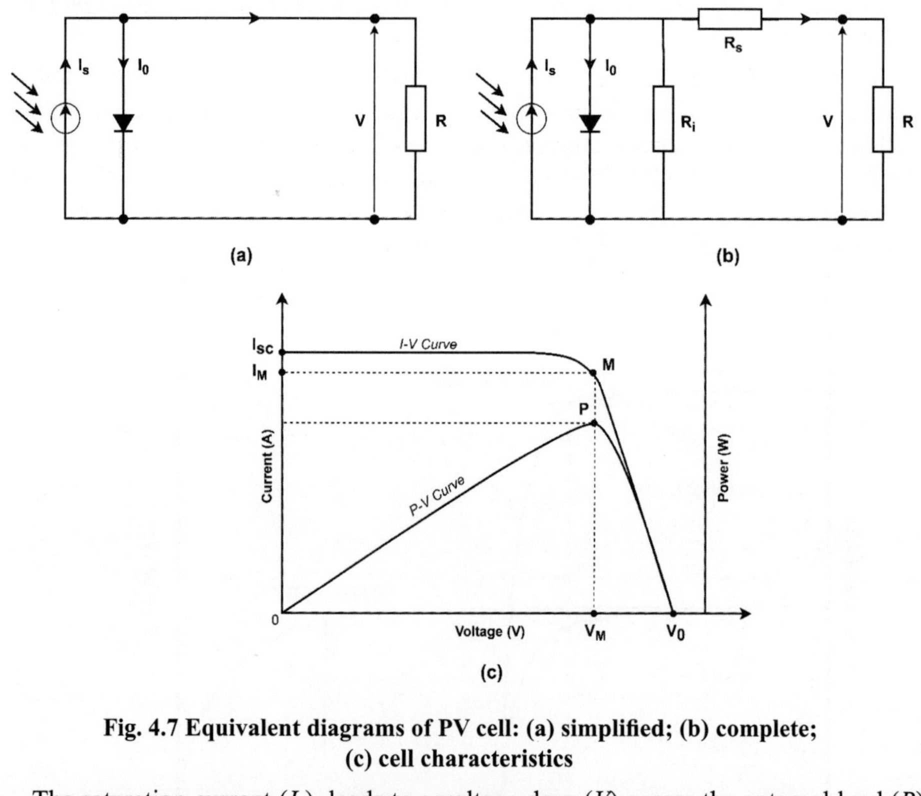

Fig. 4.7 Equivalent diagrams of PV cell: (a) simplified; (b) complete; (c) cell characteristics

The saturation current (I_0), leads to a voltage drop (V) across the external load (R) connected to the cell (Fig. 4.7). Voltage (V) generate a diode current (I_d) opposite to the direction of photovoltaic current (I_s) and it can be calculated as:

$$I_d = I_0\left[\exp\left(\frac{eV}{kT}\right)-1\right] \qquad ...(4.1)$$

Where, k – is Boltzmann's constant; T – absolute temperature and e – is the electron charge.

Therefore, in fact, the magnitude of current flowing through an external circuit (I) is nothing but the difference between photovoltaic current (I_s) and diode current (I_d), as given below:

$$I = I_s - I_d = I_0\left[\exp\left(\frac{eV}{kT}\right) - 1\right] \quad ...(4.2)$$

The basic circuit of a photovoltaic cell, illustrated in Fig. 4.7a, resembles to Eq. 4.2. With the assessment of leak resistance R_i of *p-n* junction of a PV cell and the cell base-spreading resistance R_s, one can demonstrate a comprehensive arrangement of PV cell (Fig. 4.7b). Today's cutting-edge technologies and manufacturing processes aid in achieving cells with $R_i = \infty$ and $R_s = 0$, hence the basic equivalent circuit is acceptable. Then, the electrical power applied across load (R) can be determined as follows:

$$P = VI = I\left\{I_s - I_0\left[\exp\left(\frac{eV}{kT}\right) - 1\right]\right\} \quad ...(4.3)$$

A peak power value, when power is a function of voltage [$P(V)$], is attained at point M ($I_M \times V_M$) of the current–voltage characteristic, whose coordinates are the resultant of the condition $dP/dV = 0$:

$$V_M = V_0 - V_T \ln\left(1 + \frac{V_M}{V_T}\right) \quad ...(4.4)$$

Where, $V_T = \frac{kT}{e}$ is a constant called thermal voltage.

The implicit equation above does not have a simple solution, but it converges quickly with iteration. An initial guess of $V_M = 0.9\ V_0$ gives an accurate solution in two iterations.

Also,

$$I_M = I_S\left(1 + \frac{I_0}{I_s}\right)\left(\frac{V_M}{V_M + V_T}\right) \quad ...(4.5)$$

Where, $I_s = I_{sc}$ i.e. short-circuit current.

The optimum value of the external passive load resistance (R_M) can be found by:

$$R_M = \frac{V_M}{I_M} \quad ...(4.6)$$

4.3.2 Parameters related to solar PV cells

Solar PV cells and modules come with the manufacturer's data regarding the technical parameters for standard operating conditions of the product:

(i) Global solar radiation on the cell surface, G = 800 Wm^{-2}

(ii) Air mass coefficient, AM-1.5

The air mass coefficient defines the direct optical path length through the Earth's atmosphere, expressed as a ratio relative to the path length vertically upwards, i.e., at the zenith. The Air Mass number refers to the intensity and spectral distribution resulting from a certain path length of sunlight through the atmosphere. AM-1 means the sun is directly overhead (G = 1000 Wm^{-2}). AM-1.5 applies to the inclined (45° approx.) position of sun, such that the path length is 1.5 times.

(iii) Cell temperature, Tc = 20°C

(iv) Wind velocity = 1 ms^{-1} or more

(a) *Short-circuit current:* The short circuit current is the current when the negative and positive terminals of a solar cell are circuited by shortest path without any external load resistance. At this point, the voltage across the solar cell drops to zero, thus the power output from the cell also becomes null. In the case of solar cells, $I_{sc} = I_s$

(b) *Open-circuit voltage or No-load voltage:* The open-circuit voltage (V_{oc}=V_0) is the maximum voltage available from a solar cell when no current is drawn out of the cell (I = 0) i.e., no external load is connected across the terminals of solar cell. The open-circuit voltage corresponds to the amount of forward bias on the solar cell caused by the internal electric field induced at the *p–n* junction due to incident solar radiation. It can be determined by:

$$V_{oc} = V_T \ln\left(\frac{I_s + I_0}{I_0}\right) \approx V_T \ln\frac{I_s}{I_0} \qquad ...(4.7)$$

For silicon based solar cells, thermal voltage (V_T) is about 26 mV; and I_0 is infinitesimally small as compared to I_s and the ratio, therefore V_{oc} = 0.6 V

(c) *Maximum power:* The maximum power (P_M), also referred as peak power is the product of the current and voltage at maximum power point (MPP) on the *I-V* characteristic curve of a solar PV cell.

$$P_M = I_M \times V_M \qquad ...(4.8)$$

(d) *Fill factor (FF):* It is defined as the ratio of the maximum power output of the solar cell to the product of V_{oc} and I_{sc}.

$$FF = \frac{I_M \times V_M}{I_{sc} \times V_{oc}} \qquad ...(4.9)$$

The fill factor serves as a quality index for a solar photovoltaic cell. Greater numbers are attained when a *p–n* junction exhibits excellent rectifying properties i.e., for a small saturation current (I_0), at a minimal series resistance R_S and a higher parallel resistance R_P (usually $FF > 0.7$).

(e) *Conversion efficiency of solar cell:* It is the ratio of actual power output from the solar cell (with an effective aperture area (A, m^2) operating at the maximum power point (MPP) at a specified temperature) to the incident solar radiation power (G), expressed as percentage.

$$\eta_{conv} = \left(\frac{I_M \cdot V_M}{A \cdot G}\right) \times 100 \qquad ...(4.10)$$

Crystalline silicon cells with efficiencies in the range of 13 to 25 percent, have been successfully generated in laboratory circumstances, whereas efficiencies of 12 to 14 percent is common under commercial manufacturing conditions. In laboratory circumstances, overall performance efficiency of all polycrystalline silicon cells could reach 20 per cent. The theoretical limit of efficiency for crystalline silicon cell is 37 per cent, whereas for amorphous silicon cell it is 28 per cent.

The losses in the photovoltaic conversion process are subject to:

(i) Grid coverage: To receive the electrons released by the photoelectric effect, the cell's surface must be covered by a conductive metal grid.

(ii) Reflection loss: The front surface of a cell reflects some of the incident solar energy.

(iii) False absorption: Impure atoms in the semiconductor crystal capture some of the electrons excited from their energy shell.

(iv) Sometimes solar radiations are not energetic enough to excite an electron from its low energy shell.

(v) Some of the solar radiations possess a much higher energy level than required to dislodge an electron from its shell. In the semiconductor crystal, such additional energy is dissipated in the form of heat.

(vi) According to the quantum physics, only 90 per cent of photons with the necessary energy level to excite an electron from its base shell would probably strike an electron.

(vii) Absorption does not occur at the p-n junction. Far from the junction, some photons are absorbed by the crystal. These photons produce electron hole pairs that instantly recombine, leaving a tiny amount of heat behind.

(viii) Solar cell shading: Whenever a portion of a solar cell or module comes under shade, the overall output voltage drops sharply.

(f) *Normal operating cell temperature (NOCT):* This refers to the temperature of a photovoltaic cell in no-load operating circumstances at the surrounding temperature of 20°C, global radiation of 800 Wm^{-2} and wind speed equal to or greater than 1 ms^{-1}. For typical photovoltaic cells NOCT constraint is between 42°C and 46°C. If NOCT constraint is specified by the manufacturer, cell temperature T_C for other operating conditions characterized by the ambient air temperature T_A and global radiation G could be determined.

$$T_C = T_A + \left(\frac{NOCT - 20}{800}\right) \times G \qquad ...(4.11)$$

(g) *Effect of solar radiation and temperature on I-V characteristics of solar cell:* The short-circuit current is directly proportional to the intensity of incident solar radiation. On the contrary, the open-circuit voltage depends logarithmically on solar radiation and varies only on minute scale (Fig. 4.8). For different values of solar radiation G, the corresponding short-circuit (I_{sc}) current can be determined approximately using the following equation:

$$I_{sc} = \frac{G}{G_{st}} \cdot I_{scst} \qquad ...(4.12)$$

Where I_{scst} is the short circuit current resultant to the standard global solar radiation $Gst = 1{,}000\ Wm^{-2}$.

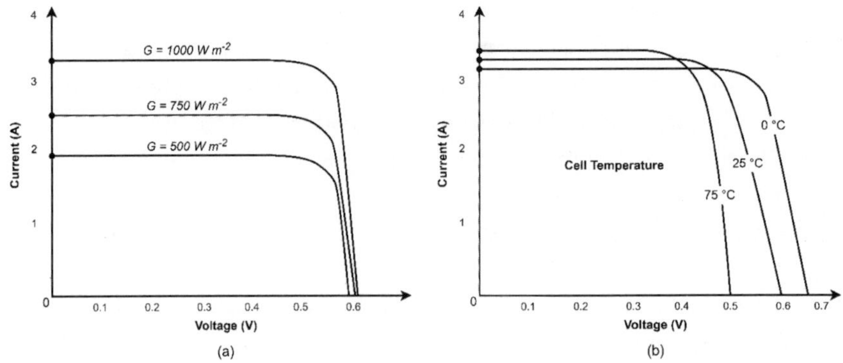

Fig. 4.8 Behaviour of I-V characteristics of a solar cell with variation in solar radiation and temperature

On the other hand, cell temperature affects the open-circuit voltage (V_0) considerably and has a little effect on short-circuit current. The open-circuit voltage falls with the rise in cell temperature. The voltage variation coefficient with respect to the cell temperature is determined using the following formula:

$$V_0 = V_{025} - 0.0023(t - 25) \qquad ...(4.13)$$

Where, 'V_{025}' is the PV cell open-circuit voltage at standard operating temperature (25°C) and 't' is the actual temperature of the cell.

(h) ***Solar cell combinations and output configurations:*** The standard PV hierarchy is shown in Fig. 4.9. A single photovoltaic cell has output voltage of around 0.5–0.6V, which is insufficient to run the general appliances. Therefore, photovoltaic cells are coupled in series within a *module* to boost the overall output voltage. The count of PV cells in a module depends on the desired operating voltage. Majority of the PV module manufacturers produce the modules compatible to 12V batteries. To ensure the reliable operation of the system, few additional volts to charge the battery and to compensate for lower output under non-standard test conditions are to be taken into account. Hence, modules usually have 33 to 36 solar cells in series to output the open-circuit voltage in the range of 18 to 21 volts.

To ramp up the output current, the series strings of solar cells are connected in parallel. To meet various *I-V* operating conditions, solar cells are connected in both parallel and series combination. The modules can then in turn be connected in series and parallel to have the desired PV system voltage and current. Such combinations of modules arranged on a common circuit board to function as a single unit are referred as *solar PV panel.* Depending on the required power output, multiple panels are connected in series and/or parallel combinations in the solar field, called as *solar PV array.*

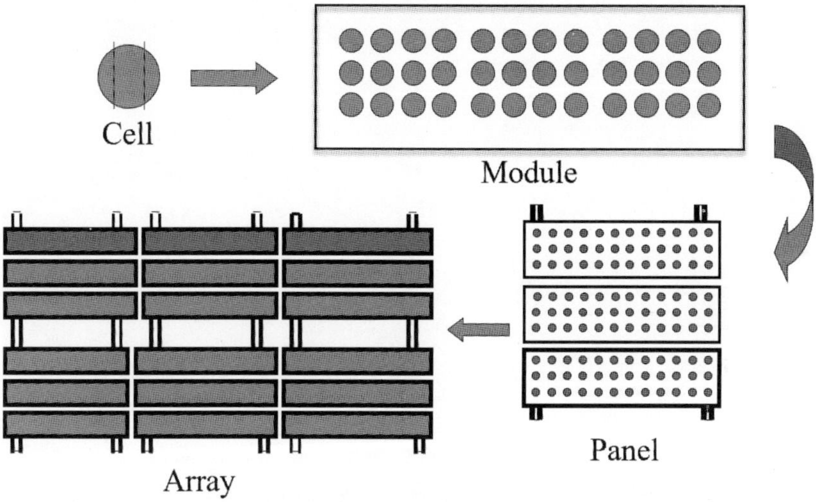

Fig. 4.9 Photovoltaic system hierarchy

4.4 Present Scenario and Future Prospects of PV Technology

Today, majority of solar cells are being commercially manufactured from silicon; with increasing efficiency and cost, as the materials range from amorphous (non-crystalline) to poly-crystalline to mono-crystalline (single crystal) silicon forms. Unlike the electrochemical batteries and fuel cells, solar cells do not utilize chemical reactions or require fuel to produce electric power. Unlike electric generators, they do not have any moving parts. The flow of electricity results from the characteristics of the semiconductors and is induced entirely by solar radiation striking the cell.

Since, the solar photovoltaic (PV) cells have no moving parts that need maintenance or fuels that would require replenishment, solar cells are widely preferred to power most of the space installations, from communications and weather satellites to space stations. Moreover, elimination of intermediate processing through the technological chains, moving components, noise and vibration along with modular construction and a service life of over 25 years, are the key points asserting that the future of decentralized energy shall belong to photovoltaic technology.

CHAPTER

5

Fundamentals of Wind Energy

The power of the wind was known to man since the early days of human civilization. Wind power has been used for propulsion of ships until the eighteenth century. Wind-powered machines have been widely used to grind grain and pump water for more than a millennium. Like most renewable energy sources, the energy of wind also comes from the sun. Out of the 173,000 TW of solar energy radiated on the earth's surface, about 2 per cent is converted into wind energy. Wind power is extracted from the energy contained in the wind using suitable converters called wind turbines. Early wind machines were used for milling of cereals and hence they were popularly called 'wind mills'. Currently, the most important use of wind energy is electricity production and the wind energy conversion system is often referred as 'aero-generator'.

5.1 Generation of Wind

Wind is nothing but air in motion. Winds are basically caused by temperature gradients created in different regions of earth's atmosphere due to differential heating by solar radiation. Obviously, the regions near the equator are receiving more solar radiation and hence heated more than the rest of the earth's surface. When the atmospheric air in contact with the surface gets heated up, it will rise up as hot air is lighter than cold air. The less dense hot air continues the upward flow until it reaches approximately 10 km high and then will move to the north and the south. The cold air from the north and south should keep flowing towards the equator to occupy the low-pressure area created by the upward movement of hot air at the equatorial region. This motion of air caused by convection is referred as 'convective circulation'.

If the earth did not rotate, the convective circulation would have driven the hot air from the equator to the poles direct. This air will get cooled and sink down there to flow towards the equator. But the earth is rotating from west to east, and the direction of flow of air gets tilted due to this rotary motion. The wind, which should have been moving straight to the north, is diverted to the right in the northern hemisphere. In the southern hemisphere it is diverted to the left. This bending force is known as the 'Coriolis force'.

The differential heating of land and water during the day is another phenomenon responsible for localised movement of air. The landmass is heated faster than the water bodies like sea due to absorption of solar radiation. This results in the lifting of warmer air in the atmosphere consequently creating a low pressure to move cooler air from the oceans towards the landmass. During the night the land cools down at a faster rate and the direction of air flow is reversed, even though at a slower rate due to the decreased temperature gradient. Another local phenomenon is the 'valley wind' due to mountain slopes. The wind we observe in a locality is likely to be the net effect of all these phenomena.

Winds are also sometimes classified as primary, secondary and tertiary winds. Primary winds, also called 'planetary' winds are those which blow in a particular direction throughout the year. Winds subjected to seasonal changes in direction are called secondary winds or seasonal winds. Local winds like valley winds are also called 'tertiary' winds.

5.2 Advantages and Disadvantages of Wind Energy Utilization

Wind energy has emerged as a promising renewable source with its major advantage of minimal adverse impact on environment. It has the potential to be widely used for electric power production to replace power plants based on fossil fuels. Wind is abundant in many regions of the world and is widely distributed. Being a clean technology there is no harm of greenhouse gas emissions during their operation. It is safer compared to other technologies like nuclear energy and land requirement is less.

Advantages of wind energy	*Disadvantages*
Renewable energy source	Wind is a dilute energy form
Non-polluting and environment friendly	Highly fluctuating
No fuel requirement	Difficulty in storing
Comparatively moderate energy cost	Restricted to potential sites
Possibility for fast installation	Noise problem

5.3 Wind Farms and Installed Power Capacities

The terminology 'wind farm' is used to refer to a group of aero generators collectively operated in the same locality. Large wind farms usually have hundreds of individual wind turbines installed over a large area. Jiuquan wind farm in China is the world's largest wind farm with a planned capacity of 20 GW. Jaisalmer (Rajasthan) wind farm in India with an installed capacity of 1600 MW developed by Suzlon Energy is the second largest in the world.

At present, the total installed capacity of the wind power projects in the world is about 750 GW. China is the leading country with about 290 GW capacity followed by USA (123 GW) and Germany (63 GW). India is in the fourth position with about 39 GW. In India, Tamil Nadu is the leading state followed by Gujarat.

There are also offshore wind farms located in water bodies like lakes and oceans. Offshore wind farms have the advantages that they operate in stronger, less turbulent wind which enable them to produce more power. But the higher costs associated with installation, operation as well as maintenance makes them expensive. Most of the offshore wind farms have been installed in UK, Germany and China.

5.4 Classification of Wind Energy Conversion Systems

Wind energy systems are usually classified on the basis of their capacities, electricity output and power utilization pattern.

5.4.1 Classification based on output capacity and nature of output

Wind energy systems up to 2 kW capacities are regarded as small scale, those in the range 2 to 100 kW as medium scale and regarded as large scale above 100 kW. This is an old classification when large systems were less in numbers. As systems with capacities greater than 100 kW have become common, the above classification does not have much relevance. Aero generators can be classified as those with DC output and AC output.

5.4.2 Classification based on operational speed

Wind turbines can be of fixed speed type or variable speed type. Fixed speed turbines work with a speed variation within one percent. Most of the modern high power wind turbines are designed as variable speed machines.

5.4.3 Based on Grid connection

(a) **Grid connected (Central Grid):** This is the most common type of energy system in which the output power is supplied to the central grid. There is no storage of electric power. These wind farms have multiple turbines with individual turbines typically having capacities above 200 kW. The total capacity can go up to several GW.

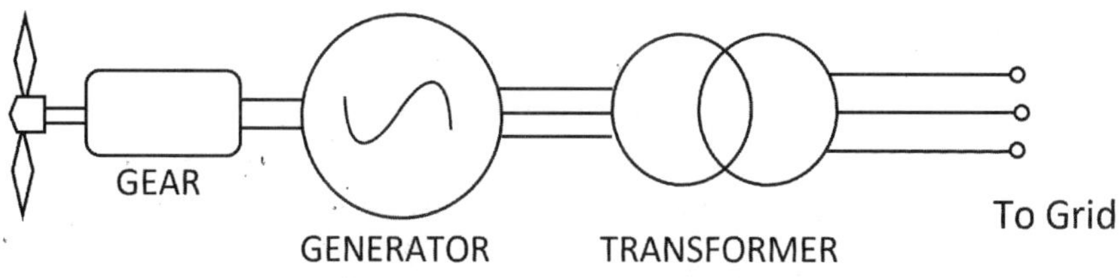

Fig. 5.1 Schematic representation of a grid connected aero generator system

Fig. 5.1 shows a simplified illustration of the grid connection scheme of a wind generator. The generator output is stepped up to required grid voltage level. In the early aero generators, ordinary squirrel cage induction generators were used.

(b) **Isolated-Grid:** In isolated grid a small area or limited users are provided with electricity from a localized grid. Medium scale turbines typically 10 to 200 kW are generally employed in these systems. These kind of systems are suitable for remote and isolated areas like islands. The electricity generation cost can be minimised in such cases and hybrid systems (e.g., wind-diesel, wind-solar) are often used (Fig. 5.2).

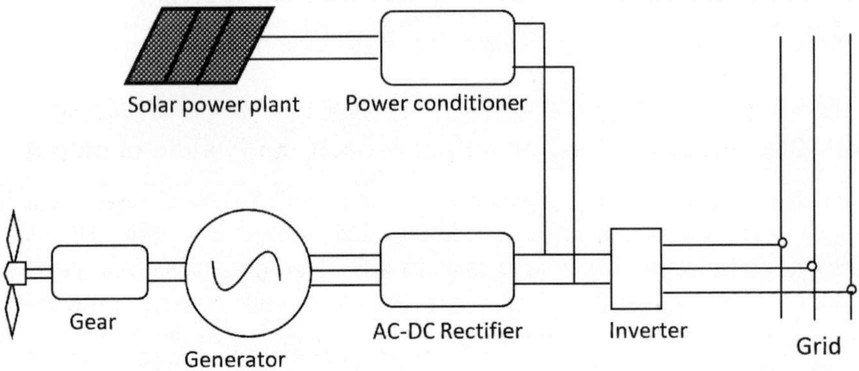

Fig. 5.2 Schematic representation of a 'Solar – Wind Hybrid System'

(c) **Off-Grid (stand-alone) systems:** Small turbines (50 W to 10 kW) are used in limited off grid systems. Very small systems are mostly used for battery charging or water pumping. Higher capacity systems with battery storage can also be connected to energise utilities (Fig. 5.3).

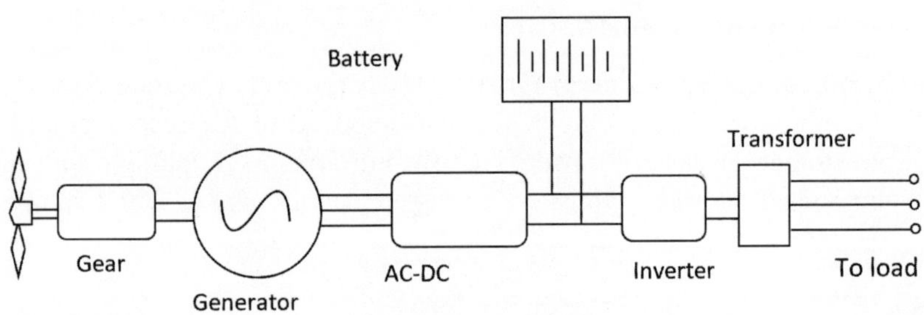

Fig. 5.3 Off-grid wind power generation system

5.5 Active Control of Power Output for Grid Connected Wind Farms

The electricity grids are required to stay balanced in order to prevent blackout. The amount of electricity fed into the grid and the consumption should be equal to avoid frequency variation in the grid. In India and Europe, the grid operates at 50 Hz frequency. The grid is set to a tolerance threshold and variation beyond this value results in partial or complete collapse. The conventional electrical grids are energised from thermal power plants which are based on turbines connected to generators. They have 'grid inertia' so as to cope up with the changes in load and power demand. Renewable energy systems have less grid inertia. When the load is not matching with the generation there can be problems in the grid and this can be more serious when aero-generators contribute a significant share of the power. Hence special arrangements are required to match the voltage and frequency of generation with that of the grid under changing load conditions. These imbalances also occur with changes in generation due to variation of wind velocity.

In most countries regulations for 'active power control requirements' have been formulated to make sure that wind farms regulate the power output to a defined level. They should also possess frequency response system so as to maintain the frequency in the permissible limits.

5.6 Forces that Drive Wind Turbines: Lift and Drag

There are basically two forces which are utilised to transform wind power into mechanical power. First one is the 'drag force' by which a body gets dragged or pushed by wind. This force drags the body in the direction of the wind. The other force is called the 'lift force' and as indicated by the word 'lift', the body is lifted by wind. But the movement of the body is in a direction perpendicular to the wind direction.

The drag forces are in the direction of wind and hence the machines to convert wind energy to mechanical power are placed in such a way to oppose the direction of wind. They utilise the force applied at right angles on an area by the wind. The commonly used instrument to measure wind velocity, the cup anemometer is an example for drag machine as shown in Fig. 5.4.

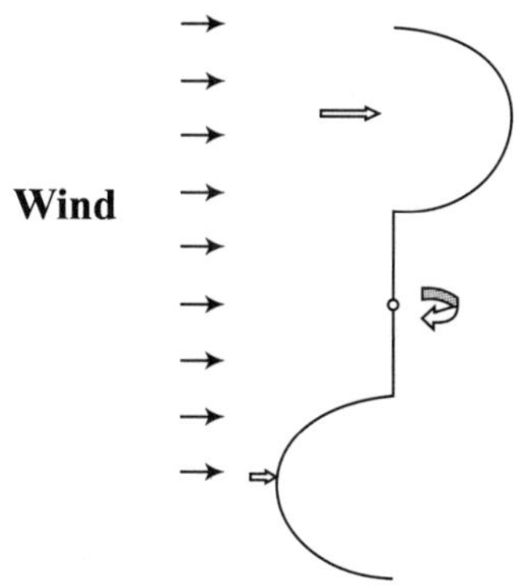

Fig. 5.4 Drag force on the cups of an anemometer

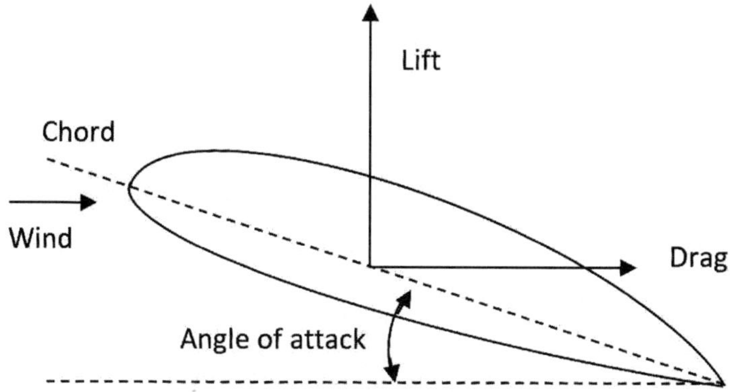

Fig. 5.5 Lift and drag forces acting on a blade aerofoil section

The aircraft wings are designed to utilise lift forces which act perpendicular to air flow (Fig. 5.5). When air at a certain speed act upon a surface kept at an angle to the direction of air flow the perpendicular lift force is produced. In the case of wind turbines, the lift force is much advantageous than the drag force, and most wind machines for electricity production make use of lift force.

5.7 Power Contained in Wind

Wind is air at motion and possesses kinetic energy by virtue of its velocity and mass. The power '*P*' contained in the wind depends mainly on the mass 'm' and on the velocity '*v*' of the wind.

The kinetic energy of the wind,

$$\text{K.E.} = \frac{1}{2}\text{mv}^2 \qquad ...(5.1)$$

The mass of air with density 'ρ' passing through an area 'A', in unit time with a velocity 'v',

$$m = \rho \cdot A \cdot v \qquad ...(5.2)$$

Power is energy per unit time which can be obtained by substituting the value of 'm' in Eq. 5.1,

$$P = \frac{1}{2}\rho \cdot A \cdot v^3 \qquad ...(5.3)$$

'*P*' is proportional to the density ρ of the air, to its area and to the cube of its velocity.

As Wind power is proportional to the cube of the wind velocity, the available power increases eightfold when the wind velocity doubles.

The area swept by the rotor is normally circular with a diameter, '*D*'. Hence,

$$P = \frac{1}{2}\rho \times \frac{\pi}{4}D^2 \cdot v^3 \qquad ...(5.4)$$

5.8 Power Extraction from Wind

Any device for extracting power from wind should invariably slow down it in order to convert it to useful work. But, the entire kinetic energy of the wind cannot be extracted from it, so also the power. The power extracted will be equal to the product of the wind speed as it passes through the rotor and the pressure drop Δp. The fraction of the power in free-flowing wind that can be extracted by a rotor is called the *power coefficient* (C_p).

Cp = (Power of wind rotor)/(Power available in the wind) ...(5.5)

The transformation of the kinetic energy of the wind to mechanical power by the rotor is achieved by slowing down the flowing air mass (Fig. 5.6). The wind velocity in front of the rotor, v_1 is reduced to the wind velocity behind the rotor v_3.

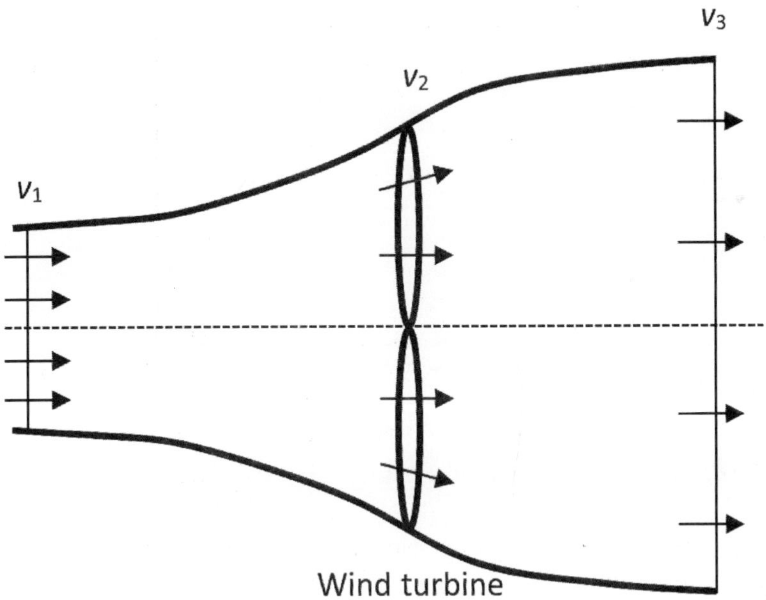

Fig. 5.6 Wind velocities when free wind is intercepted by a rotor

The energy extracted by the rotor is the difference of the energy in front of the rotor and the energy behind the rotor:

$$E = \frac{1}{2} \cdot m \cdot (v_1^2 - v_3^2) \quad \text{...(5.6)}$$

5.9 Coefficient of Performance, Betz Coefficient and Overall Efficiency

All of the wind's kinetic energy cannot be captured by rotor, or air would be completely still behind rotor and not allow more wind to pass through. The ideal or maximum theoretical efficiency of a wind turbine is the ratio of the maximum power obtained from wind to the total power available in wind. This value is equal to $^{16}/_{27}$ or 0.593 and is known as *Betz coefficient*, named after a German physicist Albert Betz, who derived that no wind turbine could convert more than 59.3 per cent of the kinetic energy of the wind into mechanical energy turning a rotor.

When the wind velocity behind the rotor (v_3) is one third of the velocity in front of the rotor (v_1), maximum efficiency can be achieved (Fig. 5.7).

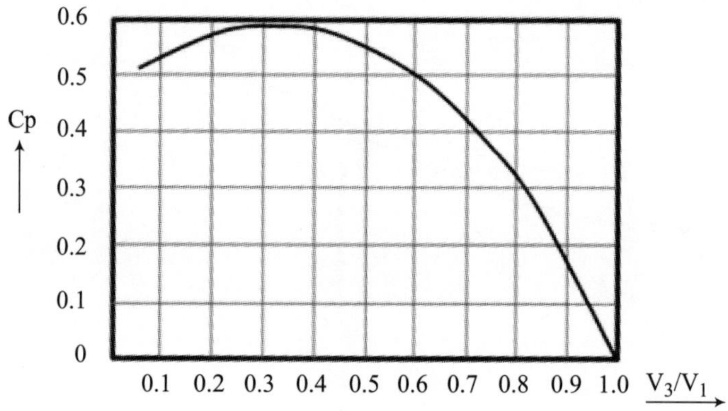

Fig. 5.7 Coefficient of performance vs. velocity ratio

Or, the condition for power coefficient to be equal to Betz limit is, $v_3 = 1/3\ v_1$ and Cp = 0.593

Overall efficiency of an aero-generator system is the product of individual efficiencies of the turbine, gearing system and the electrical generator as shown below:

$$\eta_{overall} = \eta_{turbine} \cdot \eta_{gearing} \cdot \eta_{generator} \quad ...(5.7)$$

5.10 Force and Torque Exerted by Wind on Turbines

The axial force exerted by wind on the rotor (F),

$$F = \frac{1}{2}\rho \cdot A \cdot v^2 \quad ...(5.8)$$

Hence, the maximum theoretical torque is obtained by multiplying the force (F) with radius (R) of the rotor.

$$T = \frac{1}{2}\rho \cdot A \cdot v^2 \cdot R \quad ...(5.9)$$

Actual torque, T_a will be less than this theoretical value and the ratio between the actual torque and the theoretical torque is termed as the Torque coefficient (C_T),

$$C_T = T_a / T = 2T_a / (\rho \cdot A \cdot v^2 \cdot R) \quad ...(5.10)$$

5.11 Tip Speed Ratio

The power extracted by a wind rotor is also dependant on the ratio of the linear speed at the tip of the rotor blade to the speed of the wind. This ratio is called the tip speed ratio (λ). If 'ω' is the angular velocity of turbine rotating at 'N' rotations per unit time,

$$\lambda = \omega R/v = 2\pi NR/v \quad ...(5.11)$$

The power coefficient will be maximum at a particular value of tip speed ratio. The ratio of Power coefficient, C_p to Torque coefficient C_T will be equal to λ.

$$Cp/C_T = \lambda \quad ...(5.12)$$

5.12 Solidity Ratio

The ratio of the projected area of the rotor blades (in the direction of wind) to the rotor swept area is called 'solidity ratio'. Fig. 5.8 shows a multi bladed wind turbine with 'n' blades (n=3), 'S' represents the projected area of blades and 'R' is the rotor radius.

Solidity ratio $(S_R) = n \cdot S/\pi R^2$ ...(5.13)

Most aero-generator rotors have rather thin blades (n=3) which have low solidity and they rotate faster than multi bladed high solidity turbines. The blade tip moves at a speed higher than the wind speed in low solidity wind turbines which use the lift force for their propulsion.

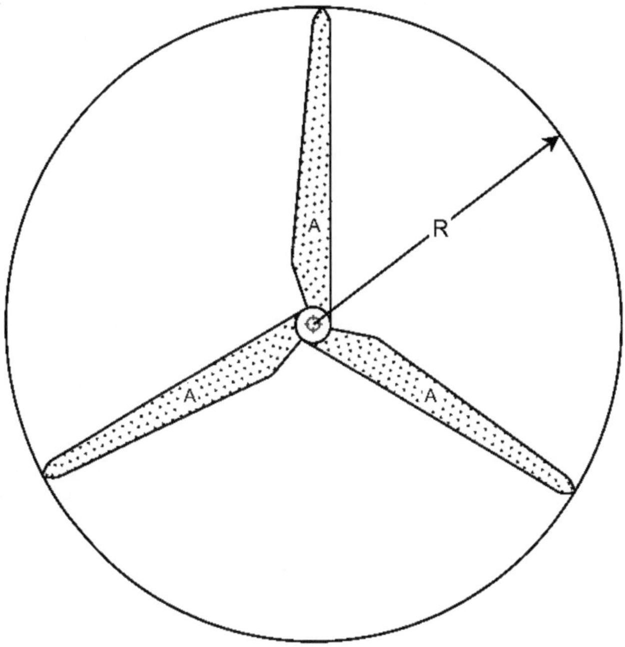

Fig. 5.8 Solidity of wind rotors

Practically, the wind turbines with high solidity have high starting toque as they operate at low values of tip speed ratio. Conversely, turbines with two or three thin blades have low solidity and hence they have very low values of starting torque. The variation of starting torque coefficient with respect to tip speed ratio for low solidity and high solidity turbines are illustrated in Fig. 5.9.

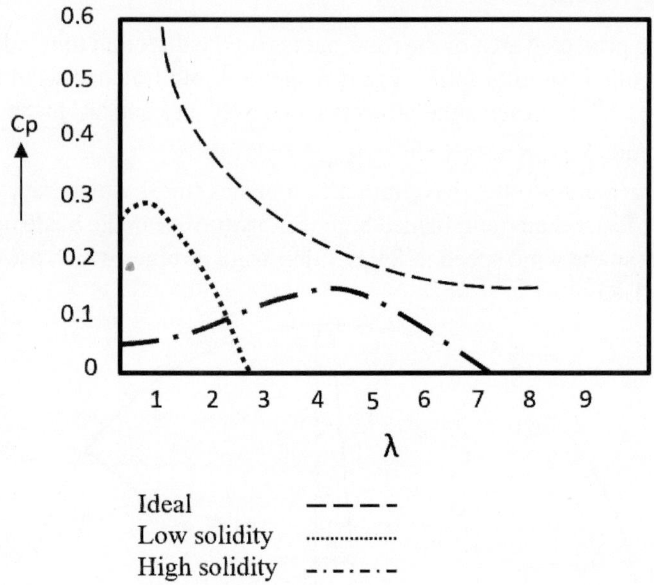

Fig. 5.9 Torque coefficient vs. Tip speed ratio for low solidity and high solidity wind turbines

CHAPTER 6

Wind Turbines

A turbine is a rotating device which convert kinetic energy of a fluid into useful work. During early days, wind driven machines for electricity generation were also referred as wind 'mills' due to their history of evolution from wind operated grain mills. Now a days, 'wind turbine' is the standard terminology for devices used to produce electricity from wind power. We know the power production from wind depends on the efficiency of the turbine at the expected operating conditions. Hence the selection of proper rotor design is very important for extracting maximum possible energy from wind. Different types of wind machines have been evolved over decades.

6.1 History of Wind Turbine Development

As early as 600 AD, humans started using wind mills for grinding cereals and for water lifting. During 7th to 9th century wind driven machines were used in Persia to grind cereals and pump water. By 1000 AD wind driven water pumps were used in China to pump sea water for making salt. During the 12th century, the English people started using wind driven mills to grind grains. The energy requirement of the Dutch people was to pump out water from inundated low-lying areas. They started utilising wind energy in the 14th century for drainage purpose. With the colonisation of America by the Dutch, the technology reached American continent also by 17th century.

In 1887 a Scottish Professor named James Blyth of Anderson's college in Glasgow built a sail type wind turbine to produce electricity. Blyth's holiday home should be the first house where wind generated electricity was used for lighting. Consequent to this, Charles Brush developed another primitive aero generator in US during 1888. In 1891, a Danish inventor called Poul la Cour developed a refined form of an aero generator which could give a more or less steady power supply. Around 2500 wind mills for various applications such as grinding cereals and pumping water was in use in 1900. Poul la Cour later started the Society for Wind Electricians in the year 1903. By the year 1908, more than seventy aero generators were operational in Denmark. A French engineer G.J.M. Darrieus patented his 'Darrieus' wind turbine in 1931.

The first large scale wind generator of Mega Watt range was built in 1941 and was connected to the electricity grid in Vermont of USA. The first wind farm in USA started giving power to about 4150 houses. It is interesting to note that a 2 MW wind generator, the world's first multi–Mega Watt wind machine, was built in 1978 by a school teacher and his students with the technical assistance of German engineers. In 1978 itself, the famous wind turbine manufacturers, 'Vestas' produced their first machine. They decided to concentrate exclusively on wind energy in the year 1987.

The first offshore wind farm was established in 1991 at Vindeby in Denmark. The decade witnessed steady developments in the manufacture and utilization of wind generators and in 2000, the global installed wind powered electricity generation reached 17,400 MW. The world's then largest wind turbine, 'Enercon E-126' having a capacity of 7 MW was installed in 2007 at Emden in Germany. The year 2020 is probably the year of maximum new wind power installations with an amazing figure of 93 GW. Currently (2021), the total wind power installations have reached about 750 GW. Vestas is currently the world's largest manufacturer of wind generators with a market share of about 16 per cent. According to Global Wind Energy Council, other leading manufacturers are GE Renewable Energy, Goldwind, Envision and Siemens Gamesa, respectively.

6.2 Types of Wind Turbines

As already we know, turbines convert kinetic energy of a fluid to mechanical energy which in turn can be utilized directly or converted to another energy form like electrical energy. There has been various turbine types and designs evolved over centuries. But the present-day wind turbines are mostly intended for production of electricity and most of these aero-generators have similar configurations.

Different turbines have been named depending on their geometry and their mode of interaction with the wind stream. The wind turbines are mostly grouped as Vertical Axis Wind Turbines (VAWT) and Horizontal Axis Wind Turbines (HAWT). As shown in Fig. 6.1, the axis of rotation of VAWT is vertical whereas, the HAWT rotors are aligned to rotate in a horizontal axis. Wind direction is perpendicular to the axis for VAWT and parallel to the axis in the case of HAWT rotors.

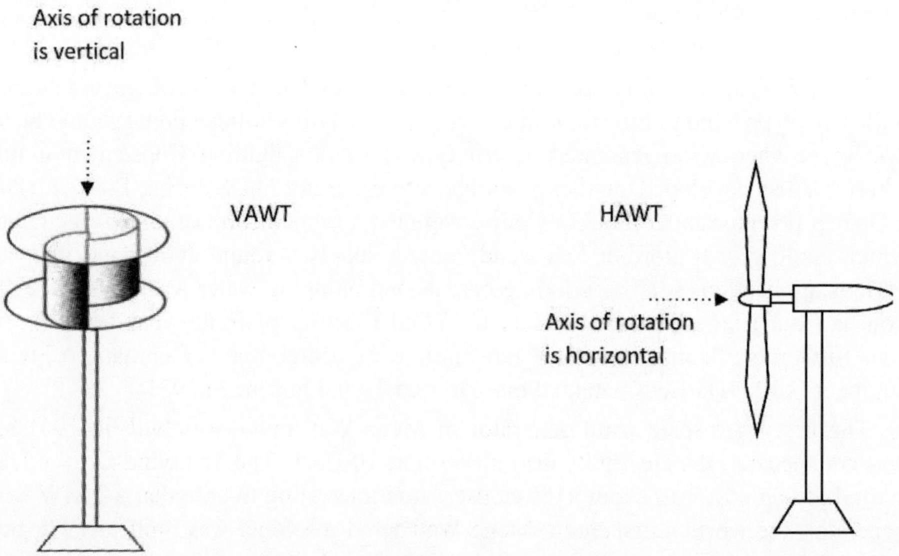

Fig. 6.1 Schematic view of Vertical and Horizontal Axis wind turbines

6.2.1 Vertical axis wind turbines

Vertical axis wind machines were known from very early days. Bishop Fausto Verancio of Venice, also an engineer, described the design of a vertical axis wind turbine in his book '*Machinae Novae*' published in 1616. As the rotor axis is vertical the wind direction is perpendicular to the axis of rotation. Hence they can utilize wind from any direction and hence called '*panemones*'. Another advantage of these rotors is that the heavy generator can be mounted at ground level. Some of the vertical axis rotors can be mounted without using a tall tower. Most common vertical axis wind turbines (exceptions are there) utilize the drag force for their working. The inherent drawback of the drag machines is that their blades cannot move faster than the wind. Fig. 6.2 depicts the principle of drag. Hence drag machines are usually not used for electricity generation purpose.

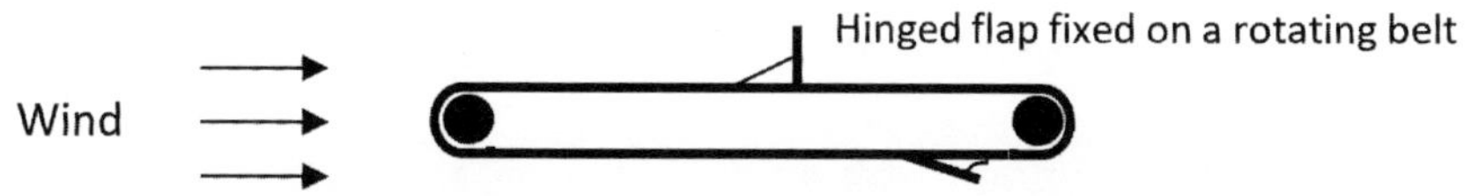

Fig. 6.2 Principle of drag force machine

(a) Cup anemometer: Cup anemometer is an example for vertical axis wind turbine. They are used for wind velocity measurements as their rotational speed has a linear relationship with the wind velocity.

(b) Savonius rotor: The Savonius rotor invented by Sigurd Johannes Savonius in the year 1922 was similar in principle to the vertical axis machine described by Bishop Fausto Verancio. If we split a cylinder and mount the halves side by side to the central axis (Fig. 6.3) we get a Savonius rotor. The major force driving the rotor is drag. Due to high solidity, it has high starting torque. Even though it can be utilized for slow speed applications such as water pumping, it is not suitable for electricity generation.

(c) Darrieus rotor: The Darrieus rotor was invented as early as 1925 by a French aeronautical engineer named Georges Jean Mario Darrieus. The 'egg beater' shaped rotor comprises of a rotating shaft on which a number of curved blades are mounted. The curved blades have an aerofoil cross section to make use of lift force and is often referred as '*troposkein*' shape. Darrieus rotors in general have low solidity and the starting torque also is less. As they are not self-starting, sometimes a Savonius rotor is coupled to it to enable starting. The 'H rotor' working on similar principle was developed by Peter Musgrove of UK in the 1980s. Thereafter many similar versions of rotors were designed including straight bladed as well as with helical blades.

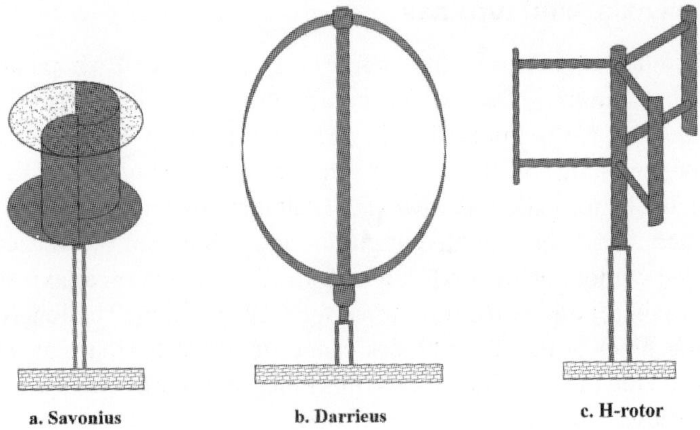

Fig. 6.3 Different Vertical axis wind turbines

6.2.2 Horizontal axis wind turbines

Early wind mills used for grinding cereals were mostly of horizontal axis type and were called Dutch wind mills; one such is shown in Fig. 6.4. These wind-driven machines used for pumping water had high solidity multi blade type rotors in contrast to aerofoil cross section blades of modern turbines.

Fig. 6.4 Traditional Dutch wind mill

As the name indicates, the axis of rotation is horizontal for these machines and the rotor is set perpendicular to the wind direction. At present, HAWTs are the most common type of wind turbines used for electricity production. They can be either mounted as upwind or down wind. The upwind type requires a tail vane to orient the turbine to face wind. Modern wind turbines depend on aerodynamic forces for their propulsion, similar to the forces created by aircraft wings. They can be single bladed, two bladed or three bladed. Single blade turbines are now a days not used due to the complexities in balancing and also due to lower efficiency compared to two bladed and three bladed designs running at same speed. For a two bladed turbine, at some instant the entire blade mass will be aligned vertically, whereas at another instant it will be horizontal. As this occur in each rotation, this creates undesirable stresses on machine components. Again, it is difficult to turn the two bladed turbine to face a changing wind direction (Yawing). In 3 bladed turbines (Fig. 6.5), the blade masses are aligned evenly in the horizontal and vertical planes at all positions. They also have appreciable aerodynamic efficiencies.

Fig. 6.5 Three bladed wind turbine

6.3 Components of a Wind Turbine Used as Aero Generator

In contrast to the traditional cereal grinding and water pumping wind mills the function of a modern wind turbine is to generate good quality electricity compatible with the grid frequency. Hence, they are envisaged as independent power plants. But, as in the case of other power plants, a modern MW capacity wind turbine is not regularly attended for its maintenance. It should run day and night throughout for 20 or more years. The development of the microprocessor has played a crucial role in enabling cost effective wind technology.

A modern wind turbine requires a number of components to accomplish its designated job of converting the kinetic energy of the wind into electrical power. Fig. 6.6 depicts the different components of a typical wind turbine. The *'nacelle'* is a common terminology used to denote the entire housing of the wind generator system mounted on the tower.

The *'hub'* holds the rotor and is a crucial part of the wind turbine. It is so designed as to withstand the stresses due to the forces imparted by the rotation of rotor at various speeds. It is the primary part that transmit the torque into the generator system. Some of the modern wind turbine manufacturers pay special attention to aerodynamic integration of hub to the rotor blades.

The *rotor blades* are the components for collecting and transforming the wind energy. The shape and mechanical properties of the rotors are crucial in the performance of the machine. Conventionally, the blades are made from fibre glass or metals, whereas many newer aerogenerators are fitted with blades made of fibre glass composites. The composite materials used for blades are comparatively low cost and they have many preferable mechanical properties such as low weight and corrosion resistance.

A *'gear system'* is required for most of the aerogenerators as the low rotational speeds of the shaft should be upscaled to higher speeds for optimum electricity production by the generator. The rotor speed may vary from 40-400 rpm depending on the wind speed whereas the generator requires speeds between 1200 to 1900 rpm. But there are also some type of DC generators used in 'direct drive wind generators' in which the generator rotor is directly coupled with the wind turbine rotor. Recently some of the major manufacturers (Enercon Ltd., Germany) have also developed special type of generators with 42 pole pairs which do not require a gear box.

'Generator' is the component for electricity production, which works on the Faraday's principle (electricity is produced by the cutting of magnetic flux by a conductor). The generators are of different types and they are selected based on the overall system configuration such as capacity, wind velocity and the turbine design aspects. Basically, the generators are of two types, alternating current (AC) and direct current (DC) generators.

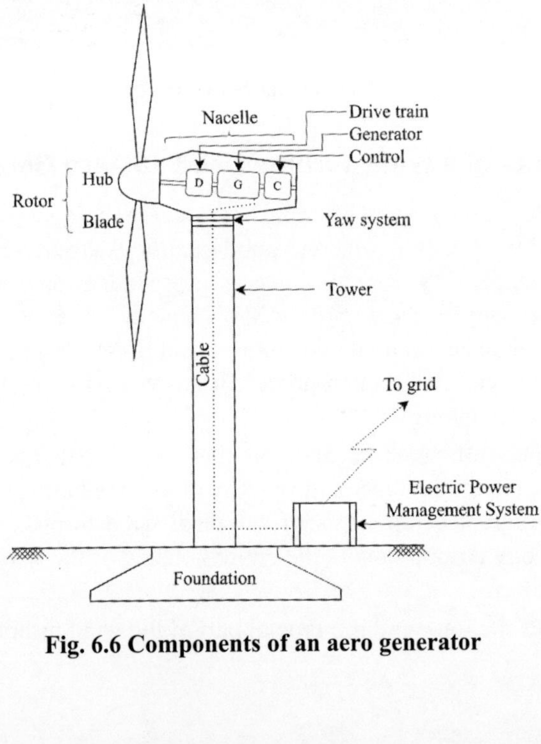

Fig. 6.6 Components of an aero generator

Modern wind turbines employ electronic *'control systems'* to regulate the entire operation of the generator and electricity transmission systems including safety aspects. Different control methods are used to optimize the power production or sometimes to limit the output as required by the grid. This is achieved by controlling the generator speed or turbine rotary speed as well as active adjustment of the blades to change the angle of attack. Maximum Power Point Tracker (MPPT) is a control system used to extract maximum power from the turbine at varying conditions.

The turbines are not designed to with stand extreme conditions when they are rotating. But once they stop rotating, they can be safe even in an extreme stormy condition. For this, a 'cut out speed' is set for the turbines beyond which they stop rotating. There are basically two methods to protect the machine from very high velocity winds. One method is *'pitch control'* in which the blades can turn in its own axis so as to vary the pitch angle. Once the pitch angle changes from optimum, the lift force gets reduced which in turn decrease the torque or speed of the turbine. Some systems use *'stall control'* in which the blades are designed in such a way that the turbine rotational speed gets decreased when wind speeds exceed certain value. Mechanical speed brakes are also sometimes used to regulate turbine speeds.

The *'yaw system'* is also a part of the control system and is used to orient the turbine with nacelle according to the change in wind direction. Yaw system for horizontal axis machines is of two types. The simplest one is *'passive yaw'* in which a tail fin is provided on the nacelle. Even though this type of yaw is enough for small wind turbines, this creates loss of power and deteriorates some of the components due to instability of yaw. Hence large machines make use of active yaw control. Instruments to measure speed and direction of wind is mounted on the nacelle. Microprocessor controlled rack and pinion mechanism rotates the hub either to face the wind for better power production or away from wind as a safety measure in stormy weather condition.

The *'tower'* supports the nacelle with the rotor so that the blades can safely rotate and there is sufficient clearance from ground. The height of the tower may vary according to the wind speeds at different heights. The increased cost for taller towers is to be economically justified and there can also be other factors in deciding the height of tower. Bigger wind turbines are generally mounted on tall towers of height more than 50 m. The giant 3 MW wind turbine of GE Renewable Energy has a rotor diameter more than 130 m and is mounted on tubular steel towers having heights above 85 m. Wind turbine tower height is site specific and may depend on several site characteristics such as nearby structures and topography. In general, it is recommended that the tower height should be such that the lowest point on the rotor is above a minimum distance of 9 metres above the tallest obstacle near the turbine.

6.4 Aerodynamic Design of Wind Turbine Blades

The three bladed rotor is the most common type of rotor used for aerogenerators and the design of blades are of utmost importance in extracting maximum power from the wind stream. Hence a rotor blade has to be aerodynamically designed similar to the design of an aircraft wing. The major parameters of a wind turbine blade are illustrated in Fig. 6.7.

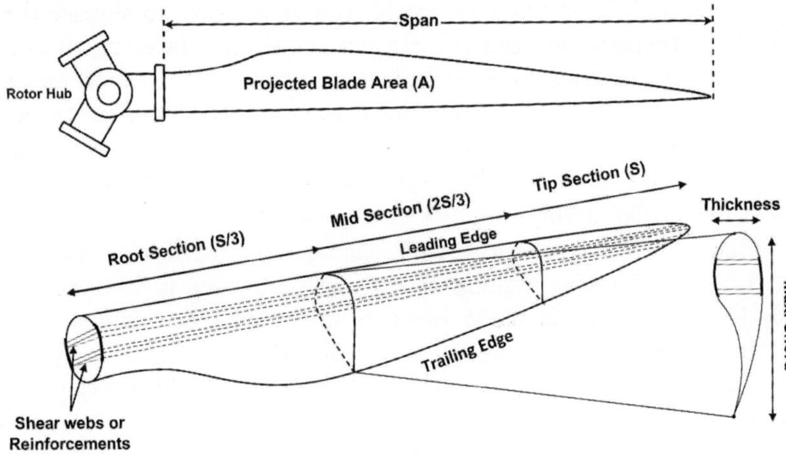

Fig. 6.7 Typical configuration of a wind turbine blade

The wind turbine rotor blade has an approximately rectangular shape with a leading edge which cuts through the air stream along with the trailing edge on the opposite side. The configuration and durability of material is most crucial for the leading edge. The leading edge is subject to corrosion by rain drops and snow eventually resulting in the reduction of power output and may ultimately cause damage of the blade. The blade root is the portion of blade fixed to the hub of rotor. The blade tip is also a critical part of the blade as the tip has the maximum speed when the turbine is rotating. The blade tip of some of the large wind turbines may be moving at a speed of 200 to 250 kmph. As already discussed, tip speed ratio is an important parameter deciding the performance of the machine. The radius of the blade is the distance from the rotor shaft to the outer edge of the blade tip. If we examine the blade cross section, it is not symmetric bilaterally or radially. The leading edge is flatter than the trailing edge, which is almost tapering. This aerodynamic profile is required to enhance lift forces. Similarly, *aspect ratio* of a given blade play an important role in generating lift and torque in wind turbine.

$$\text{Aspect Ratio} = \frac{\text{square of blade span}}{\text{projected area of the blade}} = \frac{s^2}{A} \qquad ...(6.1)$$

Wind turbines extract energy from the force of the wind on an aerofoil, in this case it is a turbine blade. The relative motion between the air flow and the turbine blade, is the same as for an aircraft wing aerofoil (see previous chapter section 5.6), but in this case the wind is in motion towards the passive turbine blades, so that the external thrust provided by the moving air flow is in the opposite direction to the thrust provided by an aircraft wing. In other words, as the wind blows, a pocket of low-pressure air forms on one side of the blade. The low-pressure air pocket then pulls the blade toward it, causing the rotor to turn. This is called lift. The force of the lift is much stronger than the wind's force against the frontal surface area of the blade, which is called drag. Turbine blades thus experience lift and drag forces similar to the aircraft wing, which set the blades in motion transferring the wind energy into the kinetic energy of the blades. The turbine blades are connected to a single rotor shaft and the force of the wind along the length of the blades creates a torque which turns the rotor.

Hence, shape of the aerodynamic profile of the blade is decisive for rotor performance. The shape is selected with care on the basis of available information gathered experimentally. The United States National Advisory Committee for Aeronautics (NACA) then; now National Aeronautics and Space Administration (NASA) developed a large number of aerofoil profiles which could be readily selected for the required characteristics of the rotor. A rotor with properly designed blade profile will start slowly but will gather speed as lift forces keeps on increasing. Fig. 6.8 illustrates NACA's wind tunnel testing of an aerofoil design i.e., Clark-Y (named after Virginius E. Clark, U.S. Army's military aviation pioneer who designed it) at an aspect ratio of 6 demonstrating variations in the characteristic lift (C_L) and drag (C_D) coefficient curves for different angles of attack (α) at constant air speed. Practically, all aerofoils have critical angles of attack, beyond which it struggles to generate lift and further increment in angle of attack results in stalling the rotor due to excessive drag force.

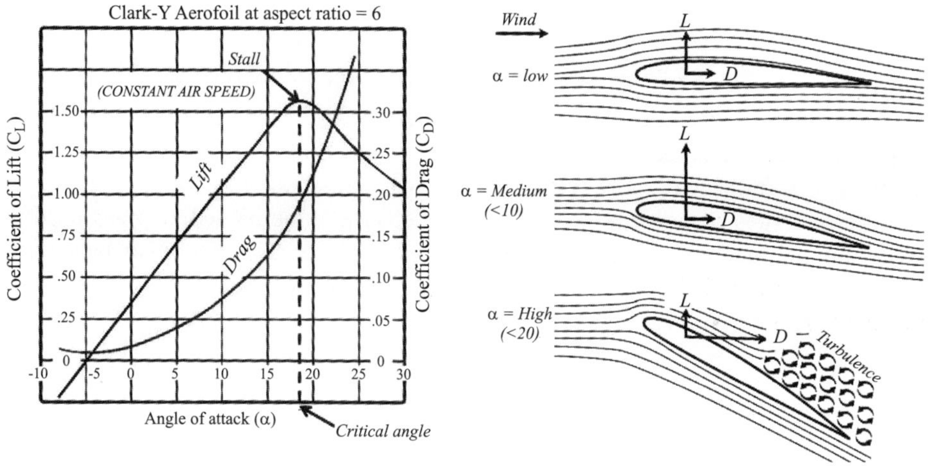

Fig. 6.8 Variations in Coefficients of Lift and Drag at constant air speed

6.5 Characteristics of Wind Turbines

Power obtainable from wind turbine is dependent on the rotor speed. There are many factors which affect the rotor speed, the major factor being wind speed. Once the turbine reaches the rated speed, the control system will come into action and the turbine speed will be regulated even though the wind speed is high. Hence the actual power produced depends on the pitch of the blades also in addition to the load of the mechanical and electrical systems coupled to the rotor shaft.

Fig. 6.9 depicts the power production characteristics of a wind turbine with respect to the power available in the wind at varying velocities. As wind velocity increases, the power of the wind stream increases proportional to the cube of velocity. But the wind turbine starts at a particular wind speed when the wind is strong enough to overcome the friction in the drive and generator. This speed at which the turbine starts to rotate is called the *'cut-in velocity'* and is in between 3 and 4 metres per second. Now, the power produced keeps on increasing along the path 1-2. At point 2, the turbine attains its rated speed and hence forth will run only at that speed even though the wind velocity keeps on increasing. At point 3, the wind velocity is catastrophically high and the turbine stops rotating beyond the *'cut-out velocity'* or *'furling velocity'*. The usual cut out wind velocity for turbines is 25 metres per second, beyond which no safe power production is possible.

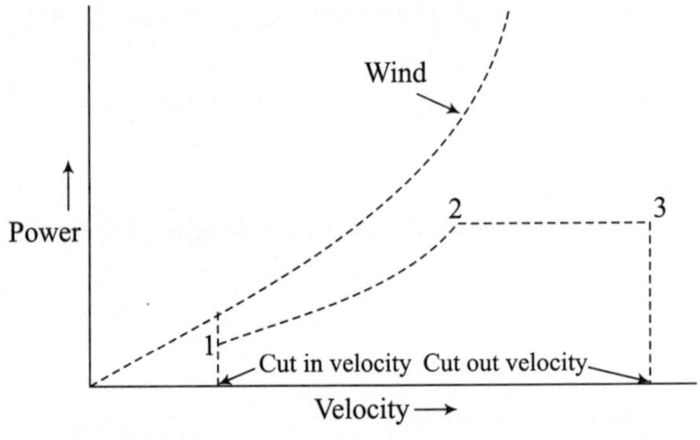

Fig. 6.9 Power-Velocity characteristics of a wind turbine

The power coefficient (Cp) of the turbine at different wind speeds is another important characteristic. The variation of power coefficient with respect to the tip speed ratio of turbine is more relevant and have been studied in the past in the case of different types of turbines. Modern propeller type three bladed turbines used for electricity production have a Cp in the range 45 to 50 per cent and it is dependent on the tip speed ratio (TSR denoted as 'λ'). In most cases, maximum Cp is observed at a tip speed ratio 5 to 6 (Fig. 6.10).

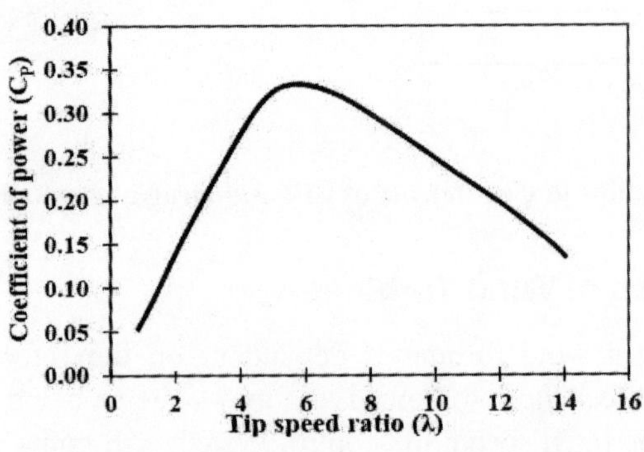

Fig. 6.10 Variation of Cp with Tip speed ratio

The variation of Cp with λ is different for different air foil sections. For other types of turbines like Savonius, Darrieus or multi bladed, the trend of variation of Cp with TSR is very much different. The multi bladed type HAWT machines generally have a very low Cp. Among vertical axis machines, Darrieus rotors show better Cp in the range of 25 to 35 per cent.

CHAPTER 7

Wind Resource Analysis

Wind is an intermittent energy resource and even small change in wind speed causes large deviation in the power output from wind turbines as the power is proportional to cube of wind velocity. Assessment and analysis of wind speed is the most crucial factor in the siting of a wind generator. The characterisation of wind resources should be done carefully as it is site-specific and is an essential prerequisite in the selection of potential sites.

7.1 Wind Velocity Data

Wind velocity is a vector quantity, specified by its magnitude and direction. The magnitude of horizontal component is important for wind turbine operation. On a regional level, surface wind data can be represented as 'Isovents' or 'Isodynes'. Isovents are contours of constant average wind velocity (m/s or km/h) and contours of constant wind power (Watts/m^2) are called Isodynes; both represent quantities averaged over a particular period. Indian Meteorology Department (IMD) have analysed and presented the tabulated wind data on hourly basis for selected locations. As per the NREL (National Renewable Energy Laboratory, USA) standards, wind speeds at heights of 40m, 25m, and 10m are measured and analysed.

Wind speeds vary with heights and also are affected by various obstructions. The reduction in speed and power due to a building of height 'H' is depicted in Fig. 7.1. Even at a distance of 10H, 6 per cent of wind speed and 17 per cent of output power gets reduced.

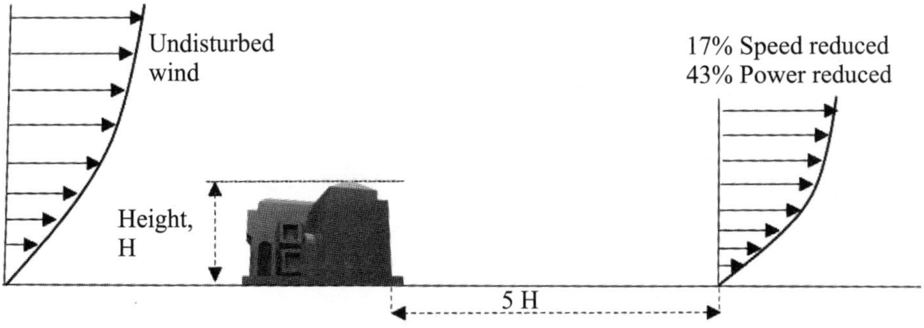

Fig. 7.1 Reduction of wind speed and power by obstruction *(Source: NREL)*

Cup anemometers are most commonly used for horizontal wind speed measurement. This instrument consists of a cup assembly of three or four cups which are connected to a rotating vertical shaft. The rotation of this shaft is linearly proportional to wind speed. The senser with a transducer converts this rotational motion of the shaft into an electrical signal to the data logging system. Wind direction can be measured and recorded using a modern wind vane. The common type has a fin connected to a vertical shaft which positions the wind vane in the least resistance position, i.e., in the wind direction. Wind vanes make use of a transducer that produces an electrical signal with respect to the position of the vane and is transmitted to a data logger.

The U.S. Department of Energy has classified wind power cites from Class-1 to Class-7 based on the power density. The different classes characterised by their wind power density is shown in Table 7.1. Based on updated wind data, gridded maps have been prepared. Grid cells designated as Class-4 or greater are generally considered to be suitable for most wind turbine applications. Grid locations categorised as Class-3 are suitable if we are installing turbines with tall towers of 50 m height or more. Class-2 sites may not be suitable in general and there can be some microsites within the area and Class-1 locations are not suitable for wind energy development programmes.

Table 7.1 Different classes of wind power sites

Class	At 30 m height		At 50 m height	
	Wind speed, $m.s^{-1}$	Power density, Wm^{-2}	Wind speed, $m.s^{-1}$	Power density, Wm^{-2}
Class 1	5.1	160	5.6	200
Class 2	5.9	240	6.4	300
Class 3	6.5	320	7.0	400
Class 4	7.0	400	7.5	500
Class 5	7.4	480	8.0	600
Class 6	8.2	640	8.8	800
Class 7	11.0	1600	11.9	2000

(Source: Wind Resource Assessment Handbook, NREL)

Global Wind Atlas (https://globalwindatlas.info), a free web-based application developed by the Technical University of Denmark is intended to help wind energy planners and investors to identify suitable locations for wind power generation anywhere in the world. The data as well as wind maps are free to be downloaded and used (Fig. 7.2).

The National Institute of Wind Energy, Chennai (NIWE) of India, has installed over 800 wind-monitoring stations all over the country and prepared wind maps at 50m, 80m, 100m and 120m above ground level (AGL) for potential sites. The map was prepared in Capacity Utilization Factor (CUF) scale and CUF more than 20 per cent was considered for potential estimation. The recent (2019) assessment indicated a gross wind power potential of 302 GW in the country at 100 m AGL. Further 120m high wind potential map was prepared by NIWE to cater to increased hub heights of wind turbines and it was estimated that 695.50 GW is the potential at 120 m AGL. The potential of major wind rich states is given in Table 7.2.

Table 7.2 Wind Energy Potential of major windy states in India

Sl. No.	State	Wind potential at 100 m (GW)	Wind potential at 120 m (GW)
1.	Gujarat	84.43	142.56
2.	Rajasthan	18.77	127.75
3.	Maharashtra	45.39	98.21
4.	Tamil Nadu	33.79	68.75
5.	Madhya Pradesh	10.48	15.40
6.	Karnataka	55.85	124.15
7.	Andhra Pradesh	44.22	74.90
	Others	9.28	43.78
Total		302.25	695.50

(Source: MNRE, 2021)

7.2 Wind Rose

Wind rose is a graphic representation which depicts the direction and speed of wind at a particular location. They are presented in a circular format and the length of each sector of the circle indicates the amount of time that the wind blows from that particular direction. The distribution of wind spread over different directions are usually marked on twelve or sixteen sectors in a circular compass. Different colours and gradients or patterns can be given to demarcate different ranges of wind speed as shown in Fig. 7.3

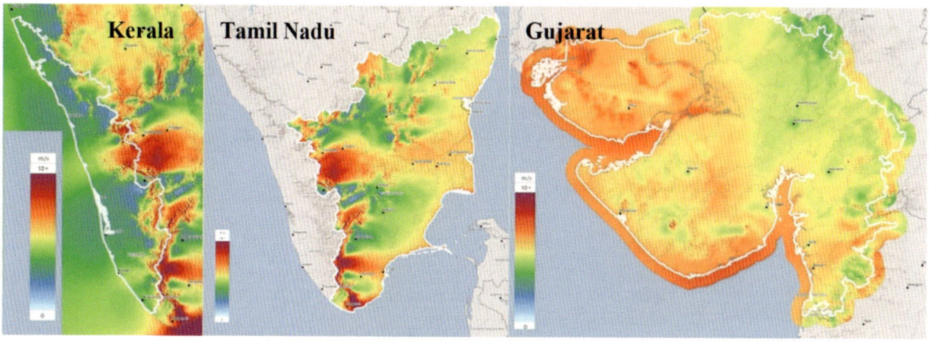

(Source: Global Wind Atlas)

Fig. 7.2 Wind map showing mean wind speed at 100 m for three states in India

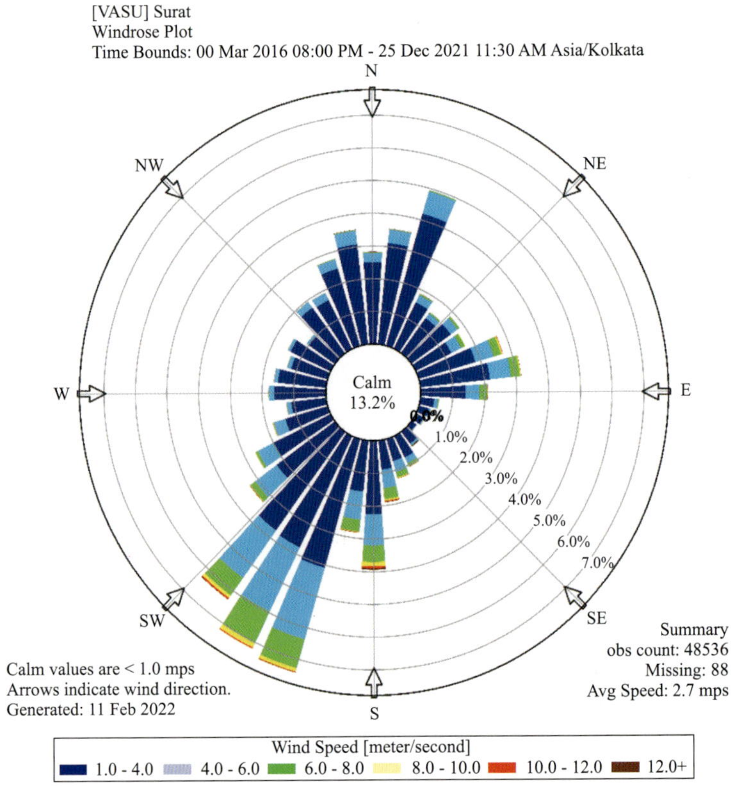

Fig. 7.3 Wind rose – Speed distribution

(Courtesy: Iowa Environmental Mesonet, Iowa State University)

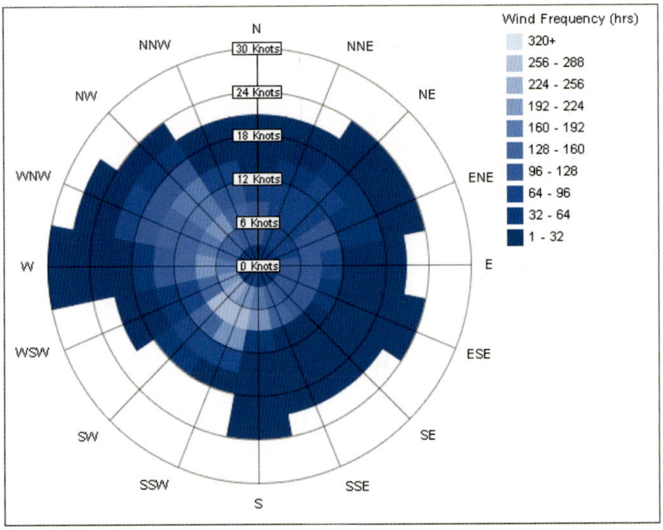

Fig. 7.4 Annual wind rose representing frequency distribution

(Courtesy: Autodesk Knowledge Network)

Wind rose can also be drawn to give more information as shown in Fig. 7.4. The radius of the outermost wedges represents the frequency of 16 wind directions. The radial scale now represents wind speed rather than percent of time. Also, the segments with colour gradient of each spoke represent hours rather than wind speed.

7.3 Energy Estimation

The preliminary estimation of wind energy at a particular site may be done using mean wind speed data over a period. The maximum values may be much different from the mean values as shown in Fig. 7.5. It is more economic to select a site with fairly high winds even though they occur occasionally, than a site where we get constant low wind speed.

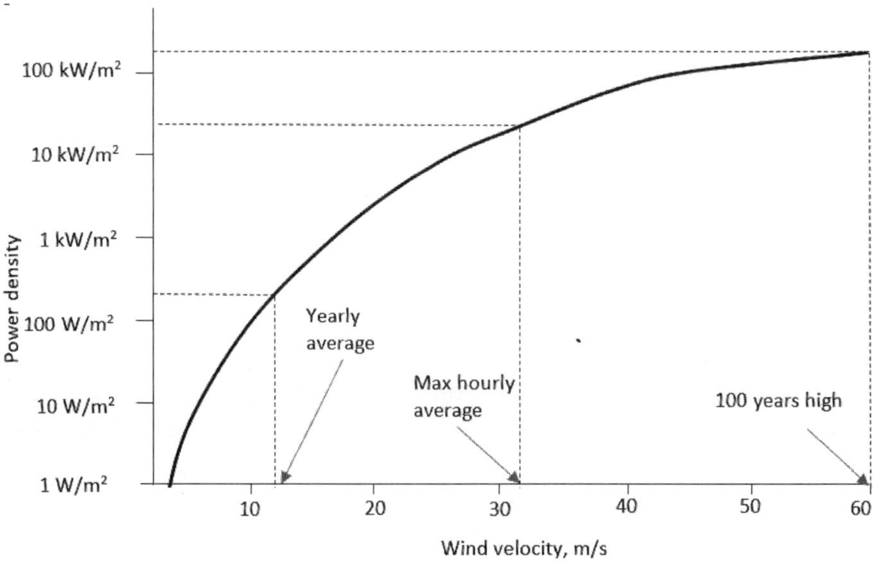

Fig. 7.5 Wind velocity and power density

7.4 Analysis of Hourly Mean Wind Velocity

Basic wind data of hourly mean wind velocity can be analysed in different ways to get a clear assessment on the power availability in a particular site.

7.4.1 Frequency-duration graph

The number of hours per year during which the wind velocity exceeds certain value is an important information which can be depicted by a frequency duration graph (Fig. 7.6). 'Weibull' statistical distribution is a commonly used statistical tool in analysing wind speed measurements.

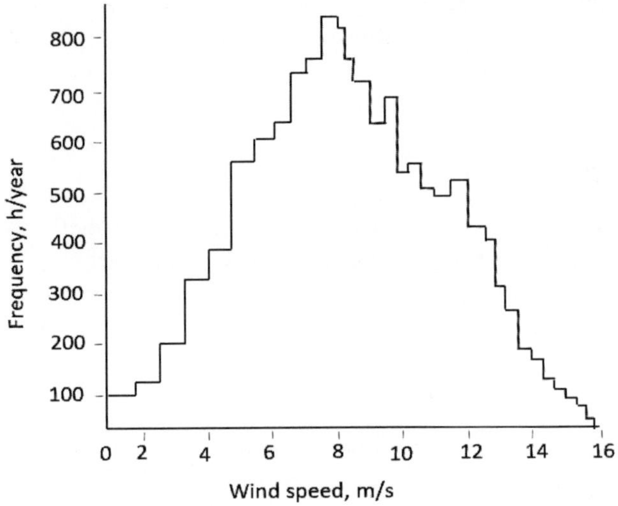

Fig. 7.6 Frequency – Duration graph

7.4.2 Speed – Duration curve

The wind speed data can further be processed into number of hours in the year for which the speed equals or exceeds a particular value. These speed values plotted against the hours in a year is called speed - duration curve (also referred as velocity-duration curve). The values on X-axis of the graph range from 0 to 8760 (Total number of hours in a year). The curve will touch X-axis indicating the point where the wind speed is above zero. The curve merge with Y-axis indicating the speed which is not achieved even for a single hour in a year. In a similar way, monthly speed – duration curve also can be prepared for detailed analysis of the wind speed data (Fig. 7.7).

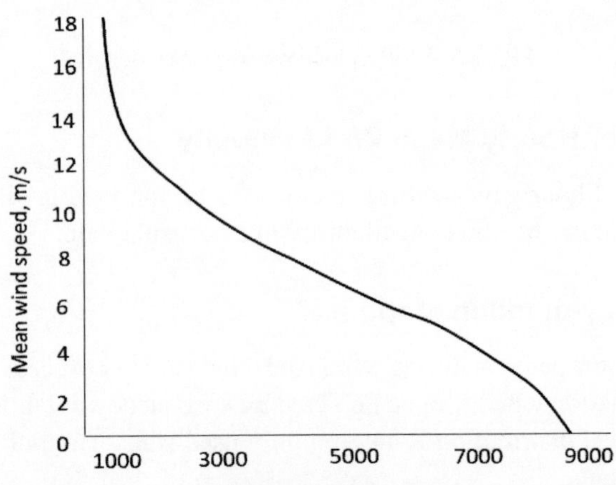

Fig. 7.7 Speed – Duration curve

7.4.3 Power- duration curve

As power is proportional to cube of wind speed, the cube of velocity (v^3) is plotted against hours to get the power- duration curve. These plots are useful in assessing the potential of a site and also to find the design wind velocity for the wind turbine to be installed at the specific site. The wind turbine design should match the wind characteristics for efficient energy production (Fig. 7.8).

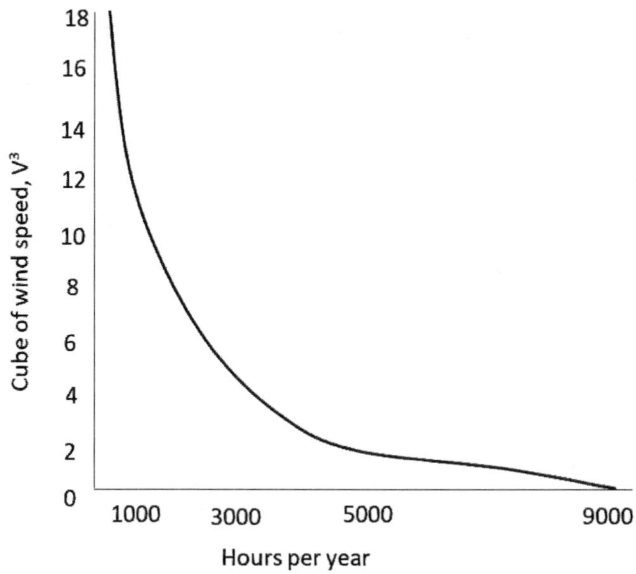

Fig. 7.8 Power–Duration curve

7.5 Site Selection for Wind Energy Conversion Systems

Wind Energy Conversion System (WECS) should be located in areas where the winds are strong and reliable with favourable frequency–duration characteristic, as the power availability in the wind increases tremendously with velocity. Hence, the basic requirement is 'high annual average wind velocity'. For precise assessment of wind power potential, anemometer data should be available for the site at the hub height. It is desirable to have average wind velocity in the range 3.5 to 4.5 metre per second (12 to16 kmph). This is the lower limit at which large WECS start operation (cut-in velocity).

Usually, three different approaches are adopted for analysing the suitability of a particular site. The first approach is to make use of the data on wind speed from the nearby available meteorological station. The power output can be assessed reasonably accurate from this and this is probably the easiest and low-cost method. More precisely, we can also make some limited measurements in the proposed location so that the available wind data from nearby stations are adjusted on the basis of the collected data to get a clear picture on actual power output that can be expected. In case the data from nearby station is not representative of the site or dependable data is not available, we need to collect wind data and analyse it to get the probable power output.

In general, the characteristics for a good site are:

- High annual average wind speed
- There should be no tall obstructions in the vicinity
- Favourable topography such as a smooth and round hill top with gentle slopes
- Mountain gap which produces wind funnelling (as in the case of Palakkad gap)
- Nearness to load centres
- Low-cost land availability

7.6 Layout of Wind Turbines in a Wind Farm

Preparing a layout for a wind farm involves locating the individual wind turbines optimally inside the given boundaries of the farm. This is very important when the land availability is limited. The expected power produced by the wind farm may get reduced by various losses. 'Wake effect' is considered the most important factor influencing the power production adversely. As wind turbines extract energy from the wind and there is a resulting 'wake' from the wind turbine downstream, where the wind speed gets reduced to some distance. Turbulence caused by upstream wind turbines along with wake effect from nearby turbines can cause problems when a number of turbines are installed closely in a wind farm. Turbulence can cause unwanted loads and resulting damages. A typical spacing as shown in Fig. 7.9 may be adopted to prevent power losses due to wake effect and turbulence from adjacent turbines.

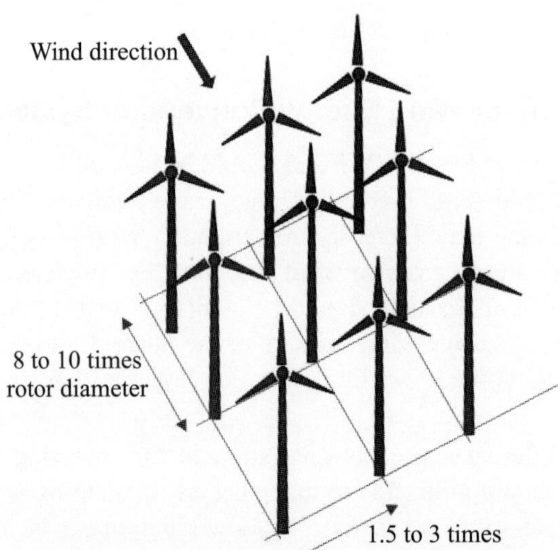

Fig. 7.9 Layout of wind turbines in a wind farm

CHAPTER 8

Energy from Biomass–Bioenergy

The most vital energy conversion process taking place spontaneously in the nature is the production of biomass by plants using sunlight. Biomass includes all contemporary biological matter, both of plant and animal origin. Solar energy is captured by plants and stored as chemical energy in biomass. Energy derived from biomass is called bioenergy and this energy can be used to generate thermal energy, electricity or run engines.

Energy conversion of wastes and biomass is inevitable to provide clean energy for the society. Bioenergy is considered to be clean energy as they do not cause pollution as in the case of fossil fuels. Systems, which utilize energies produced from biomass, are typical examples of energy recycling systems. Whatever CO_2 they produce when burned, has been trapped from the atmosphere during their life time and if you allow plants to grow, the same CO_2 will be recycled.

Bioenergy is at present the most important energy source in some of the developing countries and is capable of playing a big role in their future energy development strategies as well. Biomass in developing countries mostly includes fuel wood, animal wastes and agricultural residues, which is primarily utilized for essential activities such as cooking. Development of technologies that efficiently produce biomass, process it for better utilization and convert it to more convenient forms of energy is therefore gaining attention.

In addition, bioenergy has the potential for decentralized energy generation which is highly relevant in a developing country like India. The technological development in this area has a direct positive impact in the quality of life in rural areas. On a national level, the possibility of biomass-based power generation is around 20 gigawatts (GW).

8.1 Photosynthesis

The word '*photosynthesis*' has its origin in Greek and has the meaning 'putting together using light'. Solar radiation striking the earth on an annual basis is equivalent to 178,000 terawatts (TW), i.e., 15,000 times that of current global energy consumption. Although photosynthetic energy capture is estimated to be about 40 TW, eight times that of global annual energy consumption, only a small part of this solar radiation is used for photosynthesis. All biomass (plant, animal and microbial) originates through CO_2 fixation by photosynthesis and is the most important physiological process for life on earth. Approximately two thirds of the net global photosynthetic productivity is by terrestrial plants, the remainder being produced mainly by phytoplankton (microalgae) in the oceans which cover approximately 70 per cent of the total surface area of the earth. Biomass constituents like sugars, starch, hemi-cellulose, cellulose and lignin are formed by fixing atmospheric CO_2 in the plant cells. The maintenance of oxygen balance of the atmosphere is also achieved naturally by this phenomenon (Fig. 8.1).

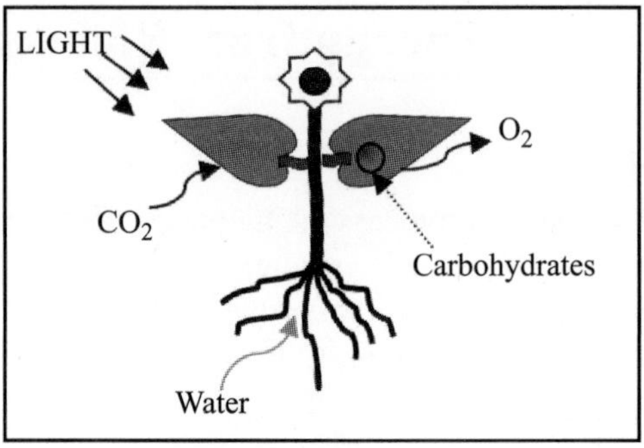

Fig. 8.1 Photosynthesis

Chemically, photosynthesis may be defined as the photochemical reduction of CO_2 and can be simply represented by the equation:

$$6CO_2 + 6H_2O \xrightarrow{\text{Sunlight + Chlorophyll}} C_6H_{12}O_6 + 6O_2 \qquad ...(8.1)$$

The chemical energy contained in the products, glucose and oxygen together is much higher than that in both water and CO_2. The source of this excess energy is solar energy. The two basic aspects of photosynthesis are:

(i) Conversion of light energy from sun.

(ii) Conversion of CO_2 into organic compounds (biomass).

The actual photosynthetic reactions are complex and are different in higher plants and lower organisms like photosynthetic bacteria. The photosynthetic process in plants differs from that in microorganisms, based on the electron source used as energy carriers. In plants (such as green algae) and cyanobacteria (blue-green algae) water is the electron source. In photosynthetic bacteria, organic or Sulphur compounds provide electron sources.

Even some plant species also differ in their process. Scientists have proved that the oxygen in the process is evolved from water and not from CO_2. In fact, the overall reaction is better represented as:

$$6CO_2 + 12H_2O \xrightarrow{\text{Sunlight + Chlorophyll}} C_6H_{12}O_6 + 6H_2O + 6O_2 \qquad ...(8.2)$$

Therefore, 12 water molecules are involved in the process and the hydrogen required for reduction of CO_2 is taken from water.

The process of photosynthesis consists of two phases, one light dependent phase (light reaction) and one light independent phase (dark reaction). The light reaction is also called 'photo stage' and the dark reaction, 'synthesis stage'. Basically, the process starts with absorption of solar energy by the so called 'reaction centres' in the plant cells where green pigments called 'chlorophylls' are present (Fig. 8.2 and Fig. 8.3).

Absorption of light by pigments occur in the grana portion of chloroplast. Here, the water is split (photolysis) and the oxygen is released, paving way for further synthesis with the freed hydrogen atoms to form intermediate compounds viz. nicotinamide adenine dinucleotide phosphate (NADPH) and adenosine triphosphate (ATP).

The next phase of reactions is light independent, leading to formation of sugars to store energy. The dark reaction is enzymatic which takes place in the stroma portion of the chloroplast. The conversion of CO_2 to carbohydrates occurs during this phase. Calvin and his associates could describe this complete biochemical process in the form of a cycle which is now called 'Calvin cycle'.

About 114 kilocalories of free energy are stored in plant biomass for every mole of CO_2 fixed during photosynthesis. Thus, biomass production has the advantage that it is a storable energy form naturally derived from solar energy. Although, photosynthesis is fundamental to the conversion of solar radiation into stored biomass energy, its theoretically achievable efficiency is limited both by the limited wavelength range applicable to photosynthesis, and the quantum requirements of the photosynthetic process.

8.2 Energy Conversion in Photosynthesis

Photosynthesis is the crucial pathway through which energy from the physical world enters the biological system. If we expose our body to sun light our body gets heated up. But this energy cannot be stored in the body nor can it be utilized to power the muscles. All the motive power of animals is derived from the chemical energy taken as food which is solar energy converted through photosynthesis. Light energy of sun is a very dilute form of energy which is concentrated by the plants as chemical energy. It is then available as the energy source for living organisms and is liberated by the reaction:

$$\text{Carbohydrate} + \text{Oxygen} \xrightarrow{\text{Metabolism}} CO_2 + 6H_2O + 690 \text{ kcal/mole of glucose} \quad ...(8.3)$$

From the above equation, we know that the energy that should be added in the reverse reaction (photosynthesis) is 690 kcal per mole ($690 \times 4.2 \times 10^3$ Joules) of glucose produced. If we want to calculate the energy used for one glucose molecule, we should convert moles to molecules and kcal to electron volts (eV).

$$\frac{\text{Energy}}{\text{mole of glucose}} = \frac{690\,\text{kcal}\cdot\text{mole}^{-1} \times 4.2\times10^{3}\,\text{J}\cdot\text{kcal}^{-1} \times 0.63\times10^{19}\,\text{eV}\cdot\text{J}^{-1}}{6\times10^{23}\ \text{molecules}\cdot\text{mole}^{-1}}$$

$$= 30\ \text{eV} \quad ...(8.4)$$

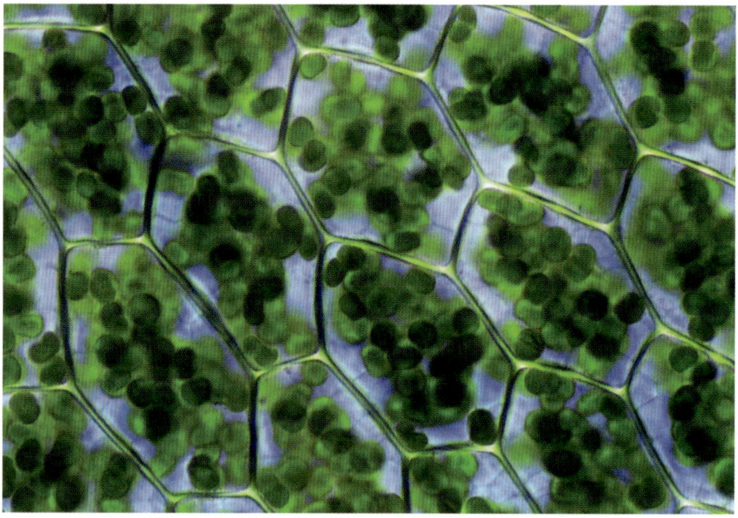

Fig. 8.2 Chlorophyll

(*Author: Kristian Peters—Fabelfroh, https://www.nationalgeographic.org/encyclopedia/chlorophyll/ , CC BY-SA 3.0 Unported via National Geographic, under GNU Free Documentation License)*

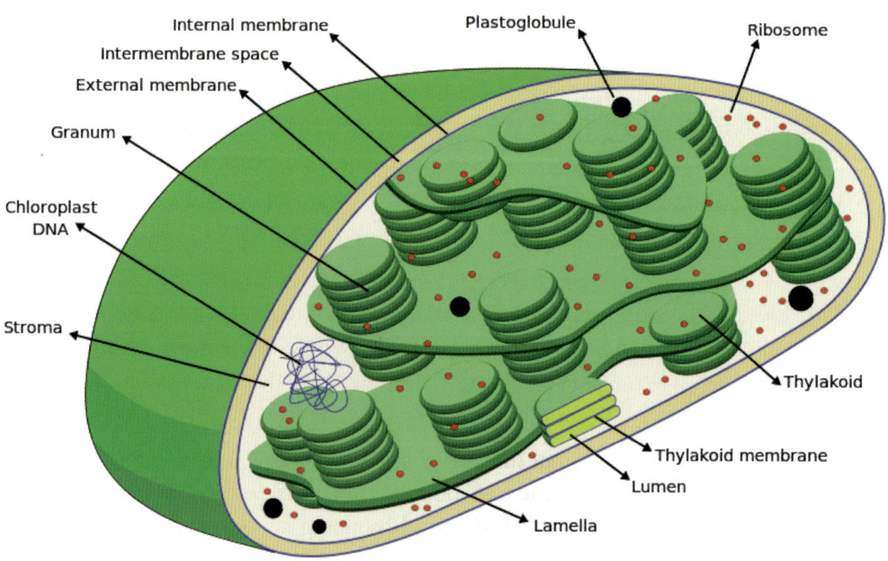

Fig. 8.3 Chloroplast configuration

(Author: Miguelsierra, https://commons.wikimedia.org/wiki/File:Scheme_Chloroplast-en.svg , CC BY-SA 4.0 International, via Wikimedia Commons, under GNU Free Documentation License)

There are 6 carbon atoms in one glucose molecule and hence this means, 5 eV ($^1/_6$ of 30 eV) of energy is required for each carbon atom fixed in photosynthesis. As we know, light is made up of photons and the energy of photon is proportional to frequency of light.

$$E = h \cdot \nu \quad ...(8.5)$$

And,

$$h = 4.2 \times 10^{-15} \text{ eV} \cdot \text{sec} \quad ...(8.6)$$

If the entire energy to fix one carbon is obtained from one photon, we see that for $E \geq 5$ eV.

The corresponding frequency, $\nu \geq 1.2 \times 10^{15}$ Hz

The wavelength, $\lambda \leq 2500$ Angstrom (0.2 μm)

This wavelength is in the UV range and is absorbed high in the atmosphere. It should be noted that wavelengths shorter than 0.3 μm are hazardous to life. The basic chemical component of living organisms is hydrocarbons with a carbon chain as shown in the molecular structure of glucose below (Fig. 8.4):

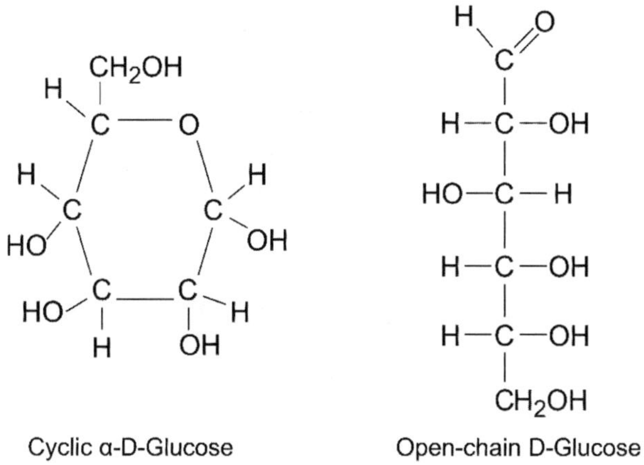

Fig. 8.4 Molecular structures of glucose

The single carbon-carbon (C-C) bond is the backbone of cell material in living organisms. The minimum energy required for rupture of this bond is 3.6 eV, and the wave length corresponding to a photon with that energy is 3500 Angstrom. That means, photons of wavelengths shorter than this (with higher energies over 3.6 eV) can make living matter unstable by breaking the C-C bond. Nature has devised photosynthesis in a safe mode without endangering this C-C bond by using several low energy photons to supply 5 eV of energy to fix one carbon atom in photosynthesis.

Let us imagine that the nature had two options, a highly efficient process in which direct solar radiation is used and a low efficiency process in which global radiation could be utilized for photosynthesis. Any device making use of beam radiation should track the sun. Hence to use the first option, the leaves of the plants should be optimally arranged without shade and should track the sun. Otherwise, the system may fail. Other option is of lower efficiency but can make use of the diffuse radiation as well and hence nature chose the second option.

8.3 Efficiency of Photosynthetic Energy Conversion

The efficiency of photosynthesis determines the area required for a specific quantity of biomass to be produced. The most important limiting factor in solar energy conversion through biomass production is the low efficiency of the process. The product of photosynthesis is basically glucose represented by the chemical formula $[CH_2O]_6$ and CH_2O is $^1/_6{}^{th}$ of a glucose molecule.

The chemical energy (ΔG) stored in the products of photosynthesis during the transformation of one mole of CO_2 to sugar is 114 kcal (477 kJ). This is the energy output in the process which can be compared with the input energy. In the process of light reactions, eight quanta of light (PAR) are utilized for photolysis of water. As efficiency of different colours in the PAR varies slightly (blue less efficient than red), PAR energy input is equivalent to that of a monochromatic light of wave length 575 nm. The quantity of energy can be calculated using the relation,

$$E = Nh\nu = \mathrm{Nh}/\lambda \qquad ...(8.7)$$

Where, 'N' is the Avogadro number, 'h' is the Plank's constant, 'v' is the frequency and 'λ' is the wavelength.

The corresponding energy is 49.75 kcal per quantum and the minimum energy input per mole of CO_2 reduced is 8×49.75=398 kcal 1665 kJ. So, the maximum efficiency:

$$\text{Output/Input} = 477/1665 = 0.286 \qquad ...(8.8)$$

PAR is only 43 per cent of the total solar radiation incident on a surface and 100 per cent of PAR cannot be absorbed by leaves. Some light is reflected and some portion is absorbed in non-photosynthetic region of the leaves. An optimally arranged canopy can absorb a maximum of 80 per cent of the available PAR. Hence, only 0.43×0.80 = 0.344 of the light energy will be utilized. Thus, the efficiency of photosynthesis drops to a value of 0.286 × 0.344 = 0.0984.

Thus, it can be understood that a photosynthetic plant making only carbohydrates and exposed to optimum sun light throughout the day can never have a photosynthetic efficiency more than 10 %. But terrestrial plants are exposed to sun light only for 12 hours in a day and they respire during night. Non-photosynthetic tissues of higher plants may respire throughout the day. Some energy is also used for bio synthesis of compounds such as proteins and fats beside carbohydrate synthesis. Considering conditions of maximum growth rate, it can be considered that approximately 1/3rd of the available energy is used for respiration and bio-synthesis.

Now, we arrive at a net efficiency, 0.0984 × 0.667 = 0.066

The overall efficiency of 6.6 per cent may be regarded as the possible maximum theoretical efficiency and is the absolute upper limit for solar energy conversion to biomass by photosynthesis. Many scientists suggest still lower values in optimum field conditions. The possible highest efficiencies in field conditions could be in the range 3 to 5 per cent, for limited periods. Typical annual conversion efficiencies are higher in tropical and sub-tropical regions (0.5 to 2.3%) compared to temperate regions (0.5 to 1.3%).

In addition to the respiration taking place during night, another oxidative process (respiration) also takes place when light is available. This phenomenon, 'Photo respiration' causes the photo-synthetically produced sugar to be converted back in a reverse process to produce CO_2 and water. Photorespiration reduces the overall photosynthetic efficiency of plants. Crops like wheat, soybean, potatoes and many trees are subject to this. But some plants, especially tropical grass species like maize and sugarcane show no photorespiration. They have an additional CO_2 fixing cycle with specialized leaf morphology. In addition to the ordinary pathway of CO_2 assimilation (C3) in the Calvin cycle, they have a special (C4) cycle and they are called C4 plants. Table 8.1 gives values of photosynthetic efficiency for a number of C3 and C4 crop plants.

Table 8.1 Maximum practical photosynthetic efficiency of crop plants

Sl. No.	Category	Plant	Efficiency, %
1.	C4 Plant	Sugarcane	2.4
2.	C4 Plant	Napier grass	2.4
3.	C4 Plant	Sorghum	3.2
4.	C4 Plant	Corn	3.2
5.	C3 Plant	Sugar beet	1.9
6.	C3 Plant	Wheat	1.7
7.	C3 Plant	Alfalfa	1.4

Hence, some argue that if photorespiration is arrested for crop plants, the production can be improved. But many others are of the opinion that photorespiration is not very significant when other agronomic factors are compared for crop growth. Scarcity of available land along with insufficient irrigation facilities in hot tropical climates may favour C4 plants for maximum biomass production. Grasses like Bermuda grass and Sudan grass, which can be harvested on a weekly basis are suitable for high biomass yield. Sugar cane is another important crop suitable for hot climates with plenty of sun shine.

8.4 Yield of Biomass as a Function of Available Energy and Conversion Efficiency

As biomass production systems require an estimation of productivity in field conditions, calculation of expected yields of plant dry matter in ideal field conditions is a matter of interest. Let us assume that the plant is converting all the synthesized sugars to storage carbohydrate (starch or sugar) or structural carbohydrate like cellulose. They are all polymers of glucose joined together by elimination of a water molecule. If Glucose with the composition $(CH_2O)_6$ only is produced, the molecular weight of the product would have been $[CH_2O] = 30$ for each CO_2 fixed. But, in actual products such as starch and cellulose, one water molecule is eliminated for each glucose molecule and the molecular weight of the final product per CO_2 fixed may be taken as that of $[CH_2O]$ minus $1/6^{th}$ of a water molecule (18/6=3). In producing 27 gram (30 minus 3 g) of dry matter, 114 kcal of chemical energy per mole of CO_2 is stored. The small quantity of energy which may be utilized for polymerization of glucose may be neglected. The dry matter production, 'M' per unit of energy is obtainable in the form of dry cellulose or starch can be calculated as:

$$M = 27/114 = 0.237 \text{ grams per kcal}$$

If we are interested to estimate the maximum rate of biomass production from a unit area the following steps can be used:

Assuming moderate sun shine of 4500 kcal per sq. metre per day, the maximum energy stored by green plants,

$$E_{max} = 0.066 \times 4500 = 297 \text{ kcal m}^{-2} \text{ day}^{-1}$$

The rate of biomass production (dry weight),

$$M_{max} = 0.237 \times 297 = 70.4 \text{ g m}^{-2} \text{ day}^{-1} = 257 \text{ tonnes ha}^{-1} \text{ per year.}$$

If a high solar intensity of 6500 kcal m^{-2} day^{-1}, as available in tropics is considered, the production will be 101.7 g m^{-2} day^{-1} equivalent to 371.1 tonnes ha^{-1} per year. But, in actual field condition the growth rates are 40 to 50 g m^{-2} day^{-1} for C4 plants and 20 to 30 g m^{-2} day^{-1} for C3 plants. The total annual solar radiation received over the entire land mass of India is about 5×10^{10} kWh. If we calculate the annual net biomass production and roughly estimate the energy equivalent, it is about 3×10^{20} Joules.

8.5 Biomass Resources in India

Energy from biomass is a major source of energy for a vast majority of the people in developing world. The biomass energy usage is 38 to 90 per cent of the total energy consumption. The word biomass is a very comprehensive term indicating of all forms of organic matter derived from living organisms. The biomass produced through biological activities can be of terrestrial origin and present either on the surface layers of soil or at different depths of water bodies such as the ocean, lakes or rivers (Table 8.2).

India generates a vast amount of biomass material through its agricultural and agro-industrial operations, estimated at over 370 million tonnes every year. On the basis of calorific value, this amount is equivalent to around 250 million tonnes of coal. But only a small portion of this material is utilized productively. Crop stalks, straws, sugar cane trash, cotton stalk, maize stalk, maize cobs, sugarcane bagasse, rice husk, groundnut shells, tapioca stems, coir pith etc. are the common crop residues. Wastes from animal husbandry, and solid and liquid wastes from agro-processing facilities and related industries are also available in India. Forest industry residues like saw dust, pine needles etc. are also available in large quantities. The biomass availability can be increased at least by 30 per cent, if energy plantations are grown only on 10 per cent of the wastelands of our country. It has been estimated that around 20 GW of power generation capacity can be sustained by making use of such waste materials.

Table 8.2 Bio Resources in India

Sl. No.	Category	Biomass resources
1.	Animal wastes	Cattle dung, poultry wastes, sheep and goat droppings, pig farm wastes
2.	Forest wastes	Saw dust, wood chips, other residues and wastes like pine needles
3.	Crop residues and byproducts	Paddy straw and other straws, crop stalks like cotton stalk and maize stalk, other crop trashes like sugar cane trash, maize cobs, ground nutshells, tapioca stems, banana stems, coconut fronts and other wastes
4.	Solid agro-industrial byproducts	Sugarcane bagasse, rice husk, coir pith, coffee wastes, textile wastes
5.	Liquid wastes and byproducts from agro-industries	Distillery wastes, Dairy wastes, Starch factory effluent, Rubber processing effluent, rice parboiling effluent
6.	Urban wastes	Municipal solid wastes (MSW), urban sewage
7.	Carbohydrate crops	Sugar cane, sugar beet, cassava, maize
8.	Oil crops	Sun flower, soy bean, oil palm, rape seed, ground nut, castor, mustard etc.
9.	Aquatic biomass	Algal biomass, aquatic weeds like Eichhornia and Salvinia
10.	Fire wood trees and shrubs	Acacia, Eucalyptus, Babul, Red gum, Gum Arabic, Subabul, Bombay black wood, Casuarina, Shisham, Agathi, Arjun,

8.6 Advantages and Disadvantages of Biomass as an Energy Source

Biomass is traditionally used as a fuel even now in many villages in developing countries. Processed forms of biomass are being used in developed countries as well. The compatibility and similarity of biomass with fossil fuels like coal still provides possibilities for their continued use. Biomass fuels like fire wood and charcoal can be stored safely for long time without much deterioration and can readily be used. A coal fired industrial boiler can easily be fueled using biomass without much modification. Unlike many other renewable energy sources, biomass can easily be used as a replacement fuel in many applications. In rural India supply of energy is a big challenge and biomass can serve the needs of a decentralized energy supply system. The advantages may be listed as follows:

(i) Renewable and environment friendly energy source.

(ii) Highly versatile enabling conversion to different forms such as thermal, mechanical or electric.

(iii) Availability of technologies for conversion into liquid or gaseous fuels.
(iv) Suitable for decentralized energy supply in rural areas.
(v) Scope for rural employment generation.
(vi) Ability for promoting sustainable, self-reliant and equitable development for the lower strata of population.
(vii) Universal availability from wide range of sources.
(viii) Wide capacity range for energy generation (very small scale to large facilities)
(ix) Possibility of safe storage over fairly long periods.
(x) Easily adaptable and no need of sophisticated technology.
(xi) Comparatively low capital investment and possibility for faster commissioning
(xii) Ensure self-reliance of the country by reducing foreign money expenditure on energy.

In spite of the above advantages, biomass has some inherent problems for its wide use arising from the natural factors. Most biomass materials like crop residues have a low bulk density as well as specific energy content. Seasonal nature of availability poses difficulties in assured fuel supply. Many crop residues become useless as a fuel once it gets wet during the rains. In humid tropical climates the safe storage of bulky biomass is difficult. The major issues can be listed as under:

(i) Low bulk density and specific energy content
(ii) Collection is difficult as dispersed in large areas
(iii) No stable market due to localized price sensitivity
(iv) No assurance of steady availability due to seasonal production
(v) Difficulties in handling, transportation and storage

8.7 Energy Farms

Farms or plantations where crops are cultivated exclusively for energy are called 'Energy farms' or 'Energy plantations'. Energy ratios from 1 to 17 can be achieved for different crops in various agricultural systems. Maize production typically has an energy ratio ranging from 2.5 to 6. The choice of crop plants depends upon the agro-climatic factors as well as the possibility or need for further refinement of raw biomass. They can be of different types and can be classified as below:

(a) Silviculture energy farm or energy plantation: More than 50% of the world's total forest biomass is produced in developing countries. But they are not evenly distributed and deforestation for fuel wood is a serious problem. It is now a common practice to grow fuel wood exclusively for energy. Many fast-growing species like eucalyptus, acacia, casuarina etc. are grown as energy plantations.

(b) Oil crops: Production of bio-diesel from vegetable oils is not a recent development. It attracted attention in the last decades as an alternative biofuel for diesel engines. Many oil crops like sun flower, soybean, rapeseed, groundnut, oil palm and many other oil seeds are being utilized for biodiesel production. Non edible oil crops like Jatropha were considered as prospective crops in India.

(c) **Hydrocarbon crops:** Alcohol production for fuel purpose is also developing in a great way. Sugar cane, cassava, maize and sugar beet are common hydrocarbon crops grown and used for fuel alcohol production.

(d) **Aquatic energy farms:** Conceptually, there are land based aquatic systems as well as 'Open-ocean system'. Production of biomass as an integrated part of a waste-water treatment system is prevalent. Sewage and other wastes would provide the required nutrients and water for aquatic plants. Production of micro algae for energy has been a topic of interest in energy research. As addition of carbon dioxide to the system was required for higher biomass yields, this could be utilized as a carbon recycling technique. Growing macro algae in ocean was also considered as a means of biomass production. Utilization of naturally grown aquatic weeds for bioenergy production is another possibility.

8.8 Biomass Energy Conversion Processes

Wastes and residues currently constitute a large source of biomass. These include solid and liquid municipal wastes, manure, agro-industrial and pulp mill wastes, and forest and agricultural residues. All these biomass have different characteristics and different processes are to be employed for their efficient utilization (Fig. 8.5). The processes are generally classified as Thermochemical and Biochemical processes. Thermochemical processes are suitable for low moisture content materials while biochemical processes are employed for high moisture wastes. In thermochemical processes, the application of temperature and sometimes pressure results in the production of energy or a readily usable fuel. In biochemical processes, living organisms (microbes) are involved. They convert the waste materials (organic) into a readily usable liquid or gaseous fuel through biochemical transformations.

The thermochemical processes include direct combustion, different types of gasification processes and different Pyrolysis techniques. Hydrogen gasification, Steam gasification, Liquefaction, Distillation etc. are all thermochemical processes which can be grouped under any of the above three processes. In combustion, the material burns in a full supply of oxygen/air to release the chemical energy as heat. Whereas, the air supply is partial in gasification so as to retain a major portion of the chemical energy in the fuel gas produced. Theoretically, pyrolysis is the breakdown (lysis) of biomass by the application of heat ('pyro' indicates heat), which occurs in the absence of air. Depending on the process parameters, there can be much variation in the quality and quantity of Pyrolysis products. The products can be mixture of combustible gases, liquids or solids.

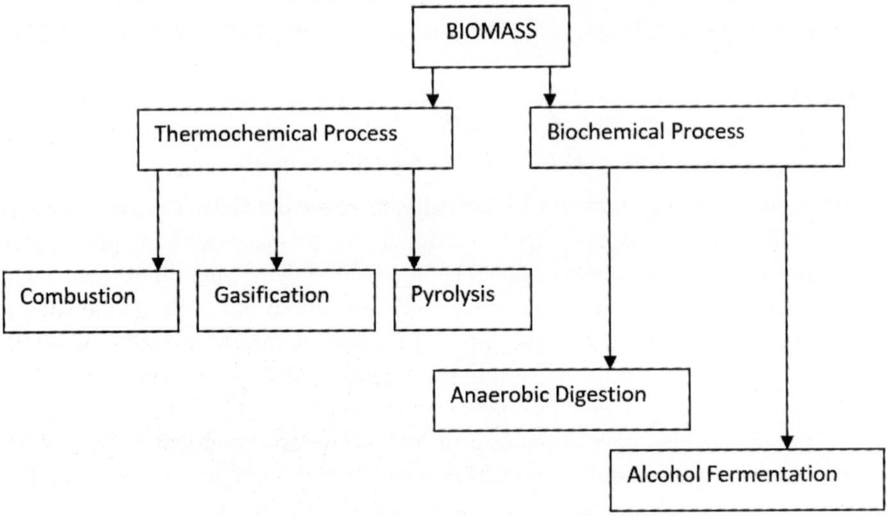

Fig. 8.5 Major biomass energy conversion processes

Major biochemical processes for the conversion of biomass to fuels are methane production by microbial consortia under anaerobic conditions and ethanol fermentation by yeast. A wide variety of organic materials from animal and human excreta to waste paper can be converted in these processes. Depending on the feed stock, operating conditions and other process parameters, the equipment and processes widely differ.

Anaerobic digestion for methane production, utilizes organic materials containing carbohydrates, lipids and proteins. Many species of microbes work cooperatively in an anaerobic digester, in which these polymeric materials are first decomposed to organic acids and then to biogas which has methane and carbon dioxide as the major constituents along with some minor ingredients like hydrogen, hydrogen-sulphide etc. A variety of raw materials which include agricultural wastes, municipal solid wastes, market garbage and organic waste waters from food and fermentation industries are useful as substrates for this process. While cow dung can be successfully converted into biogas in a domestic biogas plant, industrial effluents (organic) is converted in high-rate bioreactors like UASB reactors. Anaerobic digestion has the added bonus of solving the environmental issues related to waste disposal. Although small-scale digesters are popularly used at domestic and village levels, large-scale systems require considerable technical skill for their construction and operation. Improvements in the process and reduction of cost have been achieved in the recent past in this area making the technology suitable as an environment friendly technique useful for waste disposal as well as renewable energy production.

Wood wastes in the paper and pulp industries, willow dust in cotton industry, process waste waters in food processing industries and bagasse from the sugar-cane industry are examples of biomass likely to accumulate at a single site. Thermochemical conversion like gasification may be appropriate for wood wastes and bagasse while food processing waste waters may be successfully converted into energy by employing high-rate anaerobic bioreactors. If the requirement is alcohol fermentation, the starch as well as cellulosic biomass materials should necessarily be hydrolyzed to glucose, prior to ethanol fermentation. The net energy balance for some of these processes like alcohol production from cellulose can be problematic as the energy requirements for cellulose hydrolysis and distillation must be lower than the energy in the output ethanol.

Methane, ethanol and biodiesel can be produced from cultured micro-algal biomass through appropriate conversion techniques such as anaerobic digestion, microbial fermentation processes and physico-chemical processes respectively. Hydrogen is regarded as one of the cleanest fuels and hydrogen production by biochemical methods (Bio-photolysis) also has been demonstrated, even though at experimental levels.

Sometimes, the biomass conversion processes are also classified as dry processes and wet processes. Combustion, gasification and pyrolysis are regarded as dry processes whereas biochemical processes like anaerobic digestion and alcohol fermentation are wet processes.

8.9 Biomass Fuels

Conventionally, fuel is defined as a combustible substance available in bulk capable of economic thermal energy release on oxidation by atmospheric air. But this definition is not universal as fuels used for fuel cells and nuclear plants are exceptions. Common fuels contain carbon and hydrogen, which are easily oxidized to CO_2 and H_2O. Some fuels contain only these two elements (Hydrocarbon fuels), where as some other fuels are partly oxygenated (like biomass) with small amounts of other elements like Sulphur and Nitrogen.

Biomass fuels are fuels derived from contemporary biological matter or combustible biomass products which can be economically utilized to produce heat by burning them. Commonly available raw biomass fuels are thermally degradable solids.

(a) *Classification of biomass fuels:* Many biomass materials mostly comprising of sugars, starches, cellulose, hemicelluloses and lignin can be directly burned to produce heat or can be used to power an external combustion engine (steam engine) to produce mechanical power. There are also technological options to convert them to liquid or gaseous fuels so that they can be burned at a higher efficiency or can be used in internal combustion engines to produce mechanical power (Fig. 8.6). Basic conversion techniques as discussed in previous section as well as modified processes are used to convert them to fuels of better quality. Raw solid biomass fuels generally have low efficiency due to their inherent physical nature. When we burn solids, intimate mixing of oxygen with fuel molecules cannot happen. Hence the combustion is incomplete leading to unburnt carbon and other components in the combustion products. This results in loss of energy as well as pollution due to smoke. Liquids can be burned more efficiently by atomizing it, but still, it is not 100 per cent efficient. We observe that diesel vehicles are emitting more smoke compared to CNG vehicles. Gaseous fuels can ensure almost complete combustion because inter molecular contact between the fuel and oxygen is possible. Even if some energy is expended in transforming solid fuels to liquid or gaseous fuels, the overall efficiency can be much better. The possibility of use of biomass fuels for IC engines is another factor which prompts their conversion.

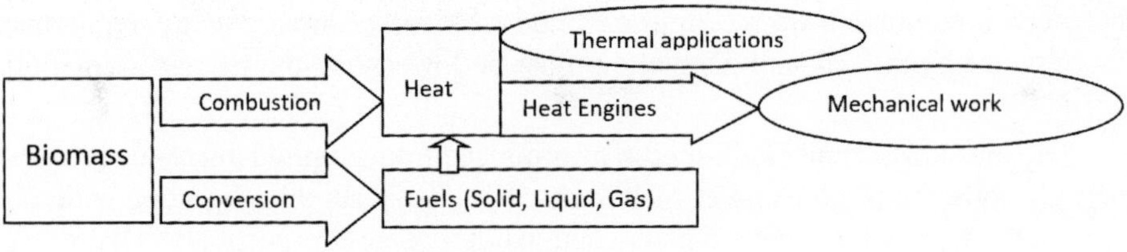

Fig. 8.6 Application of biomass fuels

8.10 Present Status of Bioenergy

Bioenergy is an area attracting global attention for clean energy options. Among different biomass conversion processes, methane and ethanol production from various wastes is economically feasible within the restraints of scale and location. Biological processes for the production of gaseous and liquid fuels have been well demonstrated in developed as well as developing countries. Although India is the pioneer in anaerobic digestion technology research has to be intensified in several areas to make the technology adaptable in different situations. An integrated approach is still lacking where these processes are integrated into a system capable of meeting basic overall efficiency requirements for converting dilute solar energy into a concentrated biofuel.

In rural sector, anaerobic digestion technology has been popularized in India as a means of providing quality cooking fuel at domestic level. Thermo-chemical gasification of crop residues is also a technological option for providing clean cooking fuel. Promotion of improved biomass stoves in rural households has much implication in reducing carbon emission, in addition to providing a pollution free indoor environment.

According to Ministry of New and Renewable Energy (MNRE) of India,

"About 32 per cent of the total primary energy use in the country is still derived from biomass and more than 70 per cent of the country's population depends upon it for its energy needs".

Realising the potential and role of biomass energy in the Indian context, MNRE has initiated a number of programmes for promotion of efficient biomass technologies. Among them, cogeneration has gained much importance in agricultural produce-based industries. Bagasse based cogeneration in sugar mills and biomass power generation are being promoted in a big way.

Chapter 9
Anaerobic Bioconversion of Biomass for Energy

Domestic production of energy from renewable sources is a major thrust area and anaerobic digestion technology for production of 'Biogas' is an environment friendly technology. Energy conversion of biomass to value added fuels through biochemical processes were basically evolved from natural phenomena called fermentation. Fermentation is a biochemical reaction actuated by micro-organisms. Anaerobic digestion is the process for fermenting organic substances in an anoxic or anaerobic atmosphere. The product is a mixture of gases which is combustible due to the presence of methane as a major component. At present this technology is regarded as one among the most environment friendly technologies for energy conversion of biomass, especially high moisture waste materials. As organic wastes can be utilised for energy production, this technology has the twin advantages of energy production as well as waste management. The production of good quality organic manure for agricultural crops as a by-product of the process is another benefit.

9.1 History of Anaerobic Digestion Technology

As a natural process, fermentation of biomass taking place in anaerobic condition was a topic of interest from olden days. The phenomenon of evolution of a combustible gas from marshy lands aroused curiosity of scientists as early as 17th century. The gas came to be called 'marsh gas' was identified as a combustible gas produced by decay of organic matter under water, where the material is not in direct contact with air. Scientists like Alessandro Volta could analyse the gas and Humphrey Davy produced it experimentally in the laboratory in 1808. It is reported that gas from a septic tank was used for street lighting in England in 1895. This led to further investigations for the successful design of biogas systems. In the early twentieth century scientists were convinced that this gas mixture comprised of CO_2, CH_4, and H_2 with traces of other gases. The anaerobic digestion technology progressed in countries like Germany and France during the first half of twentieth century, where there was increased interest to produce this gas and use it for energy production after the World War II.

In rural India, villagers used firewood and dried cow dung cakes as fuel in biomass stoves and there was shortage of cooking fuel. The rural women had to spend much time and energy in collecting fire wood from nearby forests and agricultural lands. Use of crop residues or dried cow dung cakes was the other option. The burning of these biomass fuels in traditional stoves (*chulhas*) created indoor pollution at the households leading to severe health hazards. The combustion of cow dung cakes had another adverse implication in the agricultural sector also. Cow dung, the most nutritious manure available to rural Indian farmers, were dried and used as a fuel for burning which ultimately resulted in nutrient deficiency of crops. In this juncture, biogas technology was viewed as a technology to provide clean fuel to the rural population with the simultaneous benefit of salvaging the valuable manure.

Hence, efforts to produce biogas from cow dung were intensified in India also. One of the early research activities on biogas technology was taken up at Indian Agricultural Research Institute, New Delhi (IARI) as early as 1938. The Rama Krishna Mission in Kolkata and IARI are institutions who developed early biogas plant designs. Many earlier versions of biogas plants were evolved in the 1950s including the '*Gramalaxmi*' model of J. J. Patel in 1951. He continued with refinement of his model and was the pre-curser of KVIC (Khadi and Village Industries Commission) model. Over this period, China also was interested in the technology and evolved the Chinese designs with their unique dome type gas holders. Later on, the 'Planning, Research and Action Division' of the Uttar Pradesh state (India) Government established a 'Gobar Gas Research Station' (*Gobar* means cow dung) in 1960. The Indian version of the Chinese design biogas plant was developed by them and popularised in the name 'Janata' biogas plant. There after research on domestic as well as community biogas plants were taken up at various research institutes in India like National Environment Research Institute (Nagpur), Tamil Nadu Agricultural University (Coimbatore), Indian Institute of Science (Bangalore), Murugappa Chettiar Research Centre (Chennai) etc. resulting in the evolution of a number of designs. Later on, the Indian Council of Agricultural Research launched an All India Coordinated Research Programme on Biogas Technology in 1977 with six centres and further expanded to five more centres in 1983. However, the 'floating drum' biogas plant design of Khadi and Village Industries Commission (KVIC) was the most popular design in the eighties and many designs were based on this floating gas holder design. The 'Action for Food Production', New Delhi further modified the Janata biogas plant design for reduction of construction cost and came out with the '*Deenbandhu*' model.

9.2 Salient Features of Biogas Technology

Biogas, a combustible gas mixture, is generated through a process of fermentation of organic matter in anaerobic condition. Fermentation is a biochemical process in which microorganisms are involved. Commonly, the process is referred as 'anaerobic digestion' and the resultant gas mixture is called 'biogas'. Any organic matter sufficiently biodegradable under anaerobic conditions can be used as a substrate. Waste materials such as animal excreta (cattle/ human/ birds) like cow dung, fermentable crop residues, organic waste waters from food /agro-process industries and other biodegradable organic matter are generally used as feed stock in biogas plants. Biogas production technology has the following advantages:

(i) Biogas is a clean gaseous fuel which can be produced economically.
(ii) It can be used for thermal applications like cooking as well as in IC engines for mechanical power or electricity generation.
(iii) Biogas plants can be effectively used as a waste management system for organic wastes.

(iv) The biogas slurry is enriched organic manure with better qualities compared to raw cow dung and FYM (farm yard manure).

(v) It is a renewable and carbon neutral fuel and can save trees which are being cut for firewood.

(vi) It provides pollution free indoors in rural households leading to better health of rural women and reduce the drudgery in slow cooking in traditional biomass stoves.

(vii) The time saved by rural women in their routine job of fire wood collection can be utilized for overall betterment of the family resulting in increased income and better living standards.

(viii) The technology offers decentralised fuel availability and helps in rural employment generation.

(ix) Large scale and community scale biogas systems can be used for electricity production so as to provide electricity in remote villages away from the electricity grid.

9.3 Composition and Properties of Biogas

Biogas predominantly contains the lightest hydrocarbon methane as the major combustible constituent with the incombustible CO_2 along with other minor components. The approximate composition of biogas from cow dung-based plants is given in Table 9.1.

Methane, Hydrogen and H_2S are the combustible constituents in biogas which are responsible for the heating value of biogas. Pure methane has a heating value of 9516 $kcal \cdot m^{-3}$. Neglecting minor components, the heating value of biogas with 60 per cent methane is taken as 20 $MJ \cdot m^{-3}$ (4700 $kcal \cdot m^{-3}$). Methane is a colourless and odourless gas lighter than air and hence considered safer than LPG as methane will rise to safer heights in case of gas leaks. Methane has a very low critical temperature (-82.1°C) and a critical pressure of 47.3 $kg \cdot cm^{-2}$. Critical temperature of a gas is the temperature above which the gas cannot be liquefied by the application of pressure. Critical pressure is the pressure required for liquefying it at critical temperature. As it is practically difficult to cool the gas to its critical temperature level, methane in biogas is not liquified and used for fuel purpose.

Table 9.1 Composition of biogas

Sl. No.	Component	:	Percentage
1.	Methane, CH_4	:	50-65 %
2.	Carbon dioxide, CO_2	:	30-40 %
3.	Hydrogen, H_2	:	1-5 %
4.	Nitrogen, N_2	:	0.5-1%
5.	Hydrogen sulphide, H_2S	:	0.01-0.1%
6.	Water vapour, H_2O	:	0.05-0.1%
7.	Oxygen, O_2	:	Traces
8.	Ammonia, NH_3	:	Traces

Methane has an ignition temperature of approximately 700°C and the flame temperature is 870°C. A mixture of methane with air will ignite if the methane content in the mixture is in the range 5.3 to 14 per cent. The comparison of other fuels with methane is given in Table 9.2.

Table 9.2 Heating value of various fuels and their effective thermal energy output

Sl. No.	Name of Fuel	Higher Heating Value (MJ)	Equipment	Overall thermal efficiency (%)	Effective energy output (MJ)
1.	Biogas, m^3	20	Gas stove	60	12
2.	Kerosene, litre	38	Pressure stove	50	19
3.	Firewood, kg	20	Traditional stove	12	2.4
4.	Firewood, kg	20	Improved stove	24	4.8
5.	Cow dung cake, kg	8.7	Traditional stove	12	1.04
6.	Cow dung cake, kg	8.7	Improved stove	20	1.74
7.	LPG (Butane), kg	46	Gas stove	60	23
8.	Electricity, kWh	3.6	Hot plate	70	2.52

9.4 Biochemical Process of Biogas Production

Numerous bacterial species, living in the presence of oxygen (aerobic) or without oxygen (anaerobic) are there in nature. The microorganisms involved in the production of biogas are anaerobic bacteria. They are found in decaying organic matter, depths of oceans, stagnant water bodies and marshlands as well as in the digestive tract of many herbivorous animal species. Methane (the combustible component of biogas) is produced by anaerobic decomposition (fermentation), a biochemical process in which several groups of bacteria are involved.

The biogas plant digester is in fact a bioreactor in which biochemical reactions convert the substrates to the product gas and residual spent slurry. It is now understood that the overall biochemical reactions taking place in a biogas digester (anaerobic bioreactor) are taking place in three distinct phases as shown in Fig. 9.1. The generation of biogas takes place only at the final stage of the three different phases viz., solubilisation, acidification (acidogenic phase) and methanogenesis (methanogenic phase). The bacterial groups involved are also distinct with different characteristics. The reactions taking place are consecutive; but simultaneous phases exist in the digester or bioreactor.

The first phase is the solubilisation of complex and polymeric organic compounds in the feed materials like animal excreta or other organic wastes. Carbohydrates (sugars, starches) along with cellulose, hemicelluloses etc. are the major components which may be regarded in general as ligno-cellulosic materials. They may contain some lignin also along with lipids, proteins, other nitrogen compounds, small quantities of sulphur containing compounds and some mineral matter (salts). Solubilising bacteria capable of producing cellulase (enzyme) hydrolyse cellulose to soluble carbohydrates. As the complex polymers are broken down into simple soluble sugars, this phase is regarded as the 'solubilisation phase'. The products of this phase undergo further biochemical change in the second phase called 'acidogenic phase'. The acid producing bacteria act on them to reduce the soluble organic materials to soluble simple organic acids. The heterogeneous acid forming bacterial group called 'acidogenic bacteria' are fast growing and they can survive in slightly acidic environment also. One primary acid product is acetic acid and hence this phase is sometimes referred as 'acetogenic phase' also.

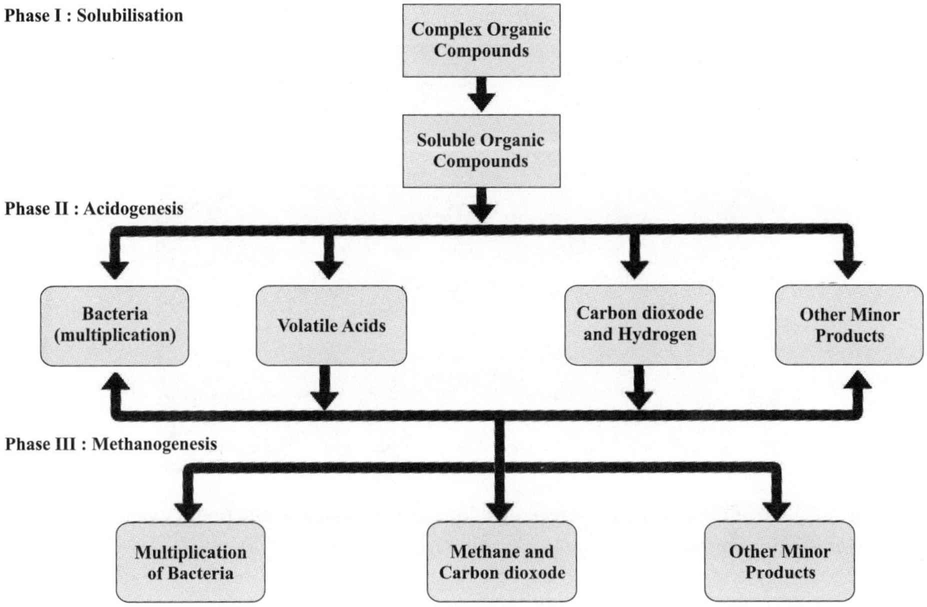

Fig. 9.1 Different phases in anaerobic digestion of organic matter

In the subsequent third step, the methane producing bacteria (methanogenic bacteria) convert the organic acids to methane and carbon dioxide. The bacteria have a specialized metabolic system which produces methane as their major product. Hydrogen and carbon dioxide produced in the second phase is converted into methane and water by some of these bacterial groups. The acetic acid route is much predominant in methane formation where it is reduced to methane with carbon dioxide as the by-product. The process may be represented by the following chemical reactions:

$$C_6H_{12}O_6 + 2H_2O \rightarrow 2CH_3COOH + 2CO_2 + 4H_2 \quad ...(9.1)$$

$$CH_3COOH \rightarrow CH_4 + CO_2 \quad ...(9.2)$$

$$4H_2 + CO_2 \rightarrow CH_4 + 2H_2O \quad ...(9.3)$$

As the production of methane actually occurs only in the third stage, a proper balance between different phases is required for uninterrupted biochemical reactions in the digester. The methanogenic bacterial groups are inherently slow growing and they are sensitive to low pH of the digester liquor. Hence any adverse digester pH will affect their growth and disrupt the balance between the *'acidogens'* and the *'methanogens'*.

Imbalances in bacterial population may result in the failure of the system due to acid accumulation. The population of acid formers should match that of the methane formers, failing which, the pH of the digester may fall to highly acidic range.

Further, the methanogenic bacteria are regarded as 'obligate anaerobes' as they cannot survive in the presence of oxygen. Their growth is drastically affected even by small amounts of O_2 and hence a highly reducing environment should be maintained in the biogas digester. Highly oxidized salts like nitrates are also unwelcome as they also can inhibit methanogens which multiply only very slowly. The methane bacteria (Fig. 9.2) include the genera such as *Methanosarcina, Methanothrix, Methanococcus, Methanobacterium, Methanospirillum* and *Methanobacillus*.

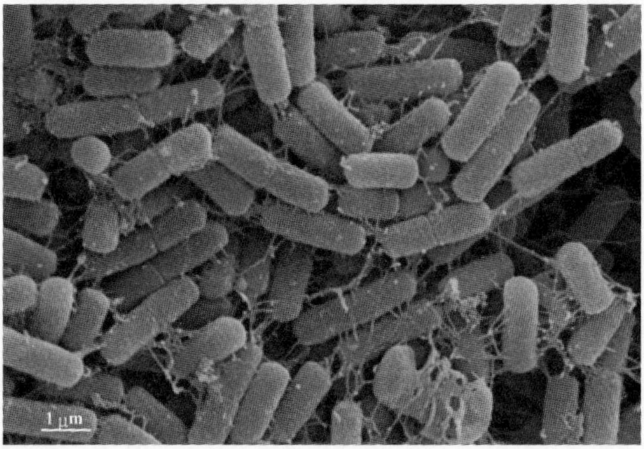

Fig. 9.2 Scanning electron micrograph image of anaerobic bacterial consortium
(Reprinted from International Biodeterioration & Biodegradation, Vol. 52:2, Ji-Dong Gu, 2003, Microbiological deterioration and degradation of synthetic polymeric materials: recent research advances, 69-91, Available at: https://www.sciencedirect.com/science/article/abs/pii/S0964830502001774#:~:text=Microbial%20degradation%20of%20polymers%20depends,others%20are%20only%20partially%20degraded. Copyright 2021, with permission from Elsevier – License No. 5183520304736, CC BY-SA 4.0)

Under balanced digester operation, the acids produced by acidogenic bacteria are broken down by methanogens into CH_4. If the balance between bacterial populations is upset, pH lowers rendering the inhibition of methanogens leading to accumulation of acids. This acidification will be further aggravated as there will be lesser number of the *methanogens* to use the acids quickly and the system fails.

This can happen due to over feeding of digesters. Bacteria in the first two phases are fast growing and hence those two phases proceed when excess feed material is available. This results in production of excess acids. Once the pH drops below 6.5, inhibition of the methanogenic bacteria occurs, but the acid production will continue as the *acidogenic* bacteria will continue to be active until the pH drops to 4.5 to 5.0. Finally, the system stops due to the death of methanogens and it is difficult to recover the system. Balanced digester conditions, can be attained by regularising the feed. In such a system, the biochemical reactions tend to maintain the pH in the proper range automatically. Although the volatile organic acids produced during the first stages of the fermentation process tend to lower the pH, the buffering capacity of the digester liquor often referred as 'bicarbonate alkalinity' can counteract the acidification effect.

The proportion of methane in the product gas decides the fuel quality and this depends on the characteristics of the feed stock. Simple sugars are easily digested by the bacteria enabling their fast multiplication. Complex sugars and other organic compounds take more time for their degradation to simple sugars which the bacteria can use. As the actual time taken by bacteria to transform the complex organic material is dependent on the biodegradability of the material, this factor decides the 'retention time'. If the feed material has predominantly ligno-cellulosic materials as components and has less proportion of easily biodegradable components, then solubilisation is the rate limiting phase in the overall anaerobic digestion process. If the material is mostly soluble as in the case of organic effluents like molasses (liquid waste from cane sugar factories), methanogenic phase is the rate limiting phase due to the slow growth rate of methanogenic bacteria.

If sulphur is present in the feed stock, hydrogen sulphide is also produced in very small quantities by some sulphur reducing anaerobic bacteria. Though the quantity of H_2S produced is negligible, it's bad odour and adverse corrosive nature makes it an undesirable component in biogas.

When the anaerobic digestion system is used as a waste processing facility, the quantity of sludge to be disposed of and its stability are also matters of concern. The left-over slurry from anaerobic digestion is a biologically stable residue with more soluble compounds compared to the original feed stock. The biogas plant slurry is rich in readily available essential plant nutrients and is a good organic manure for crops.

9.5 Substrate Parameters Relevant in Anaerobic Digestion

A proper knowledge of basic biomass properties is required to monitor digester operation. They are as follows:

(a) **Total solids (TS):** TS refer to the actual solid content of the feed material usually expressed as percent. It can also be expressed as milligrams per litre ($mg \cdot L^{-1}$) in the case of very dilute substrates like organic effluents. It is determined by oven drying method.

(b) **Volatile solids (VS):** It is the volatile fraction of the TS which is directly converted to different products during anaerobic digestion. It can be taken as TS minus ash.

(c) **Biochemical Oxygen Demand (BOD):** The BOD is a measure of the oxygen required to oxidise the organic matter in the sample of the feed material biochemically and is expressed as mg per litre ($mg \cdot L^{-1}$). Five-day BOD, commonly used to specify the organic content of a material, is estimated by fermenting the sample under specified conditions in a BOD incubator. BOD_5 is obtained by finding the dissolved oxygen in the sample before and after incubation for 5 days at 20°C. Three-day BOD is also used in some instances.

(d) **Chemical Oxygen Demand (COD):** Chemical Oxygen Demand is the amount of oxygen required for complete oxidation of the organic matter in a sample. COD is found by digestion in COD apparatus using Potassium dichromate and subsequent titration to estimate the oxygen consumed for oxidation of organic matter. COD is usually expressed as milligrams per litre ($mg{\cdot}L^{-1}$).

9.6 Environmental Factors Affecting Anaerobic Digestion

Anaerobic digestion is a complex phenomenon involving innumerable biochemical reactions carried out by several species of bacteria. Satisfactory operation of an anaerobic digestion system can be achieved only by maintaining the favourable conditions for the growth of these bacteria. Those factors can be environmental factors like ambient temperature as well as operational factors like feeding rate.

(a) **Anaerobic conditions:** The anaerobic condition inside the digester is an essential requirement for the anaerobic bacteria to survive. The methane producing bacteria are sensitive to the presence of O_2 and hence, the digester should be air tight without contact with air/oxygen. Another aspect is the danger arising due to entry of atmospheric air into the gas holder. A mixture of biogas and air can get ignited if it has a high amount of air and is a safety threat as there is a chance for explosion.

(b) **Temperature:** Anaerobic bacteria in biogas plants are very sensitive to temperature. To maintain stable biogas plant operation the operational temperature should be within the range of the optimum temperature. The optimum temperature range for the common anaerobic bacteria, which survive and grow within a temperature range of 20°C to 40°C (*mesophilic*) is between 35° and 37°C. The variation in temperature can seriously affect biogas production and care should be taken to protect the digester from sudden temperature changes. It has been observed that in places where cold winter climates make the ambient temperature to fall below 20°C, the gas production is affected badly and is likely to stop completely when temperature falls below 10°C. There are special anaerobic bacteria which survive at temperatures lower than 20°C (*psychrophilic*) and at high temperatures in the range 50 to 60°C (*thermophilic*) which can be grown by maintaining such conditions.

Minimization of the temperature fluctuations in cold climates is possible if the digester is constructed underground as is usual with some designs. Artificial provisions for heating, by way of providing a transparent cover to create greenhouse effect or using solar heated water for mixing slurry are other techniques to get an elevated temperature in the digester during cold climatic conditions.

(c) **Acidity and alkalinity of the reactor liquor (pH):** The anaerobic digestion is affected by microbial action for which a near neutral pH is congenial for their survival and growth. The optimum pH is considered to be in the range 6.8 to 7.2 as slightly acidic or alkaline conditions can be tolerated by the bacteria. Even a slightly high pH above this range does not significantly affect the gas production and the buffering capacity of the system can adjust slight lowering of pH on a temporary basis. Major changes, especially lowering to the acidic range can be drastic, as *methanogens* are very sensitive to low pH even though they can tolerate alkaline conditions up to pH 8.0.

(d) **Availability of feed materials, their biodegradability and feeding rate:** It is obvious that a biological system requires steady feed input. The proper maintenance of different groups of bacteria can be ensured by regular feeding with proper biodegradable materials. It is important that non-biodegradable materials like soil and ash are not fed to biogas plants. The feed rate should match with the digester capacity. Persistent excess feeding can result in acidification of digester liquor leading to failure. Steady supply of substrate in correct quantity and continuous operation of the system can ensure trouble free working and steady energy output.

(e) **C/N Ratio:** The micro-organisms require carbon for energy and nitrogen for their cell build up. Other nutrients also are required in limited quantities. But, the ratio of carbon: nitrogen (C/N ratio) in the feed material is very important. As many animal wastes contain sufficient nitrogen this may not be a problem for such biogas plants. But deficiency of nitrogen is observed in the feed with a higher percent of crop residues or weedy biomass. N_2 is required for bacterial multiplication and the rate of gas production will be limited by the N_2 availability in such biogas plants. If excess nitrogen is available as in the case of certain feed materials like pig manure, it will lead to ammonia formation. High ammonia concentration can create inhibition of bacteria and affect digester performance.

It has been found that C/N ratio close to 25:1 to 30:1 in the substrate is optimum for satisfactory operation of anaerobic systems. The C/N ratio of feed materials with high nitrogen content can be corrected by mixing materials with high proportion of carbon (like crop residues). Biogas plants fed with materials that are low in nitrogen can be supplemented with materials high in nitrogen to achieve the optimal C/N ratio of 30:1. In certain situations addition of urea as a nitrogen source may be required.

(f) **Solid content of feed material:** Anaerobic fermentation is a wet process which requires high moisture content. It has been observed that the digestion proceeds better if the substrate has Total Solid (TS) content in the range 8 to 12 per cent. The common feed materials like cow dung usually contain higher percentage of solids and they are diluted with water. In the case of cow dung, the recommended dilution is 1:1 (by volume) which makes a slurry of 10 to 12 per cent TS.

(g) **Presence of harmful materials:** Many materials like disinfectants and cleaning agents are toxic to the anaerobic bacteria and the entry of toxic materials in the digester may cause bacterial inhibition leading to reduction of gas production or even failure of the system. The common toxic substances are ammonia, ammonium ion, soluble sulfides and heavy metals such as copper, zinc and nickel. Higher concentrations of metallic salts of Na, K, Ca and Mg are also inhibitory. Only materials in solution have toxic effect and the threshold toxic concentration is different for different substances.

(h) Mixing: Any chemical reaction requires intimate contact of reactant molecules and this is true for biochemical reactions like anaerobic fermentation also. Stirring of slurry inside the digester is desirable to speed up the biochemical transformation by ensuring proper bacterial distribution in the feed material. Bubbling of biogas through the digester slurry creates some movement and mixing may not be essential in small digesters.

9.7 Classification of Anaerobic Systems for Energy Production

In general, the following classification is used for anaerobic systems (biogas production systems):

(a) Batch type: In batch digestion systems, the substrate is filled in the digester along with required quantity of inoculum (microbial culture). The daily gas production initiates within a short period, reaches the peak and then starts declining (Fig. 9.3). When the daily gas output reaches a low value (rated minimum gas output), the digester is unloaded and charged with fresh substrate. If we want to get steady gas supply, it may be required to operate a number of batch digesters in series so that the charging and emptying are done in succession.

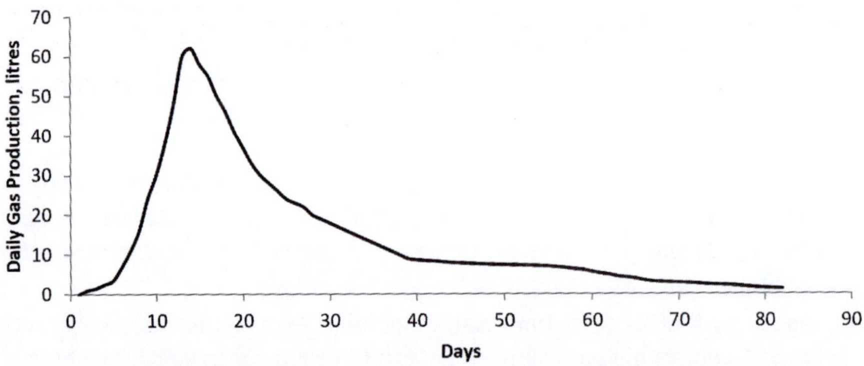

Fig. 9.3 Daily gas production in batch anaerobic digestion

(b) Semi-continuous digestion: In semi continuous digestion systems, the feed material is introduced in to the digester regularly on a particular interval. Usually domestic and community biogas plants are fed once daily preferably at the same time. This can ensure almost continuous and moderately regular gas production. As they are biological systems, changes in the composition of feed material as well as environmental factors can affect the gas production slightly. A start up period is required for these systems to reach the point of regular daily gas production. Initially the digester is filled with feed material and inoculum and run in batch mode for few days. Once the gas production peaks, daily feeding is commenced. The pattern of gas production from a semi continuous digester is shown in Fig. 9.4. The initial period of gradual increase in daily gas production indicates the start-up.

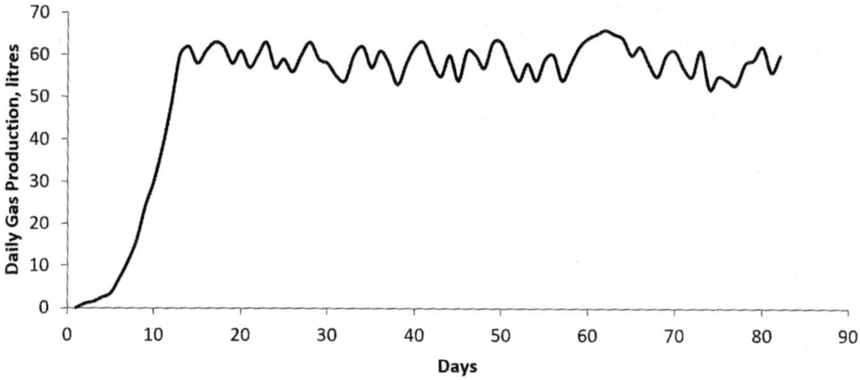

Fig. 9.4 Daily gas production in a semi-continuous anaerobic digester

(c) Continuous anaerobic digestion: In continuous feeding, the feed material is being fed throughout the operational period. This kind of feeding is often done in anaerobic systems which are treating agro-processing effluents or wastes from food processing facilities.

(d) High-rate anaerobic bioreactors: High-rate anaerobic systems have very short retention time for the feed material compared to ordinary biogas systems. They achieve this by preventing the bacteria from going out of the digester along with the outgoing effluent. These systems are usually employed for organic liquid wastes only. Specialized bioreactors are used for high-rate anaerobic digestion.

(e) High solid anaerobic fermentation: In some of the systems, high solid concentration of substrate is used, even though the optimum TS for digestion is 8 to 12 per cent.

(f) Two stage digestion: In these systems, the three phases of anaerobic digestion are carried out in two different digesters. The first two phases, solubilization and acidogenesis take place in one digester whereas methanogenesis is done in a separate digester. These types of anaerobic systems are usually employed for feed materials which needs longer time for biodegradation usually due to high percentage of ligno-cellulosic biomass.

9.8 Operational Parameters of Anaerobic Systems

There are a number of parameters which should be taken care of in operating anaerobic systems. These are also required for monitoring the digester operation so as to ensure steady gas production and prolonged trouble-free working.

(a) Hydraulic retention time (HRT): Hydraulic retention time is the most important operational parameter for an anaerobic bioreactor and is the average length of time a sample of substrate remains in the digester. For continuous (or semi continuous) digesters, fed on a daily basis, the HRT is the ratio between the volume of the digester contents to the volume of daily feed. The optimum HRTs for different types of anaerobic systems vary widely. Easily biodegradable and soluble feed stocks require shorter HRTs and less degradable materials with high proportion of ligno-cellulosic materials need long HRTs. Comparatively longer HRTs are required in colder climates. HRT for high-rate bioreactors can be as short as one day or even few hours.

(a) **Loading rates:** Loading rate is another common terminology used to indicate the daily feed quantity precisely. Feeding rate should be specified and expressed on a unit digester volume basis. *Hydraulic loading rate* (HLR) or *Volumetric Loading Rate* (VLR) is the reciprocal of HRT and is usually expressed as volume fed to unit volume of the digester per day in m^3 per m^3 digester volume.

The above terminology will not give information on how much solids are fed daily. If very dilute waste water with a TS less than 1 per cent is fed, it has no comparison to feeding cow dung slurry which as a TS of 10 per cent. In order to monitor the operation and compare the gas production potential, we need to express the feed rate scientifically. The actual feed rate of organic matter is 'Organic Loading Rate' (OLR). OLR can be expressed in terms of Total Solids (TS), Volatile Solids (VS), BOD or COD. OLR_{TS} is expressed as kg TS per cubic metre reactor volume. Similarly, $OLR_{VS,}$ OLR_{BOD} and OLR_{COD} are expressed as kg VS per cubic metre reactor volume, kg BOD per m^3 reactor volume and kg COD per m^3 reactor volume, respectively.

9.9 Performance Parameters of Anaerobic Systems

In order to assess the performance of biogas production systems, the gas production as well as reduction of organic solids are monitored.

9.9.1 Biogas production parameters

The following gas production performance parameters are used to specify the capacity of biogas systems:

(a) **Daily biogas production:** It simply tells how much biogas, in litres is produced daily in a biogas system. This actually gives little information without knowing the volume of digester or the daily feed quantity.

(b) **Volumetric biogas production:** It is the volume of biogas produced per day per unit volume of digester and is expressed as m^3 per m^3 digester volume.

(c) **Specific biogas production:** It is the volume of biogas produced per day per unit quantity of feed material and is expressed as m^3 per kg TS, m^3 per kg VS, m^3 per kg BOD or m^3 per kg COD.

(d) **Biogas productivity:** Biogas productivity of a particular material is the biogas produced from a specific quantity (Volume or weight) of the material, expressed as litre per litre or litre per kg

9.9.2 Solid degradation parameters

The reduction of solids in the digester results in gas production and these are highly related to gas production parameters.

(a) Percentage TS reduction: This is a direct indication of the reactor performance and the percentage of reduction of TS in the digester can be calculated as:

$$\text{TS Reduction, \%} = \left(\frac{TS_{\text{in}} - TS_{\text{out}}}{TS_{\text{in}}}\right) \times 100 \qquad ...(9.4)$$

Where, TS_{in} is the total solid content (TS) of the feed material and TS_{out} is the TS of outgoing slurry or effluent.

(b) Percentage VS reduction: This is a more accurate indication of the reactor performance and the percentage of reduction of VS in the digester is calculated as:

$$\text{VS Reduction,\%} = \left(\frac{VS_{\text{in}} - VS_{\text{out}}}{VS_{\text{in}}}\right) \times 100 \qquad ...(9.5)$$

Where, VS_{in} is the volatile solid content (VS) of the feed material and VS_{out} is that of outgoing slurry or effluent.

(c) Percentage BOD or COD reduction: These percentage reductions are often used when the anaerobic systems are primarily used for waste treatment, especially organic effluents. The percentage reductions of BOD or COD are also calculated in the similar way.

9.10 Biogas Plants

A biogas plant is a reaction chamber (digester) in which anaerobic digestion of feed materials takes place to produce biogas and having a storage chamber for the gas (gas holder) along with necessary accessories like feed inlet and spent slurry outlet. Basically, a biogas plant consists of the following parts:

(i) A digester tank for fermenting the feed material which can be maintained in anaerobic condition.

(ii) A suitable feed inlet through which the fermentable feed mixture (diluted with water) is introduced to the digester without disturbing the anaerobic condition in the digester.

(iii) An outlet system for the digested slurry to be conveyed out of the digester (an outlet pipe or an out let chamber as the case may be) without causing gas leakage or entry of air into the digester from outside.

(iv) A gas holder to store varying quantities of biogas with facilities to deliver the gas as and when required without causing gas leak.

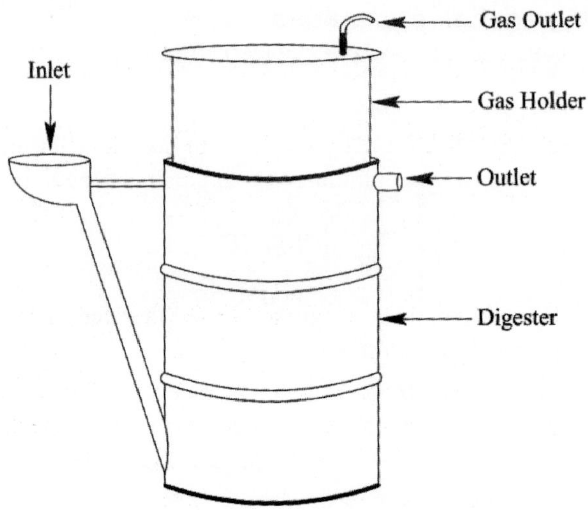

Fig. 9.5 Oil drum biogas plant

Biogas plants can be constructed in diverse designs and with various materials. There has been many such designs and models developed in many regions of the world. Some are as simple as an oil drum biogas plant as shown in Fig. 9.5. There can be flexible biogas plants (Fig. 9.6) made of materials like 'neoprene'.

The digester is underground and is supported by soil all around. As the gas holder is flexible it can expand and contract to some extend so as to contain the gas and release it when the outlet valve is open. The most common construction mode for biogas plants is masonry structures which are built with bricks and cement mortar. RCC (Reinforced Cement Concrete) structures can also be utilised in certain situations. When we construct a biogas plant as a masonry structure, we require the following components:

(i) Foundation
(ii) Digester
(iii) Gas holder or masonry dome
(iv) Mixing tank
(v) Inlet
(vi) Outlet
(vii) Gas outlet with accessories such as valves, pipe line, fittings and appliances

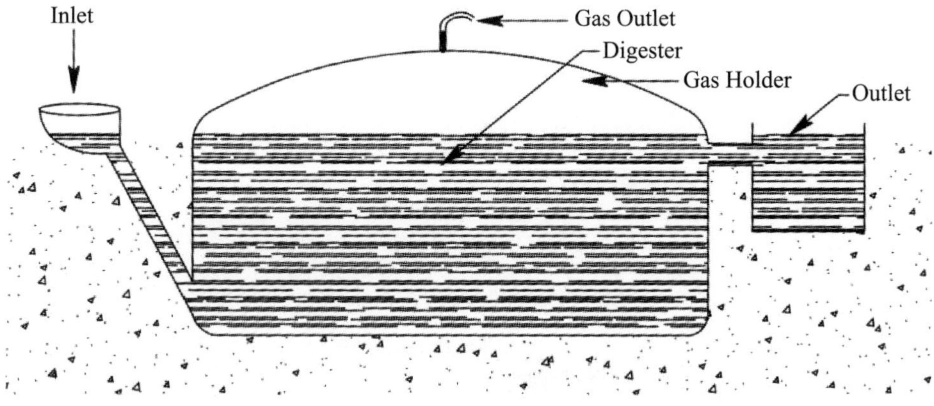

Fig. 9.6 Flexible biogas plant

Classification of Biogas plants: Biogas plants are classified based on different aspects. Based on the type of beneficiary they are grouped as 'community', 'family (domestic)' and 'institutional' biogas plants. This classification is for purposes such as giving government subsidies. Based on construction they are sometimes classified as 'vertical' and 'horizontal' biogas plants. The common biogas plants generally fall under the category vertical and the horizontally oriented digesters are suitable for shallow water table conditions. Based on design they can again be classified as 'floating gas holder' type and 'fixed dome' type. The recent class of biogas plants are 'portable', which are small capacity (daily gas output < 1 m^3) plants made of Fibre Reinforced Plastic (FRP) or other lightweight materials.

9.11 Popular Biogas Plant Models

Even though there are diverse models, basically the domestic biogas plants and to some extend the community biogas plants are either a floating gas holder type (Indian design) or a fixed dome type (Chinese design). The popular floating gas holder design is based on the KVIC design and the fixed dome designs are based on the early Janata model biogas plant.

9.11.1 Floating (Movable) gas holder type biogas plants

The characteristic feature of these Indian design plants is that they have a movable gas holder and hence popularly called 'floating drum type'. These plants are commonly known as KVIC (Khadi and Village Industries Commission) plants and were in use since early 1960s. These plants are characterised by the inverted drum type gas holder placed over the underground cylindrical digester (Fig. 9.7). The digester has an inlet pipe connected to the bottom as well as an outlet pipe placed diametrically opposite to the inlet. The other end of inlet tube is connected to a mixing chamber or inlet chamber where the cow dung (any other feed material) is mixed with water to make slurry of required solid content. In the early designs, the gas holder was made of mild steel which was considered as a major disadvantage of this model, as the mild steel easily got rusted causing gas leaks. Nowadays FRP gas holders are popular as they are corrosion resistant, light weight and durable.

In order to prevent the fresh slurry getting discharged through the outlet, a diametrical separation wall is provided at the centre up to a proper height so that the slurry can cross over from the inlet side to the outlet side through the top of the digester.

Fig. 9.7 Old KVIC model biogas plant

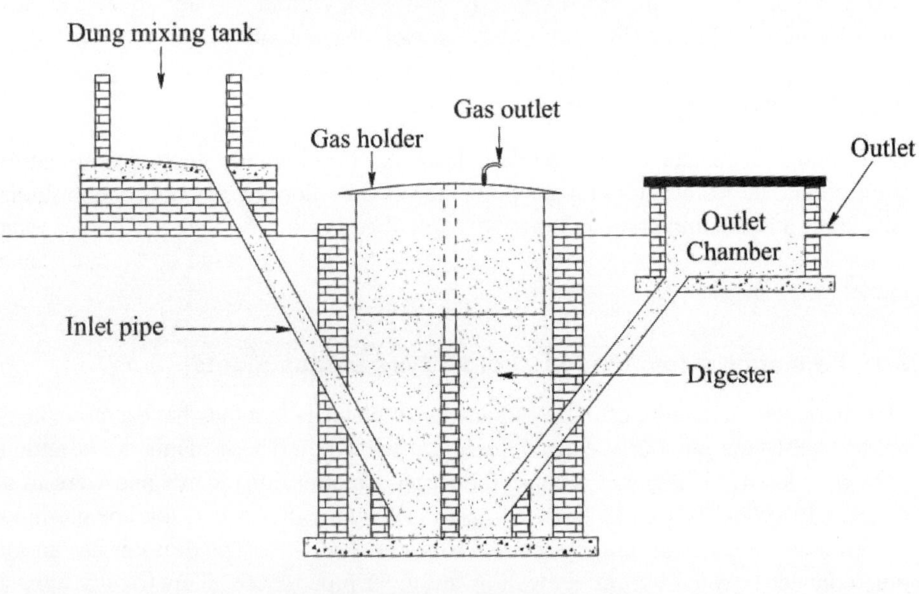

Fig. 9.8 KVIC model biogas plant

The constructional details are illustrated in Fig. 9.8. A guide frame at the top of the partition wall with a guide pipe fitted at the centre supports the gas holder. This arrangement helps the proper movement of the gas holder up (when gas accumulates) and down (when gas is being used up) along the guide pipe, as required. The gas pressure is generally about 10 cm of water column created by the weight of the gas holder. In case the pressure is lower than the requirement then additional weights are provided over the gas holder to create sufficient gas pressure to make it flow through the gas pipe to the points of use. Generally, the storage capacity of the gas holder in KVIC model plant is kept 50 per cent of the daily gas production, assuming that the gas will be used in the morning and evening for cooking purpose.

9.11.2 Fixed-dome type plants

The high cost of steel gas holders, the difficulty for in situ fabrication in rural areas along with their inherent corrosion problems paved the way for the fixed dome plants. These plants were masonry structures and could be constructed with materials available in rural area. The first popular model was the Janata model introduced by the Gobar Gas Research Station, Ajitmal (UP) during late 1970s.

(a) Janata Biogas plants: The Janata model biogas plants (Fig. 9.9) are originally Chinese design plants. The main feature of this model is that there is no metal gas holder. The digester is a shallow well, constructed with brick masonry. The gas is stored in an immovable dome shaped masonry structure above the digester as if it is the roof, often referred as 'fixed dome'. Both the fixed dome and the digester are integrated parts of the brick masonry structure. Another distinguishing feature of this model is that the lack of pipes for inlet and outlet. Large chutes are provided on diametrically opposite sides of the digester to form the inlet and outlet tanks. They serve the purpose of displacement chambers so as to contain the gas in the dome and release it when the gas valve is open. The alignment of the chute opening is carefully designed so that the slurry levels in the tanks are always kept above the level of the junction of the gas containing dome and the digester. The gas pipe is attached through a valve to the GI pipe fitted on the crown of the dome. The discharge of digested slurry takes place through an opening provided on the outer wall of the outlet tank.

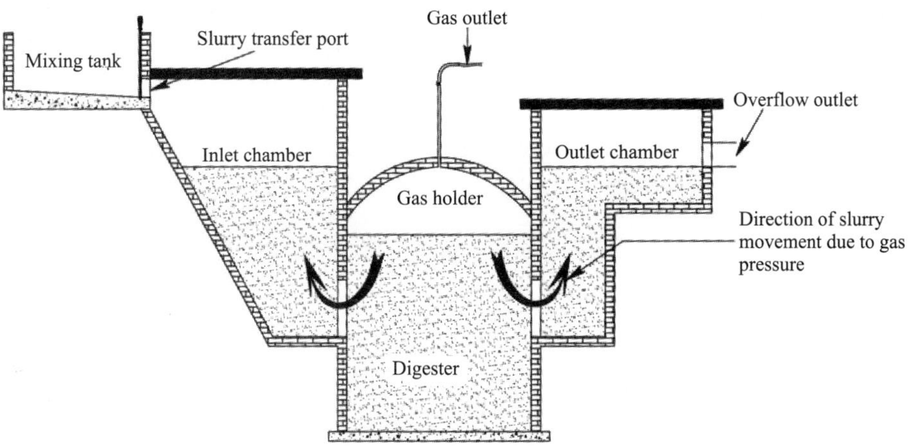

Fig. 9.9 Janata model biogas plant

Once the digester is filled with the required quantity of slurry, gas production starts and gets stored in the dome. The accumulated gas exerts pressure on the top of the slurry pushing it out through the chutes to both the displacement chambers on either side as depicted in Fig. 9.10. Now, the slurry level inside the digester is lower than that in the inlet and outlet tanks which act as displacement chambers. Once we open the gas valve, the hydraulic pressure exerted by the slurry in the tanks push the gas out. As the gas pressure decreases, the slurry flow back into the digester until there is no difference between the slurry levels in the digester and the displacement chambers. The volume of usable gas is determined by the maximum possible level difference (H) as per the design. Due to constantly changing slurry level differences, the pressure of gas in the dome is variable. Generally, the maximum level difference is limited to one metre so that the maximum pressure of biogas is safe and reasonable.

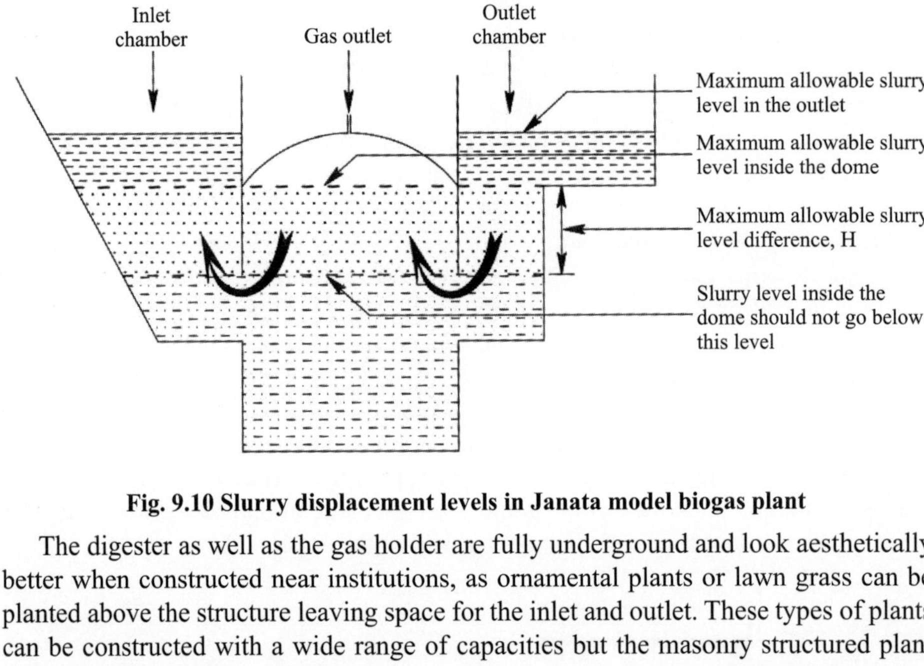

Fig. 9.10 Slurry displacement levels in Janata model biogas plant

The digester as well as the gas holder are fully underground and look aesthetically better when constructed near institutions, as ornamental plants or lawn grass can be planted above the structure leaving space for the inlet and outlet. These types of plants can be constructed with a wide range of capacities but the masonry structured plant size is limited to 15 m^3.

(b) 'Deenbandhu' model biogas plant: '*Deenbandhu*' model biogas plant (Fig. 9.11) developed by Action for Food Production (AFPRO) in the 1980s is a modified version of the Janatha biogas plant. The basic purpose of this design is cost reduction achieved by minimizing the overall surface area of the plant without sacrificing the capacity of the plant.

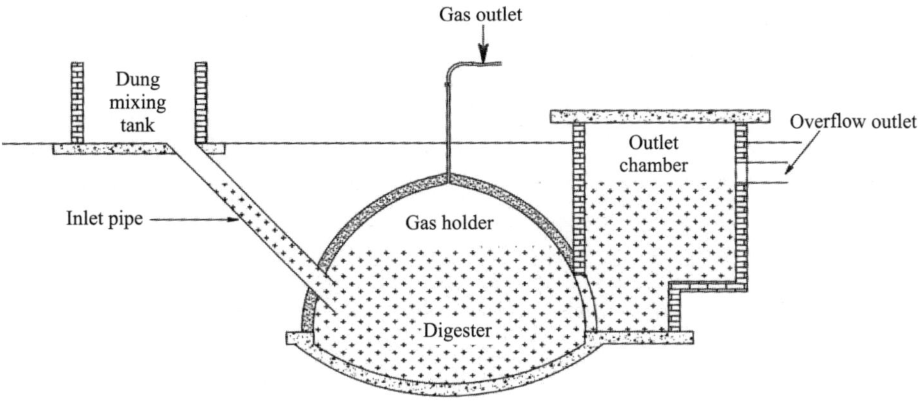

Fig. 9.11 Deenbandhu model biogas plant

The plant is conceived as a combination of segments of two spheres of different diameters, joined at their bases. The lower segment made of concrete acts as the digester and the upper part constructed as a dome with brick masonry is the gas holder. The high compressive strengths of brick masonry and concrete make the structure safe under compressive loads. The typical design makes the structure strong enough with simultaneous benefit of reduced cost. The inlet chamber of the Janata design is replaced with an inlet pipe. The outlet chamber is spacious enough to act as the displacement chamber to hold the entire displaced slurry when the gas chamber is filled to its maximum capacity. The levels of the chute connecting the outlet chamber with digester as well as the slurry outlet gate should be properly aligned to prevent gas leaks. The maximum slurry level in the dome should be safely below the gas outlet to prevent slurry from entering the gas pipe. The general recommendation for the maximum capacity of Deenbandhu model plants is up to 6 m^3.

9.12 Design of Biogas Plants

Even though the design of different models of biogas plants varies widely, there are some general principles from which the basic design parameters are arrived. The hydraulic retention time (HRT) is one of the basic parameters considered in the design as the minimum digester volume required is equal to the product of the daily feed volume and the HRT. The design can be done based on either the biogas requirement or on the availability of feed material.

(a) Selection of HRT: The HRT is dependent on the biodegradability of the material as well as the climatic condition. For areas having a warm climate throughout the year as in the case of south Indian states, the recommended HRT is 30 to 35 days for cow dung-based biogas plants. If the feed material is likely to have pathogens, as in the case of night soil, we need to give a much longer HRT for safety reasons. In areas where there are cold winters, the HRT of 40 to 45 days is recommended in India, whereas it is 50 to 55 in cool climates in the mountain region. Any biogas plant can be run on an HRT longer than the design HRT by making corresponding reduction in the daily feed volume. But the daily gas production gets decreased.

(b) Design based on requirement of gas: If there is no limitation for the feed material availability, the design will be based on requirement of biogas and the biogas productivity of the feed material. Gas availability (biogas productivity) from the feed material should be known to decide how much feed material is required to be fed daily. Ordinary cow dung has biogas productivity in the range of 0.35 to 0.40 $m^3 \cdot kg^{-1}$. Approximately 10 to15 kg of dung is available from cows depending on their body weight. The requirement of biogas for various applications are given in Table 9.3

Table 9.3 Requirement of biogas for different applications

Sl. No.	Application	Equipment/unit	Biogas required
1.	Cooking	Burner, 5 cm	0.30 m^3 per hour
		Burner, 10 cm	0.50 m^3 per hour
		Burner, 15 cm	0.65 m^3 per hour
2.	Cooking	Per person	0.30 to 0.4 m^3 per day
3.	Diesel engine	kW	0.60 to 0.68 m^3 per hour

(c) Design based on availability of feed material: If the objective is to convert all the available material into biogas, the factors to be considered are recommended HRT for the climatic condition and the maximum volume of daily feed. The daily feed volume needs to be calculated in consideration of the moisture content of the material. If it is cow dung with TS 20 per cent, then it should be mixed with equal quantity of water. Therefore, the volume of daily feed is double the volume of cow dung. If the material is having TS less than 10 per cent, then there is no need for further dilution.

9.12.1 Design procedure for a biogas digester

Standard designs with required drawings are available for the commonly used biogas plants with the state nodal agencies under MNRE. However, it is advisable to understand the basic procedure. The following example shows the design procedure for a floating gas holder type biogas plant for meeting the cooking gas requirement of a medium family:

Example 9.1: Design dimensions of a floating gas holder type biogas plant for a family of 6 members

Solution:

Gas requirement (@ 0.30 m^3 per day per person)

Gas required for cooking for 6 people = $6 \times 0.30 = 1.8$ m^3 per day ≈ 2 m^3 per day

Cow dung requirement

1 kg of wet cow dung yields (considering the minimum productivity) = 0.035 m^3

So, the requirement of cow dung = $2.0 \div 0.035 = 57$ kg ≈ 60 kg (wet)

Average cow dung yield from 1 cattle = 12 kg (wet)

So, number of animals required = $60 \div 12 = 5$ animals

Digester dimensions

Amount of slurry fed (dung water ratio of 1:1) = 60 + 60 litre per day = 120 litres per day (Sp. Gravity of slurry is assumed to be 1.0) = 0.12 m^3 per day

The plant can also be designed as per the dung availability. If 5 cows are available,

Then,

Average dung production per cow = 12 kg per day

Total dung production = 12 × 5 = 60 kg.

Assuming Sp. Gravity as 1, volume of dung = 60 litres

Daily slurry volume (after mixing with water) = 120 litres per day

Retention time = 35 days

Required minimum volume of the digester = 0.12 × 35 = 4.2 m^3

The design volume is taken as 110 per cent of the minimum volume considering the volume for bubbles during fermentation.

The design digester volume = 1.1 × 4.2 = 4.62 ≈ 4.6 m^3

Using a ratio of 1 to 1.1 for Depth (Hd) to Diameter (Dd)

(For medium size plants the Hd:Dd ratio can be 1:1 to 1: 1.2)

$4.6 = \frac{\pi}{4} Dd^2 Hd = \frac{\pi}{4} 1.1\, Dd^3$ Diameter, Dd can be found from this relation.

Dd = 1.747m and Hd = 1.92 When rounded to nearest dimensions,

Dd = 1.75 m and Hd = 1.9 m

Design of gas holder

Gas produced daily = 2 m^3

It is assumed that the gas produced during the night is used up for cooking the breakfast and lunch in the morning. The gas produced in the day time after the morning cooking is used for cooking dinner. Hence the storage requirement is 50 per cent of the daily gas production.

Gas to be stored = 1 m^3

Diameter of the gas holder will be slightly less than the diameter of the digester so as to ensure its free movement. Let us leave an annular space of 10 cm space between the digester and the gas holder. So, the actual diameter of gas holder = Dg = Dd – 0.20 m = 1.75 – 0.20 = 1.55 m

Hg can be found from the expression:

$1\ m^3 = \frac{\pi}{4} Dg^2 \mathrm{Hg} = \frac{\pi}{4} 1.55^2 \times \mathrm{Hg};\quad \mathrm{Hg} = 0.53\ \mathrm{m}.$

Considering the submergence of edges into the slurry so as to prevent gas leak an additional 20 cm may be added to the height of gas holder.

The design height of gas holder, Hg = 0.55 m

Slurry mixing tank

A slurry mixing tank is required to mix the cow dung with water. The minimum volume should be equal to the daily slurry volume. Let us assume a depth of 45 cm for the mixing tank.

Daily slurry volume = 120 litres = 0.12 m^3 = Area × Depth

Assuming square shape, Length of side = $\sqrt{\text{Area}}$ = 0.516 m

A side dimension of 55 cm with depth of 55 cm can be selected so as to leave some space at the top when slurry is prepared.

Example 9.2: Design a biogas plant (floating drum type) for a farmer having 6 cows and 6 calves. In addition to domestic cooking requirements of his 5-member family, he intends to process the milk for value added products. How many hours he will be able to use a two-burner biogas stove for milk boiling, after meeting his domestic cooking equirements?

Solution:

Design of Digester

No of cows = 6; Dung availability = 10 × 6 = 60 kg

No. of calves = 6; Dung availability = 5 × 6 = 30 kg

Total daily dung availability = 60 + 30 = 90 kg

Daily dung Volume (assuming specific weight of dung = 1 kg per litre) = 90 litres

Daily slurry volume (mixing with equal quantity of water) = 90 × 2 = 180 litres

Retention time = 40 days

Volume of digester = 180 × 40 = 7200 litres = 7.2 m^3

10% extra provided for overlap of gas holder.

Design volume of Digester = (7.2 × 110) ÷ 100

= 7.52 ≈ 7.5 m^3

Volume of digester = 7.5

= $\pi/4\ (Dd^2 \times H_d)$

Considering the Diameter: Height ratio as 1:1; Dd = H_d

7.5 = $\pi/4\ (D^3)$ and D = $3\sqrt{7.5 \times 4/\pi}$ = 2.122 m

Diameter of digester = 210 cm;

Depth of Digester = 210 cm

Design of gas holder:

Biogas productivity of cow dung = 40 litres per kg

Daily gas production = 40 × 90 = 3600 litres = 3.6 m^3

Diameter of gas holder = diameter of digester – 20 cm (For free floating, a clearance of 10 cm on sides provided) = 210 – 20 = 190 cm = 1.9 m

Volume of gas to be stored = (3.6 ÷ 2) m^3

(Assuming 50% to be used in the morning and 50% in the evening)

Volume of gas holder = 1.8 = $\pi/4\ (Dg^2 \times H_g) = \pi/4\ (1.9^2 \times H_g)$

Height of gas holder, H_g = (1.8 × 4) ÷ (π × 1.9 × 1.9) = 0.635 m

Providing a submergence height of 15 cm, Total height = 0.65 m

Dimensions of slurry mixing tank:

Daily slurry volume = 180 litres

Assuming a depth of 40 cm,

Area = (Volume ÷ Depth) = 0.18 m^3 ÷ 0.4 m = 0.45 m^2

Assuming square cross section, L = B = $\sqrt{0.45}$ = 0.67 m

Calculation of biogas usage:

Gas for domestic cooking requirement = 300 litres per day × 5 = 1500 litres = 1.5 m^3

Balance gas = 3.6 – 1.5 = 2.1 m^3

Gas requirement per hour for 1 burner = 300 litres

Gas for 2 burner stove = 300 × 2 = 600 litres = 0.6 m^3

Hours of use with balance gas = 2.1 ÷ 0.6 = 3.5 hours

9.13 Construction of Biogas Plants

The first step is the selection of a suitable site for construction.

(a) Selection of site for biogas plants: There are certain points to be considered for the selection of a suitable site for a biogas plant. Primarily, operational convenience is required and hence the plant is constructed near to the generation of feed material. Availability of water is another requirement. Sufficient space should be available for day-to-day operations like handling of digested slurry. The digester should not be a cause for water pollution and hence it should be located away from a well by at least 15 m. The site should be in such a place where the water table is sufficiently low.

(b) Construction procedure for floating gas holder type (KVIC) biogas plant: The construction is done in the following steps:

(i) **Digging the pit for digester:** After finalysing the design dimensions, the layout is marked on the ground for earth work excavation. Additional depth of 15 cm is required for the foundation.

(ii) **Laying of foundation:** The foundation is laid down at the bottom using cement mortar (cement : sand : ballast = 1:3:6). For good soil condition and for medium size plants the depth of foundation is usualy taken as 15 cm.

(iii) **Construction of digester:** The next day after laying foundation brick masonry work can be started using 1:5 cement mortar.

(iv) **Setting the inlet and outlet pipe:** Once the brick masonry reach a height of 30 cm, the inlet pipe and outlet pipes are to be placed properly aligned by giving support at the bottom.

(v) **Construction of partition wall:** A partition wall of brick masonry is simultaneosly built so as to divide the digester volume into two. In small plants this wall can be avoided, but the outlet pipe need to be placed at a height so as to prevent short circuiting of fresh slurry.

(vi) Placement of guide frame: A steel guide frame is to be placed at the top of the digester wall and the central guide pipe also is fitted vertically.

(vii) Plastering the inner surface: The inner surface of digester along with both sides of the partition wall are plastered with 1:3 cement mortar.

(viii) Slurry mixing tank: The slurry mixing tank is constructed so that mixed slurry can flow easily to the outer end of the inlet pipe. The construction may be done with brick masonry using 1:4 cement mortar. Plastering with same mix is sufficient.

(ix) Outlet tank: A spent slurry outlet tank is constructed so as to collect the output slurry for the desired number of days. The opening of the outlet tank should be properly aligned so that the slurry should not rise in the mixing tank on the opposite side. The construction is done with brick masonry.

(x) Placement of gas holder: The FRP made gas holder is fixed on the central guide pipe so as to enable telescopic movement of the central pipe of the gas holder.

(xi) Gas outlet: The gas holder has a gas pipe with valve fitted on it. Flexible pipes can be fitted with the help of necessary accessory pipe fitting components such as hose collar, hose clip etc.

(c) Construction of Deenbandhu model biogas plants: Brick masonry is used for the construction of Deenbandhu model and other fixed dome type biogas plants. The construction of dome needs to be done by a trained expert mason. The procedure for construction is more complicated and lack of expertise can lead to constructional defects. However, the brief procedure for construction is as below (Fig. 9.12 to Fig. 9.14):

(i) Digging pit: The layout of the plant is marked on the ground and the circular pit for the digester as well as the rectangular pit for outlet chamber are dug. The earth at the base of the digester pit is made to conform to sherical shape of the digester.

(ii) The foundation is laid using 1:3:6 cement concrete mix by filling the required portion to the predetermined level marked all over the bottom portion. This foundation should extend to the rectangular portion of the outlet chamber also.

(iii) The digester construction is commenced on the next day itself. At the centre of the foundation a pillar having a height equal to the level of the floor of the outlet chamber is erected. A nail is fixed on top of this pillar to mark the centre of the circular area. A straight pole (usually of bamboo) or pipe having a length slightly more than the radius of the dome is taken as a gauge. A curved hook like metal peice is fixed on one end of the pole at a distance equal to the radius of the dome. This pole is used as a gauge to lay the bricks in circular layers without altering the radius of the dome.

(iv) Once a height of 30 cm is reached, the inlet pipe is fixed by providing required supports.

(v) When the digester walls reach the required height, an arch is constructed using brick masonry at the top of the outlet chute. Necessary form work may be used to construct the arch. The outside of the digester wall is plastered with 1:3 CM.

(vi) As the dome is integrated with the digester, the placement of bricks becomes tricky after certain height is reached. It will be difficult for the bricks to be held without falling down. At this point, hooks with counter weights are used to hold the bricks in position.

(vii) When the dome is nearly finished except for a small opening of about 40 to 50 cm, a form work is required for finishing the construction of the dome. Towards the end, the diameter of the hole becomes so small to place full bricks and hence casting is done with brick pieces and cement mortar. A gas outlet pipe with necessary thread at the outer end is fixed at the centre of the crown. The top surface is finished to a proper shape using a stronger mix of cement concrete.

(viii) The outer side of the dome is strengthened by laying cement concrete (20 cm) and plastered with cement mortar.

(ix) The mixing tank is constructed as per the requirement.

(x) The annular space between the vertical soil wall of the excavated pit and the outer side of the brick masonry should be filled with soil and properly rammed.

(xi) After initial curing in a week's time, the form works inside the dome are removed and the innerside of the dome as well as the digester portion are plastered with 1:3 CM. The upper roof of the dome is plastered with a stronger mix of 1:2 CM after applying cement solution to fill the grooves on the innerside of brick masonry. A final finishing coat of cement solution is given after the plastering. This is done to prevent gas leakage through small pores, if any.

(xii) The mixing tank and outlet chamber also need plastering with 1:4 CM and the plant is kept for curing for 10 to 12 days. Regular watering of the plant is done for proper setting of the cement.

9.14 Construction of Low Cost Biogas Plants

Small size KVIC model biogas plants can be constructed with concrete rings for the digester so as to finish the construction fast. The rings can be cast in situ with the help of moulds (Fig. 9.15 and Fig. 9.16) and transported to the site. Rings made of ferrocement is used for the digester construction and FRP gas holder is readily available. The outlet tank can be constructed with ferrocement slabs so as to reduce the cost of construction to a great extend. Manufacture of FRP gas holder can be undertaken by small entreprenuers even in rural areas. Ferrocement construction requires only less raw materials and are safe for small size plants.

Cost reduction in the construction of small and medium size Deenbandhu model biogas plants is achieved by constructing the dome with ferrocement. Ferrocement construction is a technique by which cement mortar is applied on a metal mesh to get the required structure. The base frame work for the dome is made of MS rods as shown in Fig. 9.17. Then it is covered with a mesh over which cement mortar is applied after positioning the frame work over the cement concrete made bottom portion. The outlet chamber is made of ferrocement slabs and the inlet is a concrete ring readily available in the market.

Fig. 9.12 Digging of pit and laying foundation for construction of Deenbandhu model BGP

(Picture courtesy: Er. P. Jayachandran Nair, ANERT)

Fig. 9.13 Construction of dome of Deenbandhu model BGP

(Picture courtesy: Er. P. Jayachandran Nair, ANERT)

Fig. 9.14 Deenbandhu model BGP: finished view
(Picture courtesy: Er. P. Jayachandran Nair, ANERT)

Fig. 9.15 Moulds for construction of rings
(Picture courtesy: Er. P. Jayachandran Nair, ANERT)

Fig. 9.16 Finished ferrocement KVIC biogas plant
(Picture courtesy: Er. P. Jayachandran Nair, ANERT)

Fig. 9.17 Stages of construction of ferro-cement Deenbandhu biogas plant
(Picture courtesy: Er. P. Jayachandran Nair, ANERT)

9.15 Portable Biogas Plants

Portable versions of the floating gas holder biogas plants made of FRP became a popular alternative in urban and semi urban areas in Kerala. Even though only small quantity of biogas is available from these plants, they are good for domestic organic waste management. The domestic organic wastes including kitchen wastes, could be fed to the small biogas plants which are often less than one cubic metere capacity. The digested output slurry is used as a manure for kitchen gardens or poted ornamantal plants grown in the household.

FRP fabrication is a simple technology adopted widely for various applications. For constructing FRP biogas digesters, a mould is prepared first which can be dismantled into 2 or 3 pieces. After giving a chemical coating (gel coat) inside the mould, the fibre mat is spread evenly. Then the resin mixed with 'hardener' and 'accelerator' (chemical ingredients for setting) is applied using brushes over the fibre mat. One more layer of fibre mat is spread over this which is again covered with the resin mixture. Within about 12 hours the setting will be completed and the mould is split open to get the completed product (Fig. 9.18).

Another popular model of portable biogas plant has no floating gas holder. In principle it is a fixed gas holder type plant (Fig. 9.19) developed by Sri. Sunil, a private entrepreneur in Thrissur, Kerala. The hygienic problems associated with floating gas holder type plants is completely avoided in this design as it is fully sealed. The outlet and inlet are open only during the time of feeding. The design is very compact and these type of plants with a capacity of 10 kg food waste per day is available.

KAU Portable Split model: The KAU split model biogas plant was developed by Shaji James P. of Kerala Agricultural University during 2010-11 for enabling local fabrication by the rural youth theselves. This design was intended for small and marginal farmers who may have one cow and a calf only. They grow the cow mostly for their family use and the balance milk is sold to neighbours. This practice is common in small homesteads in villages of Kerala. The characteristic feature of this portable model was that the gas holder unit and the digester unit could be separated (Fig. 9.20). The digester could be kept near the cow shed and the gas holder near the kitchen. Three PVC water tanks of slightly different capacities were used for the fabrication of the biogas plant. The digester was fully sealed and could be kept huygenic. The gas holder had an inverted cylindrical drum made of PVC water tank which floated over the water in another water tank. The top portion of these two water tanks were cut. A guide pipe (PVC) was erected vertically at the centre of the base of the water filled tank. The gas holder had another PVC pipe centrally mounted with the help of supports in a leak proof manner and could move up and down as the two pipes were joined together for telescopic movement. As the fabrication could be done with readily available materials, the time required to install is minimum. A stirrer was provided in the digester to avoid scum formation. The stirrer could be operated manually by a handle easily.

9.16 Monitoring of Biogas Systems and their Maintenance

Biogas systems are biological systems which require proper monitoring and maintenance. The proper balance between different groups of bacteria is to be ensured. The proper functioning can be monitored by observing the pH of the digester contents. A simple pH meter or pH paper can be used for the purpose. A well-functioning biogas plant will generally have a pH of 7 or slightly above. If the pH is seen dropped lower than 6.6, corrective measures should be taken up urgently. The feeding should be stopped until the pH improves. Sometimes addition of alkali for pH correction may be required, especially when acidic feed materials are fed. Overfeeding is the common cause for acidification and should be strictly avoided. The recommended routine maintenance of biogas plant is done on daily, monthly and yearly basis.

(a) Daily maintenance: Feeding of biogas plant with the recommended amount of organic material after properly mixing should be done regularly. If it is a domestic portable system, it is better to keep the kitchen waste in a closed container for one day for the aerobic bacteria to act so that it can be easily solubilised in anaerobic digester. Make sure that the inorganic materials or toxic substances do not get contaminated with feed materials.

Fig. 9.18 ANERT Design - Portable FRP Biogas Plant

Fig. 9.19 Portable fixed gas holder biogas plant

Fig. 9.20 KAU Split biogas plant installed in a homestead

(b) Weekly maintenance: Stirring the digester is always good for a biogas plant. Scum formation is a common problem with cow dung fed plants in which straw and other crop residues may get mixed with dung and fed to the digester. These low-density materials tend to float in the digester and can form a thick scum. The aged scum can become a hard floating layer on the top surface and cause problems in the working. Hence a long pole can be inserted through the outlet chamber of fixed dome type biogas plants for stirring the slurry inside the digester so as to break the scum. The floating gas holder type plants usually have braces inside the gas holder which can break the scum when the gas holder is rotated by hand.

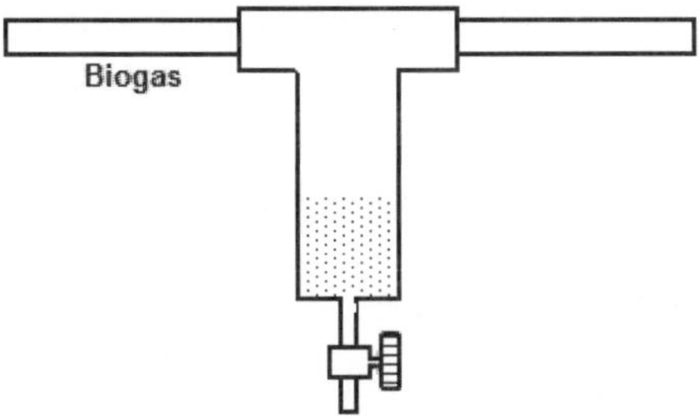

Fig. 9.21 Tee drain for water removal

A common problem with domestic biogas plants is the accumulation of condensed water in the gas pipe line. If it is not removed, the gas flow will be obstructed. During installation, the gas pipe should be aligned such that the condensed water flow back to the gas holder itself by gravity. This may not be possible always, especially when a flexible hose is used as gas pipe. In such cases a simple water removal system can be fixed at the lowest point in the gas pipe line which can trap the condensed water. A tee-drain as shown in Fig. 9.21 may be fitted for trapping water, which can be drained occasionally. A plastic bottle fitted with a tee joint in the pipe line (Fig. 9.22) can be used as a simple water trap but the gas valve should be closed to empty the bottle when it gets filled.

Fig. 9.22 Simple bottle as water trap

(c) Long term maintenance: All gas valves, gas pipe and other fittings should be thoroughly checked for gas leakage on a monthly basis. If scum build up is there, it may be removed once in three months using a hook like tool attached to a long pole.

Checking all appliances and components for defects and replacement of defective and worn items should be done annually. The gas can be fully drained once in two years and the slurry level can be checked in the fixed dome type plants. If the slurry level is below the expected level, additional slurry can be added to get the correct level. It is better to empty the entire digester once in five years so as to remove the undigested sludge which reduce the digester capacity. Once the plant is emptied, the ceiling of the dome can be checked for any minor cracks. Rectification of cracks should be done after chipping off the outer plaster and re-plastering it with leak-proof additives. After proper curing and drying the ceiling can then be painted with good quality leak-proof water-resistant paint. When all the repairs are completed the biogas plant can be refilled and started up as in the case of a new system.

9.17 Use of Biogas Slurry

Biogas slurry is a good quality manure for all crops and it is suitable for producing organic vegetables. As chemical fertilizers are manufactured from fossil fuels, any replacement is beneficial to the country both in economic and environmental aspects. The slurry is more nutritious for plants compared to cow dung or the original feed material due to many reasons (Table 9.4). Plants can absorb only soluble nutrients and the digested slurry contains more soluble components. Hence slurry is better to be used in wet condition and may be mixed with irrigation water. If the slurry is dried in shade and incorporated with bio agents like *Pseudomonas* or *Trichoderma*, it can be sold as a premium manure for organic vegetable cultivation, especially for kitchen gardens. The liquid slurry also is marketed as an organic plant nutrient in Kerala by some agencies with attractive brand names. The slurry can also be pelletized and used as fish feed or used in mushroom cultivation.

Table 9.4 Nutrient availability in organic manures and biogas slurry

Sl. No.	Manure	Available Nutrients, %		
		N	P	K
1.	Farm yard manure composted in open	0.70 – 0.80	0.50 – 0.65	0.50 – 0.60
2.	Farm yard manure composted under cover	0.80 – 0.90	0.70 – 0.80	0.7 – 0.80
3.	Biogas slurry	1.70 – 2.20	1.2 – 2.20	1.0 – 2.10

9.18 Purification of Biogas

The use of biogas in IC engines is an attractive option. One important issue with the use of biogas is the loss of power due to the presence of incombustible CO_2 and the corrosion problem due to H_2S. Hydrogen sulphide burns in the engine to produce SO_2, which creates problem with engine parts due to corrosion. In general biogas contains less than 1 per cent H_2S. If it is very negligible as in the case of biogas from materials without sulphur, this can be neglected. Hence, safe use of biogas in IC engines requires removal of CO_2 and H_2S.

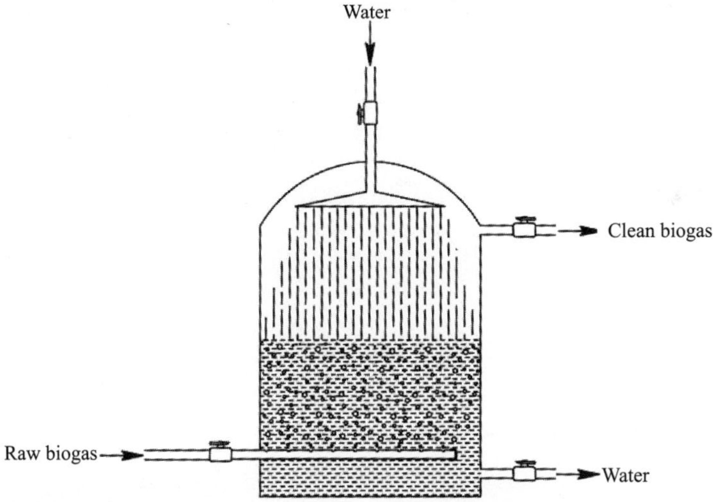

Fig. 9.23 Scrubbing of biogas for CO_2

There are methods for cleaning of biogas from both these impurities. CO_2 can be removed if the gas is liquefied by compression. As the critical temperature of CO_2 is high enough (31°C) it can be compressed and cooled to separate CO_2 and liquid CO_2 is a by-product in the process. Water scrubbing of biogas is another method for removing CO_2 (Fig. 9.23). Caustic scrubbing of biogas can remove CO_2 to a desirable level. Pressure swing adsorption (PSA) is yet another advanced technology which can be utilized for scrubbing CO_2. The basic principle is the adsorption/desorption of carbon dioxide on zeolites or activated carbon at alternating pressure levels.

Scrubbing of H_2S can be done by passing the biogas through a column packed with ferrous materials like iron scraps from lathe machines or drilling machines. The H_2S is absorbed and converted to ferrous sulphide. The absorbing material can be regenerated by exposing it to atmospheric air (Fig. 9.24).

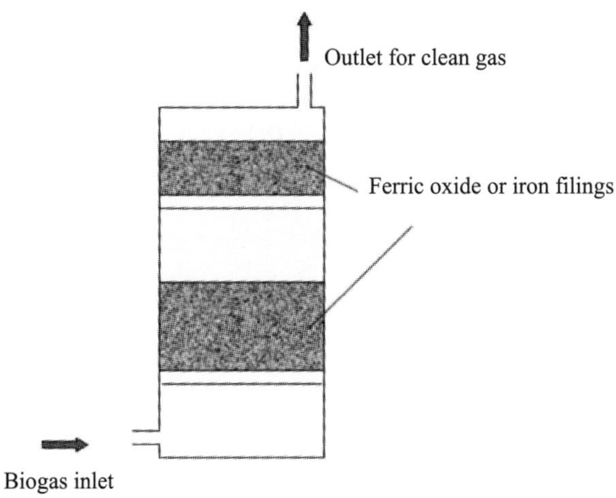

Fig. 9.24 Scrubbing of biogas for H_2S

9.19 Use of Biogas for Automobiles

Once these impurities are removed, biogas can be used as if it is natural gas, and the technology for use of compressed natural gas (CNG) is well known. Advancing of fuel injection timing may be beneficial for biogas engines as the flame speed of biogas is lower than hydrocarbon fuels. Biogas is a good fuel which can be used in IC engines. Biogas fuelled engines were already developed by some manufactures. In Auto cycle engines (Spark ignition) there is no ignition problem with biogas as fuel. But, if biogas if used in compression ignition (Diesel) engines, a pilot injection of diesel fuel is required as biogas is not a self-igniting fuel. If used in dual fuel mode, we can replace about 80 per cent of diesel fuel in diesel engines. It is also very simple to run a diesel engine on dual fuel mode. The biogas can be admitted through the air filter by providing a tee joint. The governing system in diesel engine takes care of engine speed thereby cutting the diesel intake automatically (Fig. 9.25).

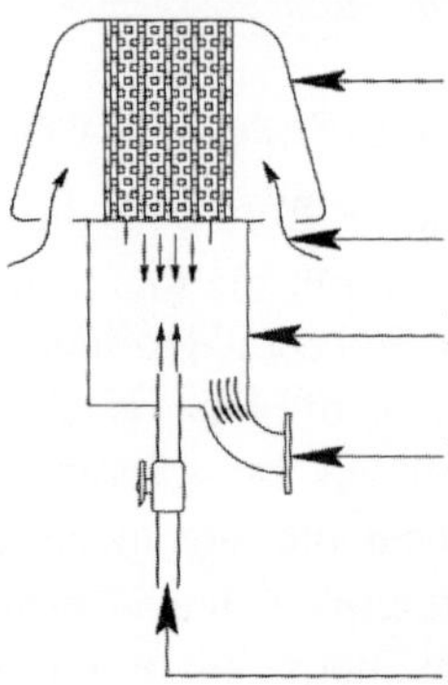

Fig. 9.25 Arrangement for use of biogas in diesel engine

In filling stations, methane can be stored at 200 atm. The vehicles should be provided with fuel selector and controlling valve along with required number of gas cylinders and a connector. Steel cylinders of 20 to 30 litres which can store methane at 180 atm., can be used for automobiles. Two to eight cylinders may be required depending on the engine power. A pressure reduction and metering device is required to be used to lower the high pressure and to deliver methane at the required rate.

CHAPTER 10

High-Rate Anaerobic Bioreactors for Energy Production

Anaerobic digestion of organic wastes is of positive environmental value since it combines waste stabilization with net fuel production and also allows the use of the residues as manure and irrigation water. Conversion of organic liquid wastes to methane by anaerobic digestion has several advantages over the conventional aerobic waste water treatment processes like a positive energy balance and reduced sludge volumes.

10.1 Need for High-Rate Digesters

The popularity of biogas technology for large scale application has been limited due to the inherent slow rate characterized by long retention time resulting in the need for very large digester volumes. The need for large digester volumes leads to escalation of installation costs as well as space requirements. Lack of assurance for steady treatment efficiency and energy supply occurring due to process instability of anaerobic digestion is another handicap in high volume waste water treatment. If these two problems in anaerobic digestion technology are solved, the well-known advantage of renewable energy generation can make anaerobic processes preferable in the organic waste water treatment sector.

Introduction of the high-rate anaerobic bioreactors like Up-flow Anaerobic Sludge Blanket Reactor have brought down the Hydraulic Retention Time (HRT) of anaerobic digesters from 35 to 40 days of conventional biogas digesters to a few hours. Such drastic reduction in HRT has produced a very favourable impact in reduction of digester sizes and subsequent minimization of installation costs. Thus, it has opened the possibility of treating high-volume low-strength wastes such as agro-processing effluents by anaerobic process.

10.2 Cell Immobilization Technology in High-Rate Anaerobic Bioreactors

The design of conventional anaerobic digesters is characterized by long HRTs due to the low specific growth rates of the methanogenic bacteria. They are generally built as a one-step continuous digester in which the bacterial biomass is continuously lost with the outgoing digested material. Usually, no technology is incorporated in them for any recycle of the active biomass or for their retention in the digester.

In aerobic waste water treatment processes such as activated sludge process, the concept of biomass (microbial) recycling is utilized. The term ‘activated’ is used to indicate the sludge which contains microbes which digest and break down the organic matter in the waste water to reduce its BOD or COD as the case may be. The large aerated tank containing raw effluent is the primary phase into which the activated sludge separated in the second phase, often called 'clarifier' is recycled (Fig. 10.1).

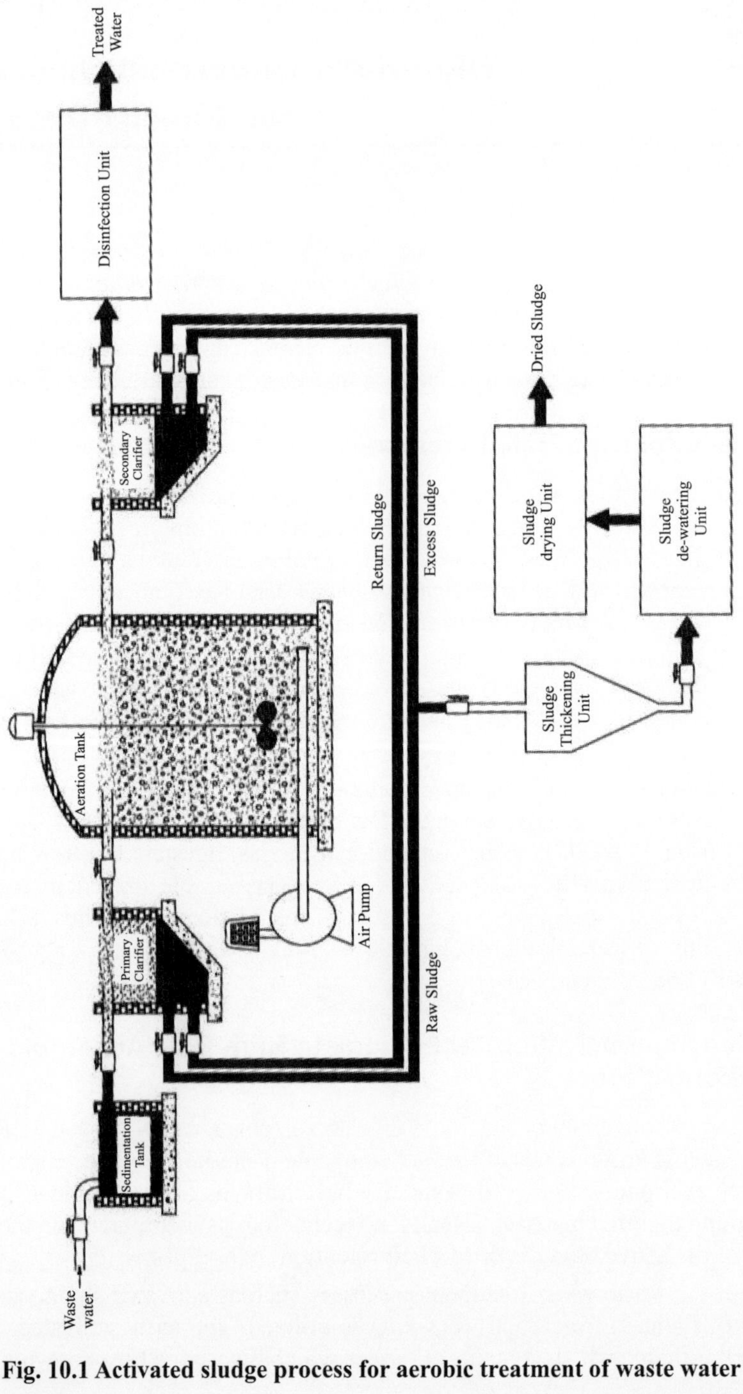

Fig. 10.1 Activated sludge process for aerobic treatment of waste water

The concept of recycling of sludge containing microbial biomass led to the introduction of the anaerobic contact or anaerobic activated sludge process and permitted a longer retention time for the active microbial biomass within the digester. This technique resulted in more efficient waste stabilization. Retention of the active microbial biomass, independent of waste water flow, could considerably reduce the HRT so that it was possible to reduce the digester volume for a given rate of waste water flow. Efficient recycle of active biological solids requires an effective separation of the bacterial cells from the effluent stream. In anaerobic systems this can be troublesome and sophisticated. As the anaerobic bacteria are slow growing, a very high recycle rate is often required to run the system so as to maintain proper treatment efficiency.

Maintenance of the active microbial biomass within the bioreactor independent of waste flow is achieved in the second generation of high-rate anaerobic bioreactors by a mechanism which avoids the costly operational problems associated with the long retention time of conventional digesters and the solids recycle system of the activated sludge process. The concept of retention and maintenance of bio-film grown on an inert support material forms the technological basis of these new generation designs. This 'immobilized cell bio-reactor' technology ensures the maintenance of a high concentration of active microbial biomass within the reactor for long periods. Different techniques such as attachment to surfaces, entrapment, containment or self-aggregation of biomass to form granular sludge are employed in different types of bioreactors. As the soluble organic matter in waste waters can easily be digested enabling multiplication as well as retention of bacterial biomass, these bioreactors can be operated at hydraulic retention times of few days to several hours. They also ensure the treatment of both high and low strength soluble wastes to more satisfactorily levels and these bioreactors can be installed and operated in economically viable conditions (Fig. 10.2).

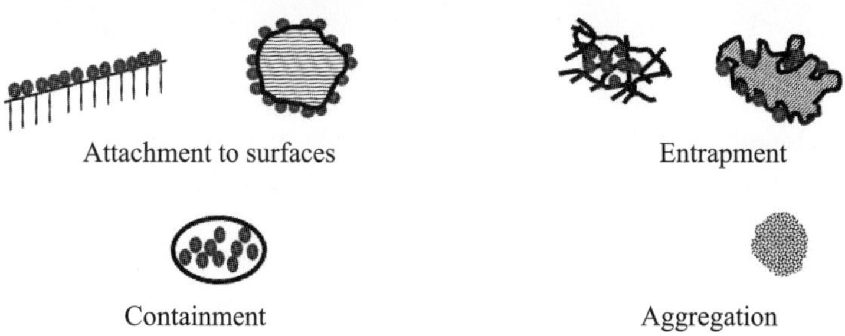

Fig. 10.2 Techniques of cell immobilisation

10.3 Up-flow Anaerobic Filter (UAF)

In fixed film anaerobic systems, the microbial biomass is retained as a bio-film attached to an inert support medium and hence can be retained in the bioreactors for much longer time duration than the hydraulic retention time at which the system operates. Anaerobic Filter is the simplest of such fixed film bioreactors, which consists of a column packed with an inert support material such as gravel, plastic, ceramic, fired clay etc. The flow of water in the filter column can be in the down ward direction or in the upward direction. Up-flow Anaerobic Filters have been found to have better process stability and they were considered superior than the down flow filters. A range of support media have been used in several investigations and the media vary widely depending on the type of reactor and the process geometry. The anaerobic filter initially developed by James C. Young in 1968 had random packed quartzite stones as medium for cell immobilization (Fig. 10.3).

At the bottom of the filter column, a distribution header is provided to create an upward flow through the submerged media bed. If the operational conditions necessitate higher up-flow velocity, a part of the outgoing treated water (effluent) is sometimes re-circulated. The initial UAF designs mostly used broken rock pieces of 25 to 65 mm diameter as media, so as to have about 50 to 60 per cent void volume. Actual liquid capacity of the bioreactor depends on the void volume. There were several investigations for assessing the performances of UAFs for the treatment of different types of waste-waters ranging from pre-treated domestic wastewater to industrial organic waste-waters. Subsequently, different types of synthetic media have been developed with high void volumes up to 95 per cent. A high proportion of the biomass (50 per cent) is not strictly attached to the support media, but is rather held in the interstitial spaces of the matrix. As long as the hydraulic flow through the filters is within safe limits, this interstitial biomass will be retained. Waste-waters which contain a high concentration of suspended solids tend to cause blockage of the interstitial spaces. Highly concentrated effluents, even though easily degradable, may also create blockage of the filter media resulting in channelling within the filter column.

UAFs can be used preferably for warm effluents with moderate organic concentrations (BOD 5000 mgL^{-1}) and low suspended solids. The highest loading rates rarely exceed 10 kg COD per cubic metre of the reactor volume per day for media with high porosity and good surface configuration for bacterial attachment. For simple UAFs with gravel or stone media, the possible upper limit for loading rates do not exceed 5 kg COD $m^{-3}d^{-1}$. Based on the experience of many researchers, about 90 per cent removal of COD is possible at low waste water loading rates (2 kg COD $m^{-3}d^{-1}$) and when the loading rates are increased the COD reduction may drop to low values. In general, UAFs can produce biogas at the standard rate of 0.5 $m^3 \cdot kg^{-1}$ COD at normal operating conditions.

As the microbial bio-film is attached to the matrix packing, the biomass remains intact even under high loading rates. Hence UAFs are stable with respect to high organic loadings and even shock loadings. UAFs are observed to withstand changes in environmental factors such as pH and temperature, which usually upset conventional anaerobic systems. On the other hand, the accumulation of suspended solids in the interstitial spaces or over growth of bio-film on the media can lead to plugging and channelling, which creates problems in reactor performance. In case blockage occurs, backwashing of the filter column may be required, which necessitates temporary shutdown of the system. The practical limitations of Up-flow anaerobic filters are due to the following:

(i) High cost of high-quality media
(ii) Limitations due to suspended solids and high strength wastewaters
(iii) Requirement of periodic back-washing

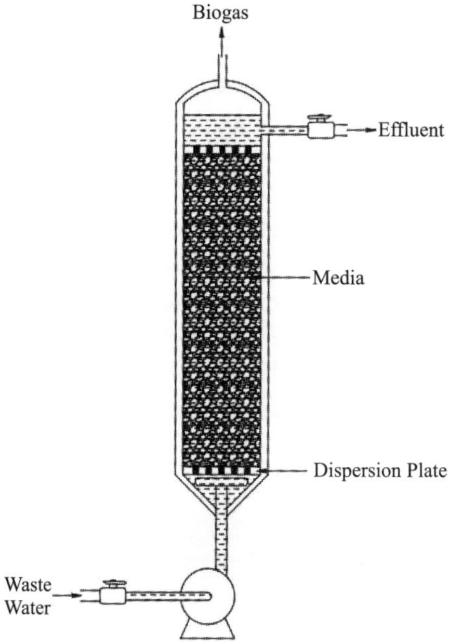

Fig. 10.3 Anaerobic Filter

10.4 Up-flow Anaerobic Sludge Blanket (UASB) Reactors

Cell immobilization in anaerobic up-flow sludge blanket reactors is achieved by formation of microbial biomass in granular form which is retained by providing appropriate hydraulic conditions in the reactor. The biomass in the UASB reactor is characterized by development of highly settleable bacterial granules of 1 to 5 mm size, which forms a blanket of sludge in a fluidized stage. The anaerobic sludge in UASB reactors is not exposed to heavy mechanical agitation and will gradually attain good settling properties. These granular sludge aggregates will be dispersed under the influence of the upward liquid flow as well as the effervescence created by biogas. UASB reactors can handle high organic loads and the agitation due to biogas production can be heavy, especially in tall reactors. The satisfactory retention of UASB granules can be achieved by separating the biogas in a gas collector system commonly known as GSS (gas-solid separator) device. It is important to separate granules from biogas bubbles as well as the liquid phase to prevent its carry-over into the treated effluent. Hence this device is sometimes referred as 'three phase separator' (Fig. 10.4).

The GSS device or three phase separator is crucial for UASB reactors and it acts as an in-built settler at the upper most part of the reactor which is usually enlarged so as to reduce the velocity of liquid. Thus, a low liquid velocity zone is created where the granules which move upwards, as if attached to gas bubbles, is disengaged and re-settle into the sludge blanket zone. Sludge aggregates bind together to form bigger granules and settle in this settling zone which eventually slide down back into the compartment beneath the GSS device.

Prof. Lettinga who was the pioneer in the development of UASB reactors along with his co-workers at the Wageningen University in Netherlands found that high loadings could be applied once the granular sludge was formed. However, the granulation process did not occur with all wastes readily. For example, slaughterhouse wastes and raw sewage did not cause granulation and rendering wastes only caused granulation at a slow rate. The slow start-up of UASB reactors is considered as one of the main drawbacks and a considerable period of time is involved in cultivating a sufficient amount of good granular sludge. Once a successful start-up leading to formulation of granular sludge is achieved, anaerobic sludge can be preserved for long periods of time even when the bioreactor is not fed. If excess anaerobic sludge cultivated in a full-scale installation is available, new installations can be started up easily by utilizing this as seed sludge.

The ease of development of granulated sludge is dependent on microbiological and biochemical factors which in turn are related to the seeding material, feedstock and reactor operating parameters. The hydraulic retention time has to be long enough to prevent washout of granular sludge. At the same time, it needs to be short enough such that filamentous and other non-settling bacterial flocks are washed out of the reactor, particularly during start-up. In most cases, a two-phase process in which the acidogenic and methanogenic phases are separated, can perform better. The reactor configuration may be such that the acidogenic phase of short HRT precedes the methanogenic UASB operated at a longer HRT. The pH of the feedstock should be greater than 5.5 and for ensuring buffering capacity, sufficient alkalinity should be there in the reactor liquor. For soluble waste waters, mixing due to the bubbling of generated biogas is sufficient. The UASB usually contains a dense sludge layer at the reactor over which the granular sludge bed remains in a fluidized stage. There is a lighter cloud of sludge usually referred as sludge blanket above this bed, with the liquid phase at the top.

Hydraulic conditions in UASB reactors are generally characterized by the superficial liquid up-flow velocity in the reactor. The up-flow velocity employed in most of the laboratory and industrial UASB reactors is about 1 mh^{-1} although higher values are possible in highly engineered systems. High up-flow velocities may cause disintegration of the granules and may get carried over with the out-flowing effluent. Very small up-flow velocities may cause creation of gas filled hollow space within the granules, leading to floatation and subsequent washout.

The following general conditions favour the successful operation of UASB reactors:

(i) Maximum organic loading rates of 12 to 15 kg COD $m^{-3}d^{-1}$ for effluents with COD less than 1000 mgL^{-1}.

(ii) A liquid up-flow velocity around 1 mh^{-1}

(iii) Strength of feed not to exceed 10000 mgL^{-1} COD

(iv) Better to have effluent recycle for high strength wastes

(v) Limit gas loadings to 200 $m^3m^{-2}d^{-1}$ to avoid too much turbulence leading to washout of granules

Preliminary treatments are required for raw waste waters. A rough screening before the pump followed by a medium screening (8 to 12 mm) can prevent entry of suspended particles which may create problems. Grit, oil and grease removal will help to reduce scum formation.

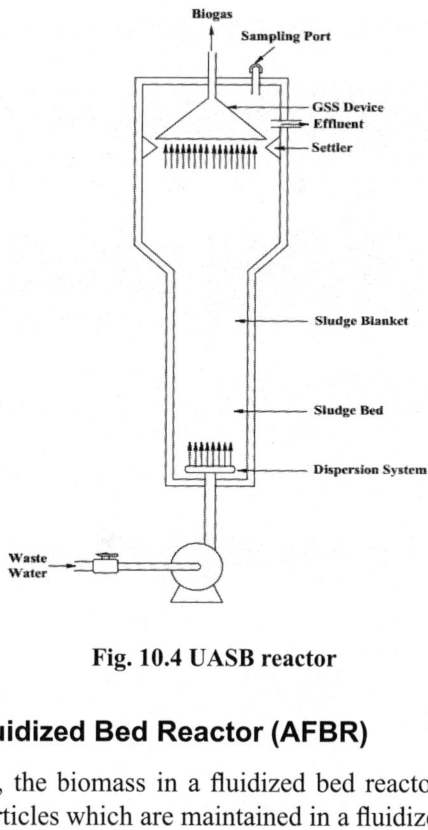

Fig. 10.4 UASB reactor

10.5 Anaerobic Fluidized Bed Reactor (AFBR)

As the name indicates, the biomass in a fluidized bed reactor grows as a bio-layer around heavy small particles which are maintained in a fluidized stage by an upwards directed flow of liquid in the reactor. This reactor configuration has many advantages over other configurations which were first recognized in aerobic effluent treatment systems. In the early development stages, the principle of expanded and fluidized beds applied to anaerobic wastewater treatment systems was found promising. The fluidization of the bed with granular and sufficiently heavy support media is achieved by providing adequate up-flow liquid velocity (ULV). In most fluidized bed reactors bed expansions are in the range 30 to 100 per cent. Fluidized beds with very low bed expansion (less than 30 per cent) are generally called expanded bed reactors (Fig. 10.5).

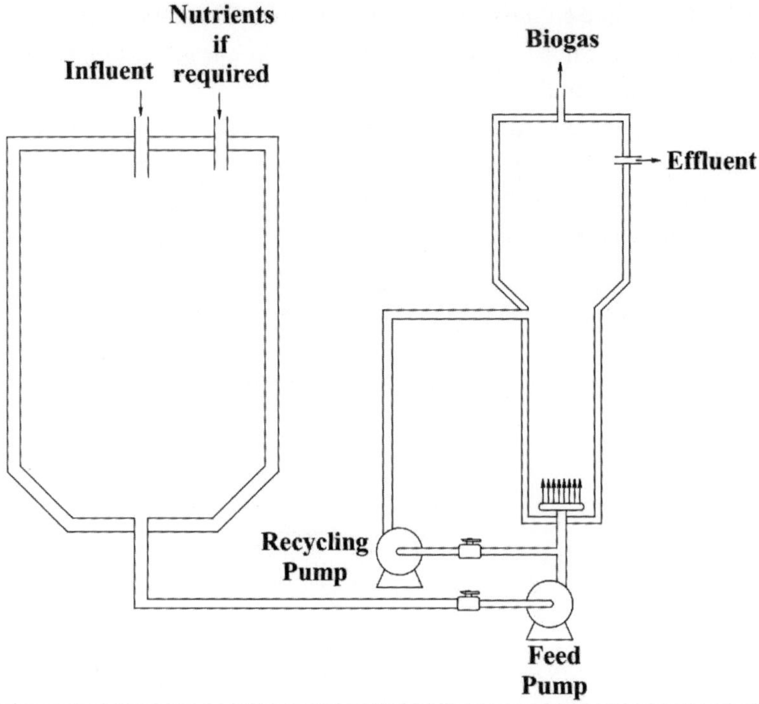

Fig. 10.5 Anaerobic fluidized bed reactor

The AFBR also is an up-flow reactor and contains approximately uniformly sized granular support medium on which the bio-films grow. The media gets expanded and remains fluidized by the velocity of the influent liquid as well as supplemented by a recycle loop. Owing to the heavy particles (sand or sand like particles having a typical diameter of 0.2 to 0.8 mm), the bed settles fast once the upward directed liquid flow is stopped. The settling velocity of these particles is high (around 50 mh^{-1}), which is more than the ULV. This enables the application of high ULVs in the range of 10 to 30 mh^{-1}. Because of this high ULV, the inert sediments or suspended particles in the waste-water do not accumulate inside the reactor. High microbial biomass concentrations are possible owing to the very high attachment surface on the small granular carriers which can provide a specific surface area of more than 2000 m^2m^{-3}. Because of the high microbial biomass concentration and their high activity, a very high treatment capacity can be obtained.

Due to the above characteristics, fluidized bed reactors are very compact and require very little space. Additional features may include a contact reactor which also acts to concentrate any sludge carryover and a bio-film reduction device to remove excess bio-film from the sand. Biogas is collected from the top of both reactors. It appears that sand of 0.3 to 0.5 mm in diameter is commonly used as the carrier in fluidized beds and that high biomass concentrations can be achieved using sand. Other materials like Al_2O_3, active carbon, synthetic resins etc. have also been used as carrier media.

Compared with other high-rate anaerobic reactors, like UAF and UASB reactors, the fluidized bed system has the potential advantages of higher purification capacity, no clogging of the reactor (as in UAF), no problems related to sludge retention (as in UASB systems) along with shorter HRT and small volume requirement. The major disadvantages of these reactors are that, they are highly engineered systems and power is required for fluidization.

10.6 Up-flow Anaerobic Hybrid Reactors (UAHR)

Hybrid anaerobic digester was a new concept evolved in anaerobic bioreactor design. The up-flow anaerobic hybrid reactor or the Up-flow Sludge Bed - Filter reactor hybridizes the UASB concept and the UAF design. Leaving the bottom portion of the up-flow reactor without any media and providing a matrix in the top third of the reactor column was found to be advantageous. Many researchers evaluated the performance of this hybrid design and reported that this configuration combined the advantages of both UASB and UAF while minimizing their limitations. The reactor was proven to be efficient in the treatment of dilute to high strength wastewater at high OLRs and short HRTs. The use of packing media only in the top portion of the reactor could minimize the channelling problem associated with UAF. Instead of a GSS device, the presence of media at the top could address the loss of biomass due to floatation in poorly performing UASB reactors. Further, the packing material enhanced the development of granular sludge in some of the difficult to granulate situations. The start-up of UAHRs were trouble free compared to UASB and had reduced operational issues in the long run (Fig. 10.6).

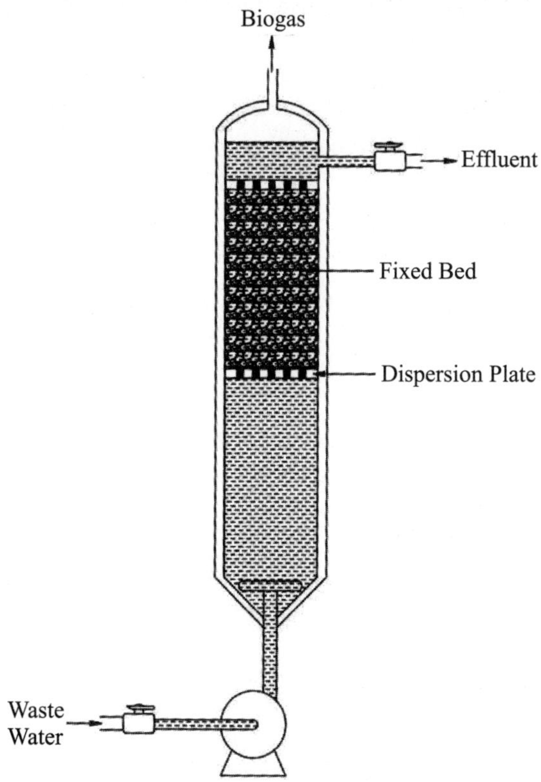

Fig. 10.6 Up-flow Anaerobic Hybrid reactor

10.7 Present Status of High-rate Bioreactors for Effluent Treatment

Anaerobic wastewater treatment can be regarded as a grown-up technology, although this is still not completely accepted everywhere. The UASB-system so far is the most widely applied reactor concept. It has found successful application for a vast number of very different waste-waters including industrial effluents. Many refined versions of the sludge blanket concept have also been developed and is being applied worldwide. Undoubtedly, various hybrid reactors have been developed which are also becoming popular. Such hybrid bioreactors with naturally occurring agricultural by-products like coconut shells as matrix have been developed. The anaerobic treatment process very likely will find increasing application for treatment of agro-processing effluents. The technology of Anaerobic hybrid bioreactors with broken coconut shell as matrix developed at Kerala Agricultural University was transferred to entrepreneurs for popularization. The technology was proven successful, especially for effluents from paddy parboiling units, natural rubber latex processing and for waste coconut water from coconut oil mills.

Even though at a limited scale in developing countries, there are also experiences with applying the process for the treatment of domestic sewage and its prospects look very promising in tropical and subtropical regions. Implementation of anaerobic treatment for this purpose can lead to tremendous savings in the investment of sewerage.

It should also be emphasized here, that for the successful application of the anaerobic process it is a prerequisite, that the implementation is pursued with the necessary care bearing in mind that we are dealing with living organisms. This is true for the design, construction as well as the operation and maintenance of the system.

CHAPTER

11

Fuel Alcohol

The development of machines for locomotion has been one of the key ingredients which accelerated global economic development. As automobiles became the crucial component of economic activities, affordability of transportation for the general public has emerged as a major driver of economies. Even though external combustion engines were used for mass transport, the real cost-effective transportation facility was realised with the inception of the internal combustion (IC) engines. Because there was no petroleum-based hydrocarbon fuel supply during the early days of IC engines, they used bio- fuels. The switch over to fossil fuels has tremendously changed the automobile industry and it is roughly estimated that about one billion IC engine powered vehicles would have been manufactured by 2010. It is also likely that an equal number of vehicles might have been manufactured in the decade 2010-2020. The development of fuel industry has also been keeping pace with this growth in automobile industry.

The alarming decline of fossil fuel reserves and the concern on carbon emission from fossil fuels have created increased interests in alternate power source for automobiles. Even though the technological improvements in IC engines could control pollution due to exhaust emission of hydrocarbons (HCs), carbon monoxide (CO) and oxides of nitrogen (NOx), the emission of carbon dioxide from IC engines of vehicles is a matter of concern. Electricity as a major alternate power source for transportation has many attractions and likely to be used increasingly in the near future. A plug-in hybrid electric vehicle (PHEV) is a hybrid vehicle driven with an IC engine as well as an electric motor. The electric motor is powered by a battery which can be recharged by plugging to an external electric power source. The Battery electric vehicles (BEVs) do not have an on-board IC engine and is propelled solely by electric motor. By the year 2030 the population of electric vehicles, both PHEVs and BEVs, may be around 20 per cent of the total automobile population. Even by 2040, it is not likely that entire electricity production will be from renewable sources. Cost effectiveness and affordability of IC engine driven vehicles is another factor and considerable number of vehicles fuelled by renewable gaseous or liquid fuels is likely to exist in many countries.

The convenience in storage and distribution of liquid fuels are factors which promote the continued use of liquid fuel powered IC engine vehicles even after 2030.

Alcohol is the most widely used alternative liquid fuel in the automobile sector. It can be produced as a renewable fuel and can easily be blended with gasoline. This is the major factor favouring its potential use in automobiles.

11.1 Suitability of Alcohol Fuels for IC Engines

Alcohols are named after the basic hydrocarbon molecules such as methanol (CH_3OH), ethanol (C_2H_5OH), propanol (C_3H_7OH), butanol (C_4H_9OH) etc. Even though any of these alcohols can be used as a fuel, Methanol and Ethanol are the simplest alcohols, which are technically and economically suitable to be used as fuels for internal combustion engines. Both methanol and ethanol are considered as renewable fuels as both can be produced from renewable sources. Both the alcohols have high octane numbers and they are suitable for spark ignition engines. Methanol can be produced by catalytic reduction of CO and H_2 produced from thermochemical gasification of biomass. But the production of methanol from biomass is not easy as in the case of ethanol, which is produced by simple fermentation of sugar. Hence ethanol (also referred as bio-ethanol) is the most important and widely used alcohol which is regarded as a fuel. It is a clean burning fuel and was tried in IC engines as early as 1870s. Ethanol had been used widely in Germany and France as early as 1894 and Brazil started using it since 1925. The blend of gasoline and alcohol (popularly called gasohol) had been used in the USA since the 1970s.

Ethanol is a high-octane fuel, where octane number is an index for a comparison of its fuel properties with isooctane (reference fuel). It is a primary alcohol and is colourless, volatile and miscible with water in all proportions. As alcohol is an oxygenated fuel its heating value is less than gasoline which is a mixture of hydrocarbons. Due to this, the specific fuel consumption is likely to be high. Energy content of alcohol is about $2/3^{rd}$ of gasoline and $2/5^{th}$ of diesel. It is used as an octane booster due to its high octane rating. But the very low cetane number makes it unsuitable for direct fuelling of compression ignition (CI) engines.

Ethanol has a higher heat of vaporisation compared to gasoline, which increase the heat requirement for vaporising the fuel. This can create cold starting problems and hence the intake air temperature should be above 10°C for easy starting. Another issue is the corrosiveness of alcohol due to its property as a good solvent. When used in engines designed for gasoline, some types of rubber and plastic components in the fuel lines may get degraded. Alcohol prepared at farm level can have a maximum proof of 190 to 191. Water cannot be removed from this product fully, as alcohol forms an azeotrope when the mixture strength reaches 95.57 per cent (weight basis). If alcohol containing water is used for blending with gasoline, water will get separated. Production of anhydrous alcohol without any water content requires special procedures to break the azeotrope.

Table 11.1. Shows the relevant fuel properties of alcohols in comparison to the reference fuel Iso-octane.

Table 11.1 Fuel properties of alcohols *(Source: FAO, 2018)*

Sl. No.	Properties	Iso-octane	Methanol	Ethanol
1.	Chemical Formula	C_8H_{18}	CH_3OH	C_2H_5OH
2.	Molecular weight	114.224	32.042	46.07
3.	Carbon (%)	84.0	37.5	52.17
4.	Hydrogen (%)	16.0	12.5	13.4
5.	Oxygen (%)	0	50.0	34.78
6.	Boiling point (at 1 atm), °C	99.239	64.5	78.40
7.	Freezing point (at 1 atm), °C	-107.378	-97.778	-80.00
8.	Density (15.5 °C), kg·L^{-1}	0.6943	0.7953	0.7944
9.	Viscosity (at 20°C,1 atm), Centipoise	0.503	0.596	1.20
10.	Specific heat (25°C,1 atm) kJ·kg^{-1}	1.163	1.396	1.396
11.	Higher heating value, MJ·kg^{-1}	47.81	22.74	29.7
12.	Lower heating value, MJ·kg^{-1}	44.35	19.99	26.9
13.	Stoichiometric air fuel ratio	15.13	6.463	9.0
14.	Research Octane Number	100	106	105
15.	Flash point temp., °C	-42.778	11.112	12.778
16.	Auto-ignition temp., °C	257.2	464	423
17.	Flammability limits	1.4 – 7.6	6.7 – 36	4.3 – 19
18.	Latent heat of vaporization at 20°C, kJ·kg^{-1}	349	1177	921.36
19.	Cetane number	--	5	8

(http://www.fao.org/3/T4470E/t4470e08.html) [Access date: 05.03.2018]

11.2 Use of Alcohol Blended Gasohol

Use of alcohol blended gasoline has become widespread and the present-day engines have components compatible to alcohol. In India 8.5 per cent ethanol is blended in gasoline as an octane booster. India's target was to achieve 10 per cent blending by 2020; but it has been extended. At present the target is to have 20 per cent alcohol blend by the year 2025.

Brazil was the pioneer in use of alcohol fuel since the seventies and since then they made it mandatory to use 10 to 22 per cent alcohol fuel blends for vehicles. In the year 1993, E22 became the standard blend. At present E25 - E27 are the fuel blends widely used in Brazil. Flexible fuel vehicles which can run on any blend are now dominating the personal passenger vehicle sector now.

The USA has different blends made available as per customer's choice. Majority of gasoline sold in US is E10 (10 per cent alcohol) and it can be used in any conventional gasoline engine. In most of the states, E10 is commonly sold so as to meet the legal requirements of oxygenated fuel sale as renewable fuel standard. E15 (alcohol content up to 15 per cent) can be used in vehicles not older than 2001. An ethanol-gasoline blend containing 51 to 83 per cent ethanol is called E85 (or flex fuel) in USA. E85 can be used in 'flexible fuel vehicles' (FFVs), whose engines are designed to run on E85, gasoline, or any blend of gasoline and ethanol up to 83 per cent. Blender pumps in fuel stations can deliver any blend like E25 or E30 depending on customer demand. Most countries in the European Union use E5 and nine countries (Belgium, Bulgaria, Estonia, Finland, France, Germany, Luxembourg, the Netherlands and Romania) have adopted E10.

11.3 Use of Straight Alcohol in Gasoline Engines

Engine modifications are required to use straight alcohol (pure alcohol) in engines designed for gasoline. Due to low energy content and low air fuel ratio of ethanol, the jets delivering fuel should be modified so as to increase the cross-sectional area to 1.5 times their original size. If any part in the fuel line is likely to corrode due to alcohol, they need to be replaced with those made of resistant material. As alcohol is a high-octane fuel, ignition advance is usually given. Due to the increased latent heat of vaporisation, an electric heating arrangement to preheat the air to a temperature above 10°C is to be incorporated. Safety against fire hazards is required as alcohol is highly inflammable. It may be better to use gasoline as the starting fuel in colder climates. An alternate solution is to use a heat exchanger to make use of the heat of exhaust gases. In this case, the intake manifolds need to be specially designed. Gasoline engines are having a low compression ratio (CR) and their thermal efficiencies can be improved if the CR is increased to 12:1. But the existing gasoline engines in general are not designed to withstand high CR.

Other pertinent aspects include a small increase in the overall thermal efficiency of the engine even though the specific fuel consumption will be high due to the reduced calorific value of alcohol. Alcohol engines run relatively cool so that the exhaust valves are less likely to get burned. Problems of soot build up also are eliminated and lubricating oil contamination is reduced. As the NOx and CO emissions are less the fuel is environmentally clean.

11.4 Use of Alcohol in CI Engines

Normally, ethanol cannot be used as a replacement fuel in a conventional diesel (CI) engine due to its low cetane number. However, specially designed, dedicated ethanol CI engine with a higher compression ratio can be made and such engines are available. But to ensure proper combustion, some additives are generally used along with ethanol. As in the case of gasoline engines, blending diesel with alcohol is another possibility. The quantity of ethanol that can be blended is limited due to issues like phase separation, improper ignition etc. Hence use of alcohol in a dual fuel mode by introducing the fuels separately into the cylinder is also tried. While the diesel is injected directly into the combustion chamber as usual, alcohol is introduced into the engine separately. The ignition of the diesel fuel readily occurs and combustion of alcohol mixed with air follows. This method is advantageous compared to other methods as higher replacement ratios are possible. The operation can also be in the flex mode, as any mixing ratio is possible. The most common method is called 'fumigation' as alcohol is mixed with the intake air stream through the inlet manifold after evaporating in the carburettor. In yet another technique, alcohol may be directly injected into the combustion chamber through a dedicated injector for alcohol fuel.

11.5 Ethanol Production

The process for fermentation of sugar containing fruit juices to make wines is an age-old technology, and was practiced even before 2500 BC. But little was known on the actual science of fermentation until 19th century. It was during mid-19th century that Louis Pasteur revealed the real science behind alcohol fermentation and the role of microorganisms in the process. Now, ethanol production for industrial use as well as for human consumption has emerged as a big industry in many countries including India. With the start of ethanol blending in gasoline, the industry is more rejuvenated.

Common raw materials for large scale ethanol production are available in the form of starch as well as sugar obtained from crops like cassava and sugar-cane (in Brazil) or maize starch (in USA) and sugar beet (Russia, France, USA and Germany). Sweet sorghum is another crop suitable for sugar extraction. Cassava, maize and cereals are used in China for alcohol production.

In countries like Brazil, ethanol is produced from cane sugar. The typical ethanol yield per tonne of sugar cane is reported to be 72.5 litres. Under good agricultural practices, more than 60 tonnes of sugar cane can be produced from one ha of land in tropical countries. In India also the major source is cane sugar, but most alcohol distilleries use molasses from sugar industry. About 500 sugar factories are there in India and more than 100 million tonnes of molasses is annually produced. From one tonne of sugar cane, about 115 kg of sugar can be produced along with 45 kg of molasses which contains unrecoverable sugar. The total content of fermentable sugars (TFS) in 45 kg molasses amounts to 18 kg, which can be fermented to yield 10.8 litres of ethanol.

Other crops can also be grown for the production of ethanol. Sugar beet and sweet sorghum are other commonly cultivated sugar producing crops. In USA, corn is the major crop for alcohol production. It is possible to produce about 300 litres of alcohol from 1 tonne of corn.

In general, there are three categories of feed stocks available for alcohol production. The sugar production routes from sugar (*sacchariferous*) materials, starchy (*amylaceous*) materials and lignocellulosic materials are shown below (Fig. 11.1):

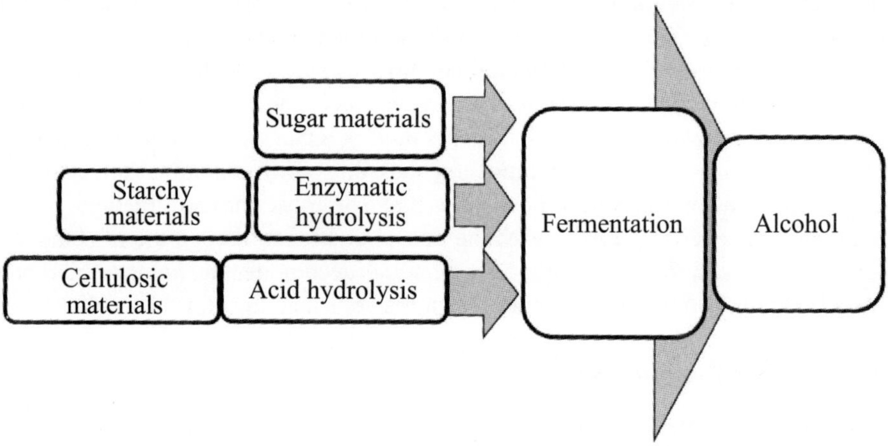

Fig. 11.1 Flow diagram of alcohol production routes

The basic chemical reaction is conversion of glucose into ethyl alcohol, as given below:

$$C_6H_{12}O_6 \rightarrow 2CH_3CH_2OH + 2CO_2 \quad ...(11.1)$$

The group of microorganisms responsible for alcohol fermentation is yeast and the specific species is *Saccharomyces cerevisiae.*

11.6 Ethanol from Sugars

Sugars can be directly fermented to alcohol, according to the following chemical equation:

$$C_{12}H_{22}O_{11} + H_2O \rightarrow 4C_2H_5OH + 4CO_2 \quad ...(11.2)$$

Commonly two methods are followed for the process. One system operates in the batch mode and the other is a continuous system in which cell separation and recycling of yeast cells are done. The process flow chart for a batch process is shown below, which is self-explanatory (Fig. 11.2).

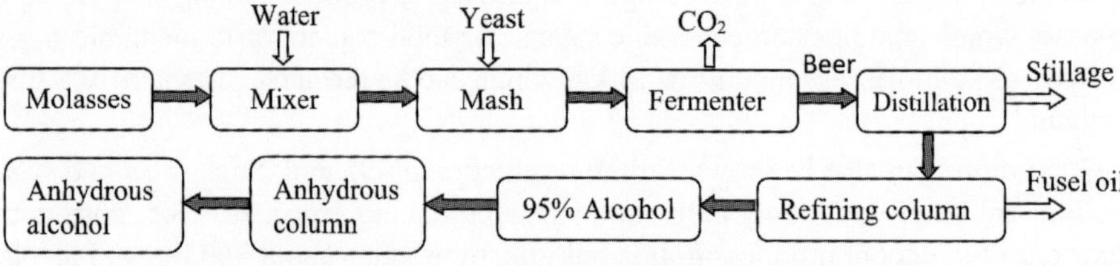

Fig. 11.2 Process flow chart of alcohol production from molasses

Molasses is the liquid by-product from sugar manufacture after the removal of possible sucrose from the cane juice. The 'Blackstrap molasses' is the left-over syrup from which no more crystalline sugar can be extracted economically. The molasses is mixed with water to form a 'mash' so that the sugar concentration is in the range 12 to

18 per cent. The pure culture of yeast strain, *Saccharomyces cerevisiae*, is multiplied and used to inoculate the mash. This commonly used yeast strain can tolerate high alcohol concentrations and hence continue to be active when the fermentation advances. Sometimes nutrients are added to mash to compensate for the deficiency in nitrogen and phosphorous. The favourable pH of the mash is 4 to 5 and the temperature is in the range 28 to 33°C. The fermentation can be completed in about 50 hours and the fermented mash is called beer. The beer is distilled to get different fractions having different alcohol contents. The fraction having very low alcohol is called 'distillery slopes'. Fusel oil is a by-product obtained in distillation of ethanol and contains some valuable chemicals.

Distillation of alcohol water mixture into anhydrous alcohol is a tricky process. When distillation proceeds, the concentration of alcohol in the liquid phase keeps on decreasing, but after some time, the amount of water vapour and amount of alcohol vapour contained in the gaseous phase above the liquid in the still reach equilibrium. After this, there will not be any change in concentration of alcohol between the liquid phase and the vapour phase. When the alcohol-water mixture reaches a concentration of 95.6 percent alcohol, the two components (alcohol and water) boil together. At this stage, the alcohol in the vapour phase is no longer more concentrated than in the liquid phase. The composition of a mixture of liquids attaining this stage is called an 'azeotropic mixture'. Sometimes a third substance is added to break the azeotrope. Different techniques are used in the anhydrous column to further concentrate the alcohol obtained from ordinary distillation.

The following methods are used to separate the azeotropic mixture of alcohol and water:

(i) *Azeotropic distillation:* This is done with the help of a reflux distillation column. The method involves the addition of a third chemical like n-hexane or benzene which breaks the azeotrope. This will change the boiling characteristics of the liquid mixture and complete separation of water is made possible to get anhydrous ethanol.

(ii) *Use of drying agents (desiccants):* Some compounds like Calcium oxide react with water and absorb it fully from the mixture without affecting the alcohol.

(iii) *Molecular sieve:* The molecular sieves make use of the larger size of ethanol molecules with respect to water molecules. In such a molecular sieve, the pores of the sieve material are smaller than ethanol molecules but larger than water. One such material is crystalline alumina silicate. Ethanol-water mixture is passed through a series of columns containing such materials in an intermittent schedule. When the mixture is passed through the first column, water is held in it until the limit reaches. Then the mixture to be separated is directed to another column. In the meantime, the water saturated column is dehydrated using hot air.

11.7 Ethanol Production from Starchy Materials

Starch is a storage carbohydrate and basically, it is a polymer of glucose having two large molecular weight compounds, viz. Amylose and Amylopectin. These are branched chains of glucose molecules which can be broken by hydrolysis. Enzymes are used for hydrolysis of starch to glucose which can be fermented to ethanol. One gram of starch can yield about 0.5 gram of ethanol (Fig. 11.3).

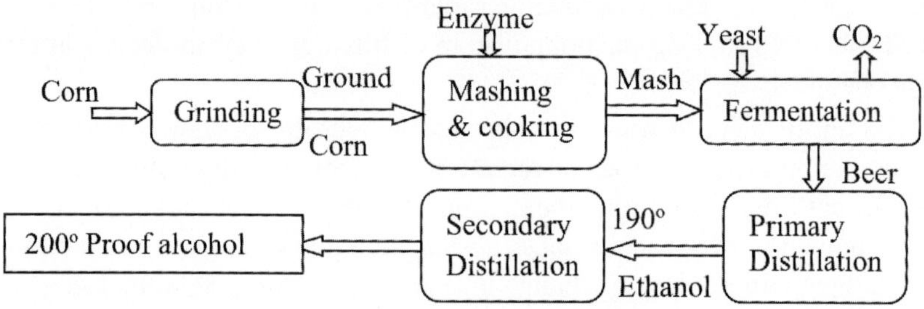

Fig. 11.3 Process flow chart of alcohol production from corn starch

11.8 Fuel Alcohol from Ligno-cellulosic Materials

Crop residues are available abundantly and efforts to produce energy from ligno-cellulosic materials in the form of alcohol have been successful to certain extend. The process becomes economically feasible only when the energy that can be produced from alcohol exceeds the energy spent for the process. These materials are actually energy-rich biopolymers in the form of plant cell wall material such as cellulose, hemicelluloses and lignin generally in the ratio 4:3:3. Cellulose is a straight polymer of glucose in which the glucose units are linked together by glucosidic bonds. These bonds are difficult to break compared to starch. The chain length varies according to the degree of polymerization. There is much variation in this for different naturally occurring materials. The chain length in paper is in the order of 1,000 whereas in cotton it is 10,000. Hemicelluloses are hetero polymers of different sugars like galactose, xylose etc. Bagasse is a commonly available ligno-cellulosic residue which has been considered for conversion to alcohol. Theoretically one litre of 95 per cent alcohol is obtainable from 6 kg of bagasse (Fig. 11.4).

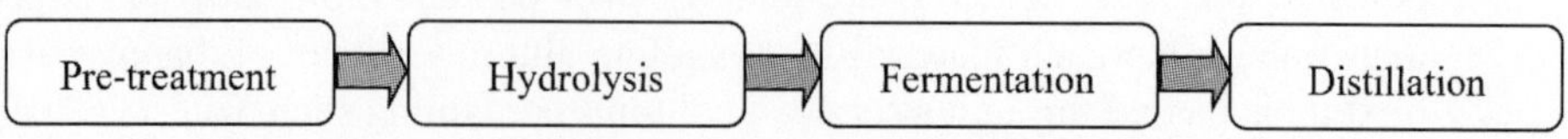

Fig. 11.4 Conversion process of ligno-cellulosic biomass to alcohol

11.8.1 Pre-treatment

The fibrous components of ligno-cellulosic biomass are mostly hemicelluloses and cellulose which can be broken down by enzymatic hydrolysis. But the lignin should be removed first so that the enzymes can act on these materials. For this, pretreatments are done in which the cellulose structure also is disrupted. Pre-treatments may comprise of physical, chemical and biological stages which may be sometimes combined. In the physical process, the particle size is reduced by shredding. In chemical treatment, cellulose disrupting acids or alkali agents are used. This method has the disadvantage that the chemicals are corrosive and they are inhibitory to microbes which will be acting on the material subsequently. Steam explosion is another method in which the ligno-cellulosic materials are saturated with water under high pressures (20 to 35 kg·cm^{-2}) and temperatures above 200°C. The pressure is then reduced rapidly forcing the water at high temperature to vaporize instantly breaking the fibrous materials. Some acids formed during the process act as a catalyst in the hydrolysis of hemicelluloses. The disadvantage is that large quantity of energy is spent in the process (Fig. 11.5).

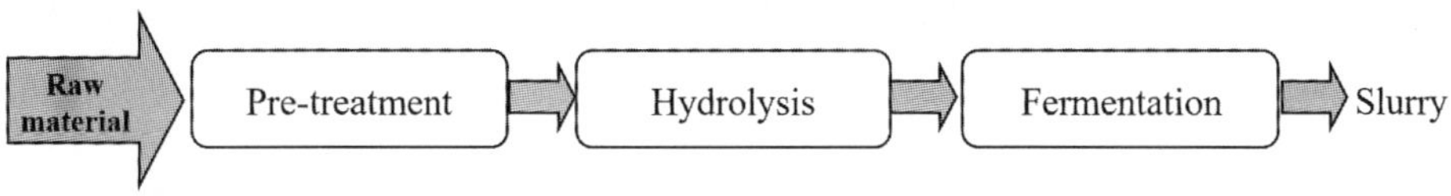

Fig. 11.5 Pre-treatment process

11.8.2 Hydrolysis

Hydrolysis or saccharification is the conversion of polymers into glucose (Fig. 11.6). This is carried out by the enzyme 'cellulase'. During this process lignin is initially attacked by lignin peroxidases whereas the degradation of cellulose requires at least three different enzyme activities (exo- and endo-β-1,4-glucanase and cellobiase). Most of these enzyme activities are present in fungi like *Trichoderma, Aspergillus* and *Fusarium*. Some bacteria such as *Clostridium thermocellum* can also accomplish these activities. Once the saccharification is over, cellulase enzyme should be recovered. The resulting sugar solution can be concentrated by reverse osmosis.

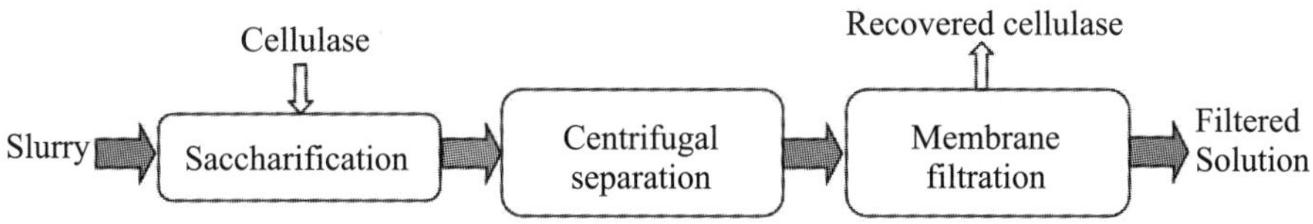

Fig. 11.6 Hydrolysis

11.8.3 Fermentation

The sugar solution obtained is then fermented using different techniques. It can be the usual batch fermentation or advanced methods such as use of immobilised yeast cells. This is achieved by holding the yeast cells in a gel (calcium alginate). Flash fermentation is a method in which the concentration of alcohol is maintained at a low level by removing the alcohol by simultaneous distillation at very low pressures. Simultaneous saccharification fermentation (SSF) is a process in which both the steps are combined.

CHAPTER

12

Biological Hydrogen Production

The global energy demand is increasing abruptly, while the fossil fuel availability is diminishing every day. The greatest concern is the fact that fossil fuel use leads to enormous emission of CO_2 threatening the very existence of future generations on the earth. These facts tempt global scientific community to explore new and sustainable fuel alternatives. As hydrogen is a clean fuel with zero CO_2 emission, it is regarded as a clean energy source. The energy yield of hydrogen (122 $kJ \cdot g^{-1}$) is about 2.5 times compared to hydrocarbon fuels. The heating value of hydrogen (142 $MJ \cdot kg^{-1}$) is comparable well with that of other conventional fuels as well. The possibility of direct conversion to electricity by the use of fuel cells has opened up tremendous possibilities for use of hydrogen as a versatile source for energy. The Department of Energy, USA expect that by the year 2040, hydrogen transport and power systems will be approachable in all provinces of the United States.

The production of hydrogen gas can be achieved by different procedures including the electrolysis of water, photo-catalytic processing, thermochemical processing, photo-electro-chemical processing and bioprocess. Even though hydrogen can be produced through several methods, the most common methods are natural gas reforming (a thermal process) and electrolysis. The major limitations of these methods are the high energy requirement and the requirement of elevated temperatures (<850°C). Steam reforming is a high-temperature process in which steam reacts with a hydrocarbon fuel like natural gas, diesel etc., to produce hydrogen. Today, most of the hydrogen is produced from steam reforming of natural gas.

12.1 Biohydrogen

Water can be considered as an inexhaustible source of hydrogen. By electrolysis of water, hydrogen gas can be produced and can be considered as a viable replacement fuel to fossil fuels. Production of hydrogen from renewable sources has been widely regarded as a 'Green technology' capable of addressing environmental issues related to fossil fuels. There are also methods by which renewable biomass fuels can be reformed to produce hydrogen, including liquid and gaseous biomass fuels. Other approaches for hydrogen production include solar-powered and biological processes. Biological processes are regarded as a radical approach for the production of hydrogen, which is often referred as bio hydrogen.

Production of hydrogen through biochemical processes has attracted attention of scientists for many decades. Biohydrogen production can be done by the use of simple solar bioreactors and require less energy input. Even though the conversion efficiencies in biological systems are low, the low energy use and reduced initial investment are favourable factors. Biohydrogen is defined as the hydrogen produced biologically, most commonly by algae and bacteria. It is a clean bio-fuel procurable from both cultivation as well as recoverable from organic waste materials. Biomass can be used as an ample renewable resource for biochemical production of fuel hydrogen.

12.2 Biohydrogen Production Processes

The different approaches of bioprocess for hydrogen production are depicted in Fig.12.1. The processes can be broadly divided into two; (i) bacterial fermentation and (ii) bio-photolysis.

12.2.1 Bacterial fermentation processes

The bacterial fermentation processes can occur in the presence of light (photo fermentation as well as in the absence of light (dark fermentation). In combined fermentation both the above processes are integrated.

(a) *Photo Fermentation:* Photo fermentative production of hydrogen includes complex biochemical processes taking place in the presence of light. The organism involved in the biochemical process is a bacteria called 'purple non-sulphur' (PNS) photosynthetic bacteria. Several strains of these bacteria have been tested for their hydrogen production capability by scientists. The process is catalysed by nitrogenase enzyme under nitrogen-deficient conditions. Under anaerobic situations with illumination, photosynthetic bacteria are capable of converting organic acids or volatile fatty acids to carbon dioxide and biohydrogen. The theoretical yield of hydrogen for photo fermentation is 4 mol/mol of acetate when acetate is used as a carbon source (Eq. 12.1).

$$2CH_3COOH + 4H_2O \xrightarrow{\text{light energy}} 8H_2 + 4CO_2 \qquad ...(12.1)$$

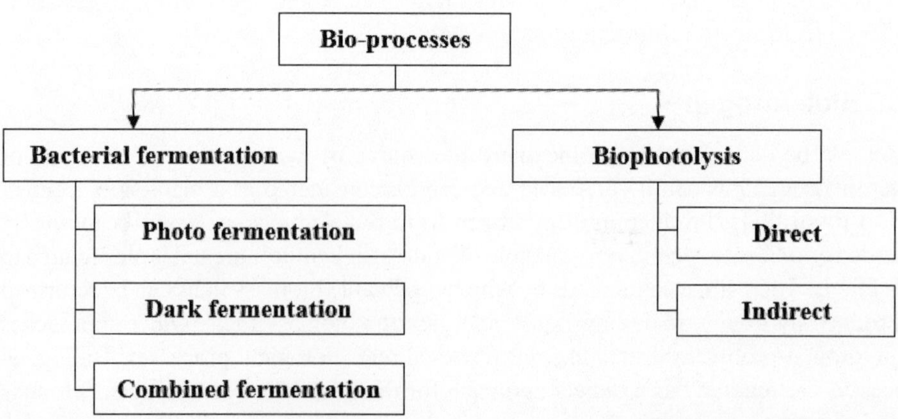

Fig. 12.1 Bioprocesses for hydrogen production

(b) *Dark fermentation:* Dark fermentation in anaerobic condition convert sugar to hydrogen. In this process, the carbohydrate substrates are decomposed to hydrogen by anaerobic bacteria. The production of hydrogen by dark fermentation includes intermediate steps such as hydrolysis, acidogenesis and acetogenesis. Acid hydrolysis is the most used pre-treatment, followed by enzymatic and physical treatments to prepare the feed stock. By hydrolysis, soluble organic compounds are produced from non-soluble biopolymers. The soluble organic compounds are then transformed to CO_2 and volatile fatty acids through a process called acidogenesis. Further, volatile fatty acids are transformed to H_2 and acetate at the end of a bio process called acetogenesis.

Theoretical yield per mol of glucose is explained as follows:

$$C_6H_{12}O_6 + 2H_2O \rightarrow 2CH_3COOH + 2CO_2 + 4H_2 \quad ...(12.2)$$

Advantages of dark fermentation over photo fermentation:

(i) Higher H_2 production rate

(ii) Energy saving

(iii) Choice of a variety of feedstock materials

(iv) Better and easy process control

(v) Reduced operational and maintenance costs

(c) *Combination of photo and dark fermentation:* The dark fermentation results in the formation of alcohols and ample amount of short chain organic acids. The photo fermentation bacteria can transform the acids to produce additional hydrogen. The lighter organic acids like acetic acid produced by the anaerobic dark fermentation of sugar molecules are then further decomposed into hydrogen by the action of photosynthetic bacteria through photo-fermentation. This results in the overall maximisation of hydrogen yield and thus maximise the conversion efficiency of a single substrate. Hence, the combination of photo-fermentation and dark-fermentation can improve the efficiency of overall hydrogen production. The combination of these two bio-processes could theoretically yield 12 mol of hydrogen per mol of glucose as given in equations 12.3 and 12.4.

Dark fermentation stage:

$$C_6H_{12}O_6 + 2H_2O \rightarrow 2CH_3COOH + 2CO_2 + 4H_2 \quad ...(12.3)$$

Photo-fermentation stage:

$$2CH_3COOH + 4H_2O \rightarrow 4CO_2 + 8H_2 \quad ...(12.4)$$

(d) *Bio-photolysis:* Photoautotrophic microorganisms such as eukaryotic green microalgae or prokaryotic cyanobacteria (green algae), which possess Chlorophyll are capable of carrying out photosynthesis like plants. They capture sunlight to generate carbon dioxide and hydrogen via photosynthesis. In the bio-photolysis process, sunlight act as the energy source while CO_2 is utilized as the source to split water into H_2.

12.3 Bio-photolysis of Water

It has been observed that few groups of micro algae and cyanobacteria are capable of producing hydrogen. They are both photosynthetic organisms which can convert

CO_2 and water to carbohydrate. The photosynthetic process takes place in two steps, a biochemical energy conversion process and a CO_2 reduction process. Under certain conditions, some species of micro algae and cyanobacteria utilize the biochemical energy to produce hydrogen instead of reducing CO_2. These microalgae and cyanobacteria have the capability to split water into O_2 and H_2 and the process is commonly referred as 'Bio-photolysis'. They utilize solar energy in extracting protons and electrons out of water as depicted in Fig. 12.2.

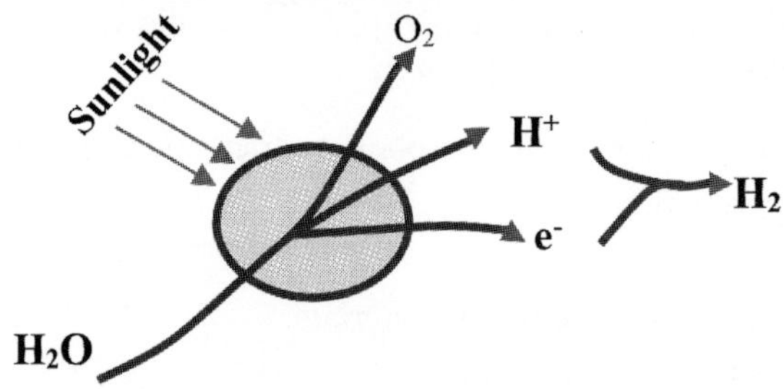

Fig. 12.2 Bio-photolysis of water

The absorbed sun light provides necessary energy for the water-splitting reactions in which electrons are transferred to hydrogenases and nitrogenases. While molecular hydrogen is evolved by these organisms, some others use hydrogen. Both these types of organisms make use of enzymes called hydrogenases. Some of them release electrons with a hydrogenase enzyme, in anaerobic condition or when solar energy is absorbed by them. The process eventually leads to conversion of intermediate hydrogen ions to hydrogen gas as electrons and protons liberated through the water-splitting process recombine into hydrogen molecules. This is described as Hydrogenase dependent hydrogen production. Nitrogenase dependent hydrogen production is yet another route. Nitrogenases are enzymes that catalyses the reduction of N_2 to NH_3, with concurrent evolution of H_2.

(a) *Direct bio-photolysis:* In this process, cyanobacteria capture solar energy to split water molecules into hydrogen ion and oxygen under anaerobic conditions. The action of hydrogenase enzyme on the generated hydrogen ions results in the formation of hydrogen gas. One of the important hydrogen-producing algae is *Chlamydomonas reinhardtii*. The important drawback of the process is the strong obstruction on hydrogenase enzyme by the generated oxygen.

(b) *Indirect bio-photolysis:* Indirect bio-photolysis and H_2 production are performed by filamentous cyanobacteria. A nitrogen fixing cyanobacterium, *Anabaena cylindrica*, was observed to be capable of producing hydrogen. During the N_2 fixation process, catalysed by the nitrogenase enzyme, H_2 is also produced, as explained in the following biochemical reaction:

$$N_2 + 8H^+ + 8e^- + 16ATP \rightarrow 2NH_3 + H_2 + 16ADP \quad ...(12.5)$$

12.4 Bioreactors for Biohydrogen Production

As biological hydrogen production using microorganisms is the best alternative to the existing expensive industrial hydrogen production, bioreactors have been developed for commercial bio-hydrogen production. In general, bioreactor is an enclosed chamber in which biochemical reactions takes place due to the growth of microbial or plant cells under controlled conditions. Commercial bioreactor systems for hydrogen production are available in different sizes.

12.4.1 Photo-bioreactors

They are used for the hydrogen production using microalgae. It has been designed to provide suitable environment for rapid multiplication and maintain high density of the microalgae culture economically. The culture is allowed to flow through an array of transparent tubes using a pump. The transparent tubes are provided with a light receiving system. The productivity of the system depends on different factors like availability of light or artificial illumination, large surface to volume ratio, pH, thermal control and flow rates. Photo-bioreactor designs include flat plate, tubular and pond (pool) types.

12.4.2 Dark fermentation bioreactors

Dark fermentation can be performed in continuous, semi-continuous and batch type bioreactors. The biohydrogen yield from the batch type bioreactors are comparatively less than the other two bioreactors. Batch fermentation is applicable for experiments on process optimization. For large-scale operations, continuous or semi-continuous types are preferred for dark fermentation.

(a) *Continuous stirred tank reactor (CSTR):* CSTRs are the widely accepted bioreactor systems for continuous hydrogen production, which are easy to operate and simple to construct. They provide effective and uniform mixing. These bioreactors are operated at high concentrations of volatile suspended solids (10 to 30 $g{\cdot}L^{-1}$) with a maximum hydrogen yield of 10.4 to 11.4 m^3/m^3/day. In a CSTR, feeding and discharge are performed continuously. Generally, suspended bioreactors have cylindrical shapes provided with mechanical turbines for agitation. Since the CSTR has shorter HRT, the size is considerably smaller compared to the ordinary methane digester (Fig. 12.3).

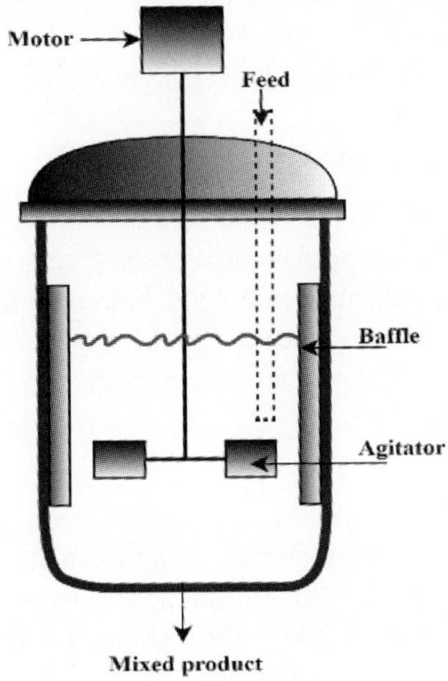

Fig. 12.3 CSTR type bioreactor

(b) *Fixed bed bioreactor:* Fixed bed bioreactors are popular reactors due to their simple construction. The reactions are carried out by a fixed bed of solid particles which contain biocatalyst on or within the bed. The solid structure may be rigid or a compressible structure made of non-porous or porous materials. A nutrient mixture is allowed to flow continuously through the bed and the product formed after the reaction is removed. The flow can be either upward or downward, however the flow under gravity (downward) is most preferred (Fig. 12.4).

(c) *Fluidized bed bioreactor:* Fluidized bed bioreactor is having a cylindrical shape with expanded top and narrow bottom. The reactions occur at the narrow bottom part called reaction column. This design help to retain the solids in the reactor while the liquid flows out. The reactions are carried out by the biocatalysts suspended in the fluids. Microbial flocs, immobilised enzymes and immobilised cells can be used as the biocatalysts (Fig. 12.5).

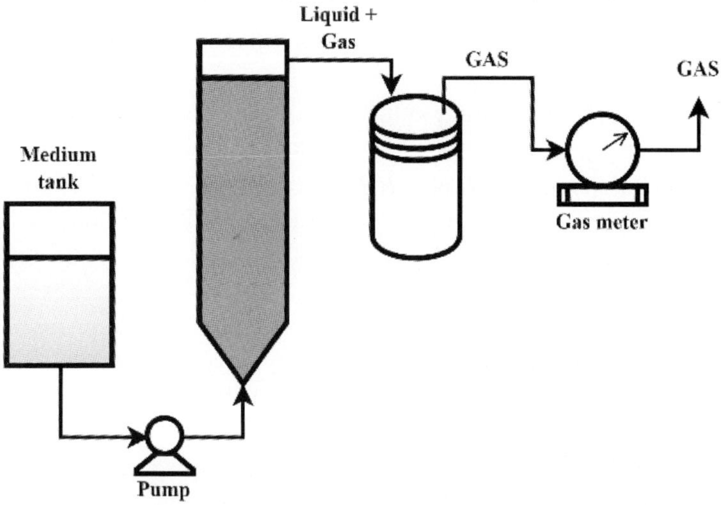

Fig. 12.4 Fixed bed bioreactor

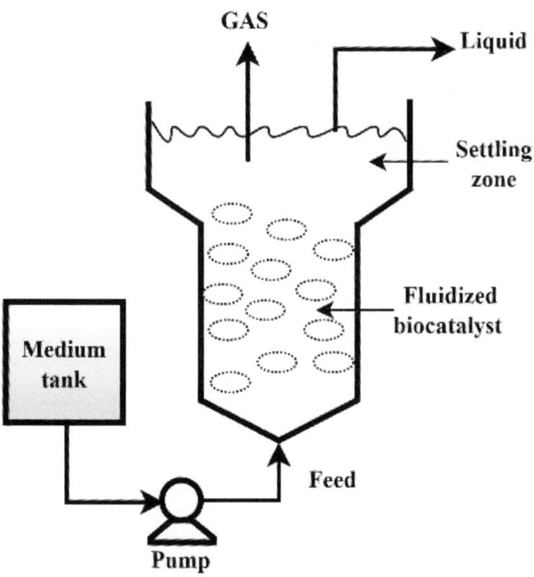

Fig. 12.5 Fluidised bed bioreactor

12.5 Factors Influencing H_2 Production in the Bioreactors

The biohydrogen yield and performance of bioreactors are greatly depending upon various factors discussed below:

(i) ***pH:*** The pH of the culture affects the metabolic pathways and yield. The optimum range of pH for maximum yield was observed as 5.2 to 6.0.

(ii) ***Hydraulic Retention Time (HRT):*** HRT can affect the population of microorganisms. The ideal range of HRT is 8 to 12 h for liquid substrates. In treating solid-type feedstock, the need of hydrolysis makes the ideal range of HRT longer.

(iii) ***Partial pressure of H_2:*** Hydrogen dissolved in the liquid phase may lead to the hydrogen partial pressure, which may affect the microbial pathway. It has been noted that, increase in the partial pressure creates an unfavourable condition for H_2 production.

(iv) ***Nutrients:*** In the fermentation process, iron, phosphate, nitrogen, magnesium, zinc and other inorganic trace minerals are necessary for the optimal cell process and H_2 production.

(v) ***Temperature:*** The microorganisms are capable of producing hydrogen in the temperature range 15 to 85°C. The optimal temperature was found to be around 35°C. Increasing the temperature in the mesophilic range is having a positive impact on H_2 yield.

(vi) ***Substrate concentration:*** The substrate concentration for consistent H_2 production varies with different substrates and at different process conditions like HRT.

(vii) ***Inoculum***: Two most widely used inoculums (Seed culture) for fermentative hydrogen production are *Enterobacter* and *Clostridium*. Mixed cultures of bacteria from compost, municipal sewage sludge, anaerobic sludge and soil had been widely used as inoculums. Mixed cultures are more practical than pure cultures.

(viii) ***Feedstock:*** Carbohydrates are one of the main sources of hydrogen. Hence, feed materials like biomass and wastes containing carbohydrates are suitable as feedstock for H_2 production. Complex solid wastes including municipal waste, mixed waste, food processing waste, organic residues, and digester sludge can also be used as feedstocks.

12.6 Limitations in Biological Hydrogen Production

(i) Direct bio-photolysis processes have problems with the oxygen sensitivity and light conversion efficiencies. The nitrogenase enzyme in indirect bio-photolysis requires high energy and it has low solar energy conversion efficiency.

(ii) The biological processes for hydrogen production still need to be integrated with a system, which is capable of satisfying the overall efficiency requirement in producing fuels from solar energy.

(iii) The processing of certain biomass feedstock is expensive which add to the overall cost.

(iii) Lack of an industrially capable and persistent microorganism that can be used for profitable hydrogen production.

(iv) Issues in separation and purification of hydrogen and difficulty in obtaining a steady hydrogen production rate for an extended period are also regarded as major limitations.

(v) Problems in design of bioreactors and their scaling-up.

CHAPTER 13

Fuel Cells

It is believed that investigations on fuel cell technology started in the early nineteenth century. The primitive model of fuel cell was discovered by Sir William Grove in 1839 and the first fully operational fuel cell was demonstrated by Francis T. Bacon in 1959. Thereafter, NASA used proton exchange fuel cells in their Gemini and Apollo space programs which led to the increased interest in fuel cell technology. The recent explorations in the field enabled the use of fuel cells for a wide range of applications.

13.1 The Fuel Cell

Fuel cell is an electrochemical device that converts the chemical energy of a fuel into electrical energy without any intermediate mechanical or thermal process. The output of the fuel cell is heat and low voltage direct current (DC) electricity, which can be used directly. The DC electrical output can be converted into AC using an inverter.

A fuel cell is also called voltaic cell or galvanic cell, and can be operated with a wide variety of fuels and oxides. Hydrogen is identified as the most appropriate fuel for the practical application as it has higher electrochemical activity compared to other fuels.

13.2 Working of Fuel Cell

The fuel cell is similar to other electrical cells, as it contains an anode, cathode and an electrolyte between them. The anode and cathode represent fuel and oxidant, respectively. Hence the fuel cell can generate electricity as long as we supply fuel and oxidant. The energy conversion process of fuel cells involves the transfer of electrons from fuel electrode to oxidiser. The most commonly used oxidiser is air or oxygen (Fig. 13.1).

The hydrogen fuel is supplied to the anode where oxidation (removal of electrons) reaction occurs. At the anode, supplied hydrogen loses its electron to become H^+. The H^+ ions and electrons thus formed, start to move towards the cathode through the electrolyte and external circuit respectively. Oxygen or air is supplied to the cathode, where it absorbs the electrons coming through the external circuit to form oxygen ion (O^{2-}). This O^{2-} ion then combines with the H^+ ions coming through the electrolyte to form water, which can be drained out from the cell. In some other fuel cells, the oxygen after reduction will move towards the anode through the electrolyte to form water. It is the type of electrolyte that decides the movement of ions through it.

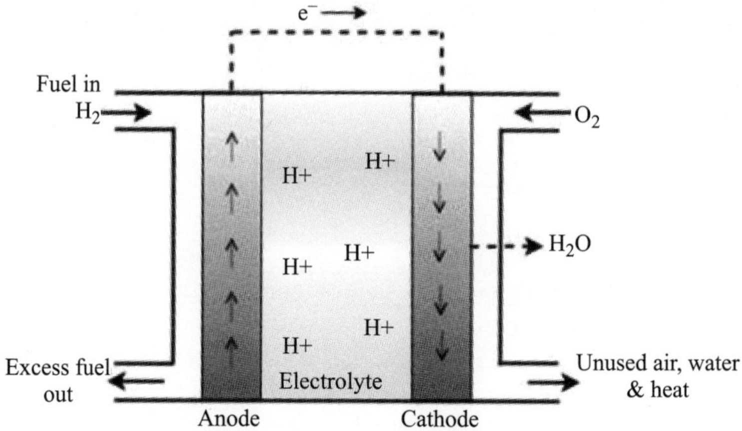

Fig. 13.1 Principle of working of fuel cell

The maximum potential difference and the maximum electrical energy output from a fuel can be achieved when it is operated under thermodynamically reversible conditions. The overall electrochemical reactions inside the fuel cell are described in equation 13.1.

$$\text{Fuel } (H_2) + \text{Oxidant } (O_2) \rightarrow W + H + \text{Product } H_2O \qquad ...(13.1)$$

Where, 'W' is the rate of electrical work done by the fuel cell system and 'H' is the rate of heat transfer into the system from the surrounding at constant temperature and pressure.

13.3 Types of Fuel Cells

Fuel cells can be classified based on the type of electrolyte employed. The operating temperature of a fuel cell greatly depends on the type of electrolyte. High temperature fuel cells have operating temperatures higher than 600°C and low temperature fuel cells operate below 250°C. Different types of fuel cells based on the type of electrolyte used are discussed below:

13.3.1 Proton Exchange Membrane Fuel Cell (PEMFC)

Proton exchange membrane fuel cell is also called Polymer Electrolyte Membrane (PEM) fuel cell. The electrolyte used in PEM fuel cells is a water based acidic polymer membrane, which is permeable to H^+ ions. Hence, hydrogen ions or protons move from anode to cathode to form water, and the product (water) can be drained from the cathode side. Platinum-based catalyst is used on both electrodes. The operating temperature is below 100°C. High-temperature variants, which can operate up to 200°C use a mineral acid-based electrolyte. One single cell can produce 1.1 DC voltage.

Reactions at anode:

$$H_2 \rightarrow 2H^+ + 2e^- \qquad ...(13.2)$$

Electrons formed creates a negative potential and move towards cathode through an externally connected circuit.

Reactions at cathode:

$$\frac{1}{2}O_2 + 2H^+ + 2e^- \rightarrow H_2O \quad ...(13.3)$$

Overall reaction:

$$2H_2 + O_2 \rightarrow 2H_2O \quad ...(13.4)$$

13.3.2 Direct Methanol Fuel Cell (DMFC)

The direct methanol fuel cell (Fig. 13.2) operates between temperatures ranging from 60 to 130°C. The DMFC generally uses a platinum catalyst at the cathode and platinum – ruthenium catalyst at the anode. The fuel used is methane as indicated in the name. The catalyst used is capable of drawing hydrogen ions from the methane supplied. The hydrogen ions diffuse through the electrolyte to combine with oxygen ions at the cathode to form water. DMFC are convenient for portable power applications with outputs generally less than 250 W.

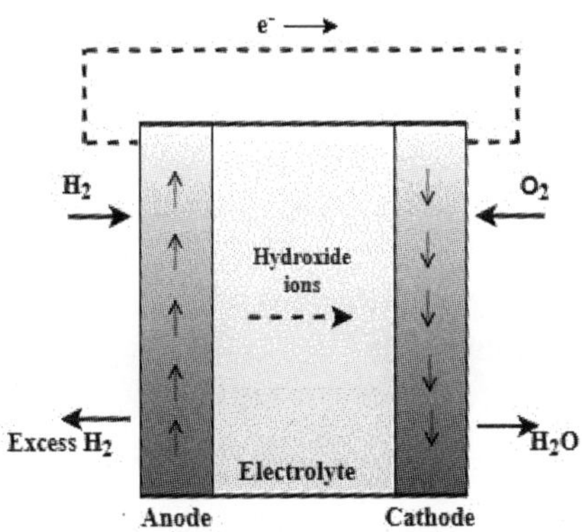

Fig. 13.2 Direct methanol fuel cell

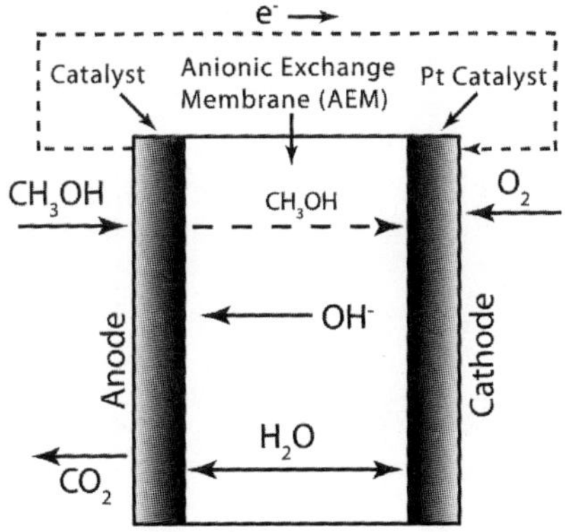

Fig. 13.3 Phosphoric acid fuel cell

13.3.3 Phosphoric Acid Fuel Cell (PAFC)

Fig. 13.3 shows a phosphoric acid fuel cell. The working of PAFC is similar to that of PEM fuel cell. The electrolyte used in this type of fuel cell is liquid phosphoric acid in a bonded silicon carbide matrix. The operating temperature is around 180°C. The overall electrical efficiency is relatively low. This type of fuel cells can be used in stationary power generators of capacity 100 to 400 kW.

13.3.4 Alkaline Fuel Cell (AFC)

This type of fuel cell uses hydrogen as fuel and oxygen as oxidant. The electrolyte is an alkaline solution such as potassium hydroxide in water (20 to 40 per cent), which is capable of diffusing hydroxide ions. The reactions at anode and cathode are described as follows (Fig. 13.4):

Reactions at anode:

$$KOH \rightarrow K^- + OH^- \quad ...(13.5)$$

$$H_2 + 2(OH^-) \rightarrow 2H_2O + 2e^- \quad ...(13.6)$$

$$1/2O_2 + H_2O + 2e^- \rightarrow 2H^+ + 2e^- + 2(OH^-) \quad ...(13.7)$$

Reactions at cathode:

Overall reaction:

$$H^+ + OH^- \rightarrow H_2O \quad ...(13.8)$$

$$2H_2 + O_2 \rightarrow 2H_2O \quad ...(13.9)$$

AFC fuel cells have relatively large efficiencies, and hence used by NASA shuttles throughout their space programmes.

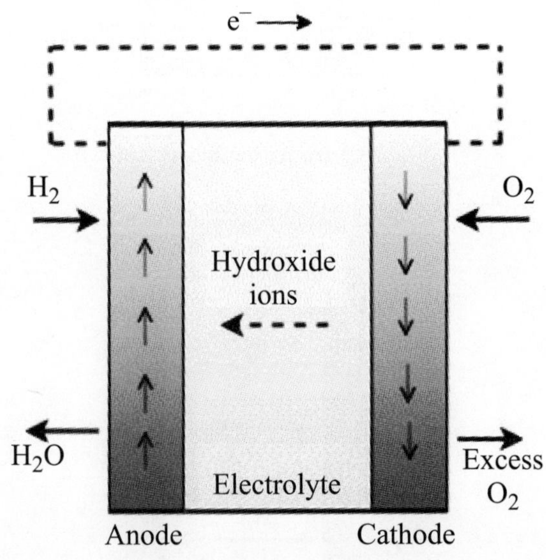

Fig. 13.4 Alkaline fuel cell

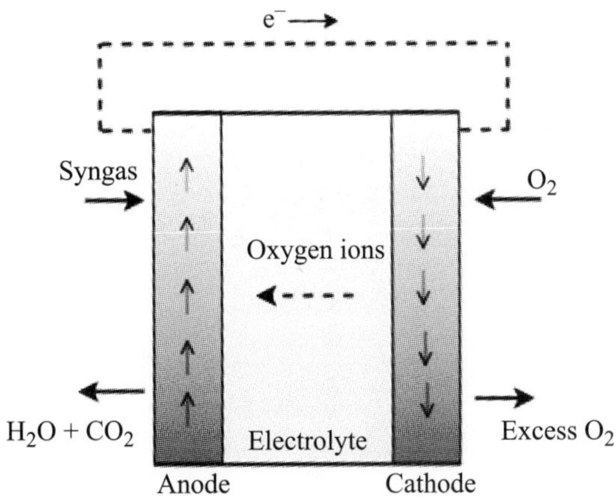

Fig. 13.5 Solid oxide fuel cell

13.3.5 Solid Oxide Fuel Cell (SOFC)

The fuels used in solid oxide fuel cells are hydrocarbons (Methane, syngas etc.). It uses solid ceramic like stabilized zirconium oxide as electrolyte, which is permeable to oxygen ions. The electrolyte carries oxygen from cathode to anode to combine with CO and hydrogen ions to form CO_2 and water respectively. SOFC generally operates at high temperatures ranging from 800 to 1000°C. This type of fuel cell is popularly used in stationary power generation (Fig. 13.5).

13.3.6 Molten Carbonate Fuel Cell (MCFC)

This type of fuel cell uses hydrocarbon fuels like methane. It uses a molten carbonate salt suspended in a porous ceramic matrix as electrolyte and does not require precious catalysts. The electrolyte carries carbonate ions from cathode to anode to form CO_2 and water respectively. The operating temperature of MCFC is around 650°C. This type of fuel cell is popularly used in powerplants of megawatt capacity. The working principle of MCFC is illustrated in Fig. 13.6.

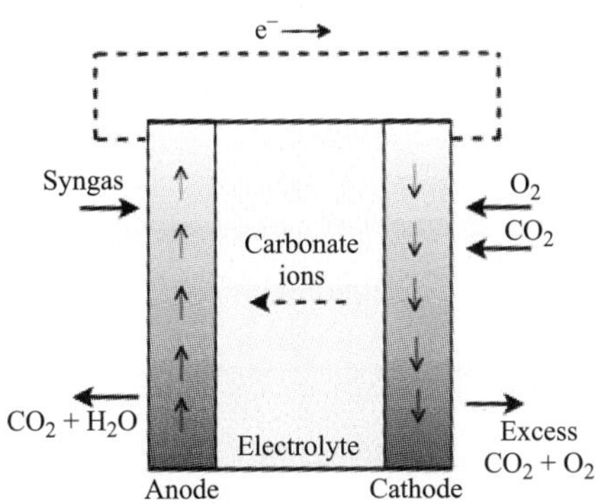

Fig. 13.6 Molten carbonate fuel cell

13.4 Fuel Cell Power Plants

Under typical working conditions, fuel cells can produce a voltage in the range 0.6 to 0.8 V only. As the voltage is low it does not serve as a practical working voltage for most applications. Hence, fuel cells are stacked in series to form a module and several modules are connected together to form a power producing unit.

A fuel cell power plant constitutes three major sub systems:

(i) Fuel processing unit

(ii) Power production unit

(iii) Power conditioning unit

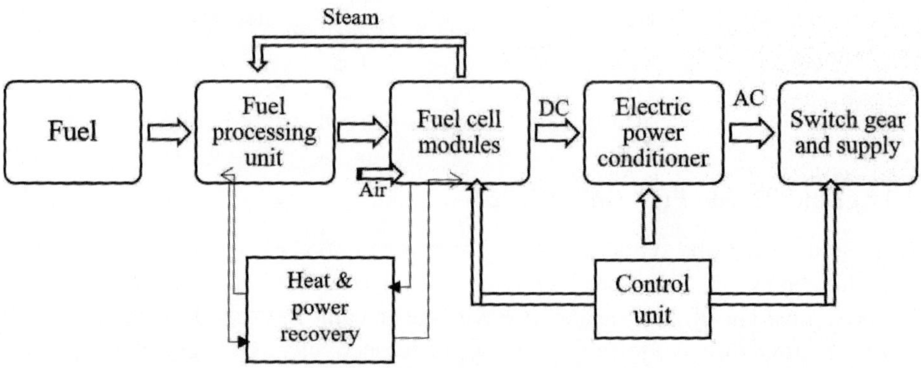

Fig. 13.7 Fuel cell power plant

There is also an energy recovery system to recover heat and power for heating and gas compression requirements. Fig. 13.7 depicts the flow diagram of a fuel cell power plant. Hydrogen rich fuel from the fuel processing unit is fed to the fuel cell power section. The DC electric output from the power section is converted to AC by the power conditioning unit, from where it can be taken out for useful work.

The electrical efficiency for PAFE systems and MCFC systems are above 45 per cent and 60 per cent respectively. Whereas the modern steam power plants have maximum efficiencies around 50 per cent.

13.5 Use of Biogas in Fuel Cell

Fuel cell power plants are one of the promising technologies for the future energy needs as they help to reduce emission of greenhouse gases. The combination of fuel cell and biogas technology can provide a renewable and efficient energy conversion system. Among all types of the fuel cells, molten carbonate fuel (MCFC) is most suited to run on biogas with a minimum expected efficiency of 49 per cent. On an experimental basis biogas have been tried as a fuel in solid oxide fuel cells (SOFC) also. It has been observed that the SOFC-biogas combination can offer high efficiency in the range of 50 to 60 per cent with lower emissions (Fig. 13.8).

Biogas can be produced by the anaerobic digestion of organic matter such as agricultural wastes, municipal solid wastes, wastewater etc. in a suitable digestion chamber or anaerobic bioreactor. Biogas typically consists of methane (65%), CO_2 (35%) with traces of hydrogen, hydrogen sulphide and ammonia. Fuel cell is sensitive to the traces of impurities like H_2S in biogas. Higher temperature fuel cells are sensitive to H_2S, whereas lower temperature fuel cells are sensitive to CO_2 and H_2S. Hence, biogas should be upgraded by scrubbing to remove impurities to have a high methane content (>97%) for use in fuel cells. There are different technologies like water scrubbing, alkali scrubbing, pressure swing process, cryogenic separation, membrane separation etc. for biogas cleaning.

The installation cost of biogas fuel cells is relatively high. The estimated cost of biogas fuel cell of 250 to 300 kW capacity is INR 5.8 lakh·kW^{-1}.

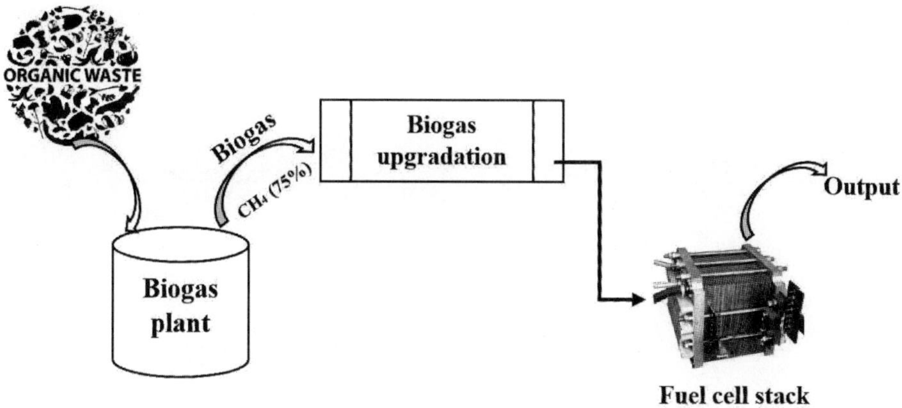

Fig. 13.8 Conceptual picturization of biogas fuel cell

13.6 Applications of Fuel Cells

Fuel cells have found their application in various sectors from automobiles to space crafts.

13.6.1 Stationary power plants

Stationary fuel cell power plants are capable of providing efficient service and reliable power with practically zero emissions. They require less space compared to other cleaner technologies. Hence, they are extensively used as primary as well as backup power by many institutions and organizations across the world including Microsoft, Apple, Adobe, Google, Honda, Walmart and many more.

13.6.2 Portable fuel cell systems

The light weight - long lasting portable fuel cell system has a wide range of applications including military electrical systems, laptops, cell phones, unmanned aerial vehicles, underwater vehicles, battery chargers etc.

13.6.3 Applications in transportation sector

Fuel cell has applications in the field of transportation including automobiles, utility vehicles, bikes etc. as described below:

(a) *Passenger cars:* Cars are one of the major markets of fuel cells as they are the most common transportation means across the globe and a major contributor to air pollution. Nissan, Ford, General Motors, Hyundai, Toyota, Honda, Subaru, Mazda are the major manufacturers involved in the manufacturing of fuel cell powered cars. Many of their fuel cell equipped models are designed to run with or without supplementary batteries. The fuels used in these models are either reformate (product of hydrocarbon reforming) or pure hydrogen in gaseous or liquid form (Fig. 13.9).

Fig. 13.9 Hyundai NEXO Hydrogen Fuel Cell Car

(Author: Dr. Artur Braun (Arturbraun), https://commons.wikimedia.org/wiki/File:Hyundai_NEXO_Hydrogen_Fuel_Cell_Car_beim_Autosalon_Genf_2018_y.jpg)

(b) *Public transport buses:* Many bus manufacturers started their demonstrations on fuel cell powered buses in the early years of 1990s. Fuel cell buses have been running successfully in many major cities including London, Amsterdam, Stockholm, British Columbia, California, Madrid, Hamburg, Luxembourg etc. Large space availability in buses makes the hydrogen storage easier than other automobiles (Fig. 13.10).

(c) *Submarines:* Fuel cells systems are functioning with low noise and infrared signature and hence can be considered as the most appropriate substitute for the existing batteries used for military submarine applications. The hot water obtained as byproduct of stationary fuel cell power plants can be utilized for the domestic uses of personnel in submarines.

Fig. 13.10 First generation fuel cell bus by Hyundai
(Author: Minseong Kim, https://commons.wikimedia.org/wiki/File:Hyundai_fuel_cell_bus_1st_gen.jpg)

(d) *Other applications:* The application of fuel cells for space crafts is well known. Three 12 kW AFCs were used in the space shuttle Orbiter. Due to their modular design, they can also be used in small portable units ranging in capacity from even 5 W.

CHAPTER 14

Biomass Combustion

The discovery of fire by the early man was the most significant step in the human history. It is believed that human species like *Homo erectus* and *Neanderthals* (forefathers of *Homo sapiens*) knew how to use fire. In addition to the use of fire as a protection against wild beasts, they might have used it for light and warmth also. Later on, it is most likely that they intentionally burned the surroundings (not to say forests, as all land was forest) to savour fried tubers and meat.

14.1 Historical Use of Biomass as Fuel

In fact, the discovery of cooking made tremendous changes in the life of early man as they could eat many food items which could not be digested in the raw form. While the apes spend much time and energy in chewing the food, men could save these two resources, which could be spent for other activities, as he started eating cooked food. Ability to use fire was the most distinguishing feature of the life style of humans compared to other animals during those days. It is reported that evidences for human use of fire, in the form of a hearth, was found at Israel's Palaeolithic archaeological site, Qesem Cave (inhabited about 3,00,000 – 4,00,000 years ago). Some archaeologists believe that humans used fire even 1.9 million years before. Even though exact dates are not known, cooking food might have been crucial in the development of the human brain, which eventually paved the way for his evolution into *Homo sapiens* (which means 'the wise man'). Humans started cooking meat in the beginning and then discovered 'stone boiling' technique by which liquids could be boiled. By the passage of time, they discovered the 'open firing' methodology for making clay utensils by heating clay to a temperature in the range 600 to 900°C, even though regional diversity in the pottery techniques existed. The advancement of technology in pottery lead to the development of the kiln which could make available high temperatures required for ceramic and porcelain pottery. During the early centuries of CE, food was generally cooked in a compatible vessel over an open flame produced by burning biomass in a hearth.

While biomass production is the most vital energy conversion process taking place in the nature, burning of biomass for thermal energy was probably the most important and primary stepping stone for the human civilization. It may be interesting to note that man started agriculture very recently (approximately 12,000 years ago) but started eating fire cooked food probably more than a million years before. As civilization progressed, the mode of cooking changed from an open fire to wood stoves. They also found better means to heat their dwelling places. In the eighteenth century, skilfully fabricated wood stoves were used. One spectacular stove design was from Benjamin Franklin in 1742. The Franklin stove had its fireplace lined with metal and had a chimney placed at a certain gap from the hearth, which enabled better heat exchange to the room air and produced less smoke compared to a traditional fireplace. Gas oven and electric cooking ranges were invented later in the 19th century. Fire places in cold climates gave way to today's advanced space heating means only very recently.

Even today, different forms of biomass are used as a fuel, especially in the developing countries of Asia and Africa. It may be interesting to note that biomass pellets are used for various thermal applications like room heating even today in many developed countries also. Biomass fired drying chambers are very common for drying various agricultural commodities in Asia. Even developed countries produce and use biomass fuels for various purposes including industrial heating. United States of America is the largest exporter of wood pellets to European markets with an export volume more than 5.5 gigatonnes (GT) in 2017. The growth of wood pellet industry has been largely influenced by the climate regulations in EU. Even though biomass fuels include gaseous and liquid fuels, the technology used for them are very similar to that for petroleum fuels. Biomass based liquid fuels (ethanol and biodiesel) are used for IC engines and seldom burned for direct thermal energy needs. Biomass based gaseous fuels (producer gas and biogas) are commonly utilised for direct burning in gas stoves specially designed for the gas. The following sections are basically intended to explain basic aspects of solid biomass combustion.

14.2 Biomass Combustion Process

Combustion is a process in which heat is released by the exothermic oxidation reactions by which the combustible constituents of biomass are converted to their oxides. Combustion process involves interaction of fuel particles, energy and environment. Direct combustion of a fuel is a very complex process influenced by the fuel characteristics as well as other process factors like supply of air. When a solid is delivered to a combustion chamber, the fuel particles do not get readily oxidised. There are three consecutive stages in biomass combustion:

(i) **Evaporation of moisture:** Most biomass materials contain moisture and due to the high temperature, vaporization of water occurs first.

(ii) **Distillation and burning of volatile matter:** Part of the biomass material is volatile at temperatures above 100°C and they are converted to gaseous form. This volatile fraction burns with flames.

(iii) **Combustion of fixed carbon:** The fraction of biomass left after volatilisation is the carbonaceous char. Burning of char do not produce flames as the volatiles have already been removed.

Pyrolysis of biomass after the drying phase and subsequent combustion will proceed in two different routes. The first route proceeds at high temperature during which thermal decomposition of biomass results in the production of combustible gases which mix with the air and results in flaming combustion. The distillation produces different fractions, viz. the volatile matter in gaseous phase and the fixed carbon in the form of solid char, which burns with different combustion characteristics. The volatile fraction burns with a flaming combustion over the fuel bed.

The other route operates at still lower temperature during which the fixed carbon in char burns with a smouldering combustion as the combustion products (incombustible gases viz. CO_2 and H_2O) remains above the fuel bed. Due to escape of volatiles, the char becomes highly porous and reactive. Consequent to this, char continues its oxidation with a glowing combustion and finally becomes ash. The two modes of combustions have entirely different reaction kinetics and depend on the physico-chemical characteristics of the fuel.

The carbon and hydrogen in the biomass are oxidised according to the following basic combustion reactions:

$$C + O_2 \rightarrow CO_2 \quad \text{...(14.1)}$$

$$H_2 + \frac{1}{2}O_2 \rightarrow H_2O \quad \text{...(14.2)}$$

The above equations represent the overall combustion process where there is no oxygen limitation. The actual reactions are very complex and many intermediate products are formed which further reacts among themselves. Combustion of biomass like wood begins with pyrolysis on the surface. When exposed to heat, volatiles move out from this layer forming a carbonaceous char layer on the outer surface. As the heat zone penetrates inside more volatiles are produced and the char layer thickens. At the same time the bed temperature is raised by the exothermic combustion reactions of volatile matter. The product gases of pyrolysis may react with carbon in the char layer. Water vapour reacts with carbon as shown below:

$$H_2O + C \rightarrow CO + H_2 \quad \text{...(14.3)}$$

Many intermediate products evolve which may subsequently react to form different products. Production of elemental carbon results in soot formation at high temperatures.

$$2CO \rightarrow CO_2 + C \quad \text{...(14.4)}$$

The CO and C are then further oxidised in the course of combustion reactions. The charred biomass after removal of volatiles becomes highly porous and reactive to form activated charcoal. If combustion proceeds, then the fixed carbon also will be oxidised. If oxygen is deficient some amount of CO may escape to surroundings leading in pollution. Proper combustion is also affected by the availability of air and removal of combustion products from the hearth which is largely decided by the design of the combustion equipment.

14.3 Combustion Intensity and the Power Rating of Combustion Equipment

Combustion reactions are very complex that the actual rate of fuel oxidation may vary. But once the process stabilises and the fuel supply is properly regulated, it could reach a moderate steady state. In that stage, the heat evolved will be proportional to the rate of fuel mass oxidised. This can be expressed as shown below:

$$\text{Combustion Intensity } (I_{com}) = \Delta H \frac{-dm}{dt} \qquad ...(14.5)$$

Where, ΔH is the heat of combustion and *dm* is the mass reduction in the time interval *dt*.

Power rating of combustion equipment can be found by knowing the energy released per unit time.

$$\text{Thermal Power Rating } (P_{thermal}) = [m(LHV)] \times \frac{\eta}{100} \qquad ...(14.6)$$

Where, m is the mass of fuel burned per second (fuel feed rate per second) and η is the overall efficiency of the equipment in percentage.

Example 14.1: Find the fuel feed rate per hour required for a wood fired furnace of 20 kW thermal power capacity. The overall thermal efficiency is 35% and LHV of fire wood is 18 $MJ \cdot kg^{-1}$

Solution:

Power (thermal) = 20 kW = m (18000 kJ × 0.35)

Fuel feed rate per hour, M = m × 3600 = 3600 × [20 ÷ (18000 × 0.35)]

= 11.43 kg per hour

14.4 Biomass Properties Affecting Combustion

The physico-chemical characteristics which govern biomass combustion can be summarised as below:

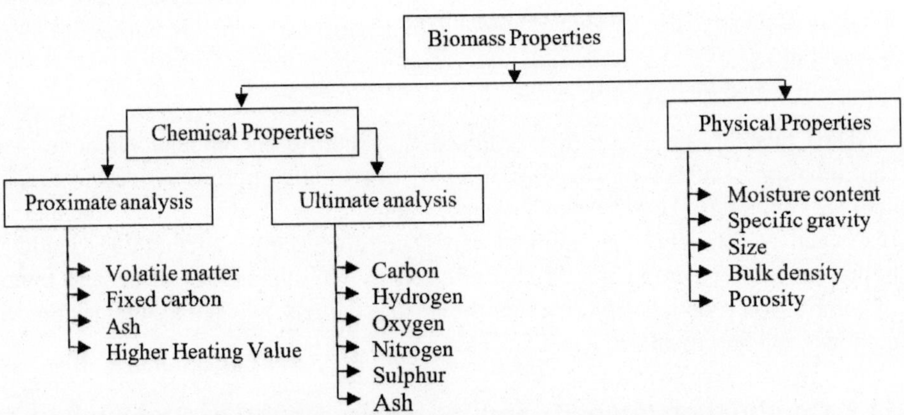

Fig. 14.1 Biomass properties affecting combustion

(a) *Moisture content:* Among physical properties, moisture content is an important parameter. Biomass materials usually contains about 30 to 50 per cent moisture in raw stage. If sun dried properly, this can be reduced to 12 to 20 per cent, depending on the nature of the material. Common biomass materials are composed of dead tissues of plants which contain water in the cavities and cell walls.

Moisture content (MC) is expressed in two forms, wet basis (wb) and dry basis (db). The MC is on wet basis when the quantity of water is expressed as a fraction of the wet weight of biomass and is on dry basis and when reported as a fraction of the dry weight. MC is very important as energy is to be spent for vaporizing water from biomass. If the MC is 50 per cent (wb), about 13 per cent of the energy available in the biomass is required just for vaporizing the moisture. Thus, the net heat release from the biomass fuel is considerably decreased. In general, about 2.8 MJ of energy per kg of moisture is required to dry the biomass in the combustion process.

(b) *Other physical parameters:* Other physical parameters viz. specific gravity, size, porosity and bulk density are also important in designing efficient combustion equipment. The energy released depends primarily on the quantity of carbon and hydrogen in unit volume of biomass for which the specific gravity is a measure of the material content. The ease of burning of the combustible constituents depends upon the mixing of air through the material if it is more porous. The removal of combustion products from the zone is also equally important. Size reduction increases the surface area of the fuel particles which in turn improves the contact of oxygen molecules with the fuel molecules. Bulk density is to be considered in the equipment design as the space occupied by the fuel in the combustion zone or fuel bed is important in deciding the rate of energy release by the system.

(c) *Chemical properties:* Most biomass materials used for combustion are classified as 'ligno-cellulosic' and can be considered as a mixture of complex organic substances. The main components are cellulose, hemicelluloses and lignin with other extractable compounds such as sugars, lipids and proteins. Cellulose is the most abundant natural polymer and the major chemical component of biomass fibres, contributing 40 to 45 per cent of the dry weight of wood. Cellulose is composed of linear chains of glucose linked by glycosidic bonds with the degree of polymerization from 10,000 to 1,000. Different types of hemicelluloses are present in biomass and they have a lesser degree of polymerisation. Lignin accounts for 10 to 25 per cent of ligno-cellulosic biomass and is the second most abundant natural polymer. Lignin is insoluble in water and stable in nature and acts as a binder between cellulose and hemi-cellulose. The extractable compounds are generally classified into water soluble and solvent soluble. The combustion kinetics depends on the chemical reactions taking place with the constituents. Two types of analyses are done to get a clear indication of the behaviour of the biomass fuel.

14.5 Proximate and Ultimate Analyses for Biomass

The overall chemical nature of the biomass is determined by the '*Proximate analysis*' in which the total solid (TS) content is fractionated to volatile matter, fixed carbon and ash. The TS content of the sample is determined by heating the sample at 105±3°C in a hot air oven for 24 hours. The dry sample is weighed and again heated for 1 hour. The weight is checked for confirmation. Volatile fraction is easily vapourised at high temperatures whereas fixed carbon remains as solid to form char. Volatile matter (VM) can be determined experimentally by heating the dried biomass sample in a closed airtight crucible at 900ºC for 6 minutes. The remaining mass is the fixed carbon along with ash. Ash content is determined by igniting the dried samples in a muffle furnace at 550±30°C for 30 minutes or till concordant weights are recorded. The ash content is the mineral matter in the material which does not contribute to the heating value. Ash contents vary widely for different biomass fuels. The actual mineral content in the ash also is relevant since some inorganic compounds like silica interferes with smooth combustion. Total Volatile Solids (VS) is the sum of VM and FC and can be obtained by subtracting ash content from TS.

Ultimate analysis refers to the elemental analysis in which the basic composition is estimated as the fraction of each element. The significant elements in biomass are Carbon, Hydrogen, Oxygen, Nitrogen and Sulphur. The heat release during combustion of different types of biomasses like wood, foliage, crop residues etc., is dependent on their chemical composition. Carbon, hydrogen, oxygen and nitrogen are the predominant elements which can be estimated experimentally using a C-H-N-O analyser. As the system is based on combustion of sample and analysis of the combustion products, it is often referred as 'Combustion analyser' also. A capsule with sample is injected into a high temperature (1000°C) furnace and completely combusted with oxygen at experimental conditions. The resulting combustion products are absorbed in columns packed with specialized reagents to estimate carbon dioxide (CO_2), water vapour (H_2O), nitrogen (N_2) and oxides of nitrogen. From these, the elemental composition can be determined.

14.6 Calorific Value of Biomass

Calorific value is the most important property of a biomass fuel which is the measure of total energy released by unit mass of the material by the exothermic reactions during complete combustion. *Higher Heating Value* (HHV) or Gross calorific value is the thermal energy released when unit weight of the fuel is completely burned and the products of combustion are cooled to 15°C to condense the water vapour. Here the heat of condensation released by the water vapour is also taken into account. The standard method for the determination of the gross calorific value of a solid biomass fuel at constant volume and at the reference temperature of 25°C in a bomb calorimeter (calibrated by combustion of benzoic acid) is specified in ISO 18125:2017.

The HHV of a fuel can be calculated by considering that the heat released by the constituent elements in the fuel is equal to the heat released when they are burned in the free (not combined) state. This is under the assumption that no energy expenditure is required to free the elements from the compounds constituting the fuel.

The energy released in the combustion reactions can be represented by the following equations:

$$12g[C] + 32g[O_2] \rightarrow 44g[CO_2] + 408.54 \text{ kJ} \quad (\text{at } 15°C) \qquad ...(14.7)$$

This means 1g of carbon releases $408.54 \div 12 = 34.045$ kJ of energy.

Similarly,

$2g[H_2] + 16g\ [O_2] \rightarrow 18g[H_2O] + 288.70\ kJ \quad (at\ 15°C)$ ...(14.8)

Therefore, 1 g of hydrogen releases 288.70 ÷ 2 = 144.35 kJ (provided the water vapour is condensed)

The HHV of any fuel containing carbon and hydrogen only can be obtained from the following equation:

$$HHV = \frac{[(C \times 34.045) + (H \times 144.35)]}{100}\ MJ.kg^{-1} \quad ...(14.9)$$

Where, 'C' and 'H' are percentages of carbon and hydrogen in the fuel.

Most biomass fuels are having oxygen also in addition to hydrogen and oxygen. Oxygen content reduces the heating value and this means that fuel is partly oxidised. Hence, HHV of a biomass fuel is given by:

$$HHV = \frac{\left[(34.045C) + 144.35\left(H - \frac{O}{8}\right)\right]}{100}\ MJ.kg^{-1} \quad ...(14.10)$$

Where, 'O' is in percentage.

HHV can be determined experimentally using a 'Bomb calorimeter' by burning the biomass in a sealed chamber. The carbon and hydrogen in biomass are converted to CO_2 and H_2O when ignited by oxygen in the calorimeter. The total heat evolved per unit mass is measured and regarded as the HHV.

The Lower heating Value (LHV) or the Net calorific value is obtained by subtracting the heat of condensation of water vapour (at 15°C) from HHV. The LHV of a sample differs from the HHV as it considers the energy required to vaporise the water generated when the hydrogen in biomass burns to produce water. This implies that the quantity of hydrogen in the biomass (or any other fuel) becomes a factor to be subtracted in the overall heat released in the combustion reactions. This difference is equivalent to the amount of latent heat of vaporization of water vapour that could be recovered in a condenser. Practically, LHV is more relevant and it can be calculated as:

$LHV,\ kJ\ kg^{-1} = (HHV\ kJ\ kg^{-1}) - 212[H] - 24.5[M] - 8[O]$ ...(14.11)

Where, [H], [M] and [O] are percentages of hydrogen, moisture and oxygen.

14.7 Thermo Gravimetric Analysis of Biomass

The study of biomass behaviour when subjected to increased temperatures is called Thermo-gravimetric analysis (TGA). The rate of weight reduction with respect to temperature and time is observed for this analysis. The weight loss can be recorded at a fixed temperature and plotted against time (isothermal TGA). Sometimes the analysis is done for a fixed heating rate. Both records are useful in understanding the thermal degradation behaviour of biomass. TGA for different types of biomasses can be prepared for comparative evaluation.

When a solid biomass fuel like fire wood is fed to a furnace, the free moisture in the material starts evaporating. This drying process continues and all the free water gets evaporated as the temperature reach 100°C. Thereafter, the bound water also is removed in the temperature range of 100 to 150°C. Beyond that, volatiles start leaving the solid biomass with the simultaneous production of char in the temperature range of 250 to 350°C. Volatile matter burns with a flame (flaming combustion) and results in further increase of temperature in the combustion zone. Depending on the composition of material, initial condition, rate of heating, nature of volatile matter, the temperature range for volatilization may vary. At about 400 to 450°C the pyrolysis (volatilisation) is completed and entire biomass material gets charred. Further heating causes the combustion reactions to proceed and fixed carbon also gets combusted without any flame. At about 800°C all carbon will be oxidised and only ash will remain. The thermal degradation of biomass can be elucidated graphically in a thermal degradation curve shown in Fig. 14.2.

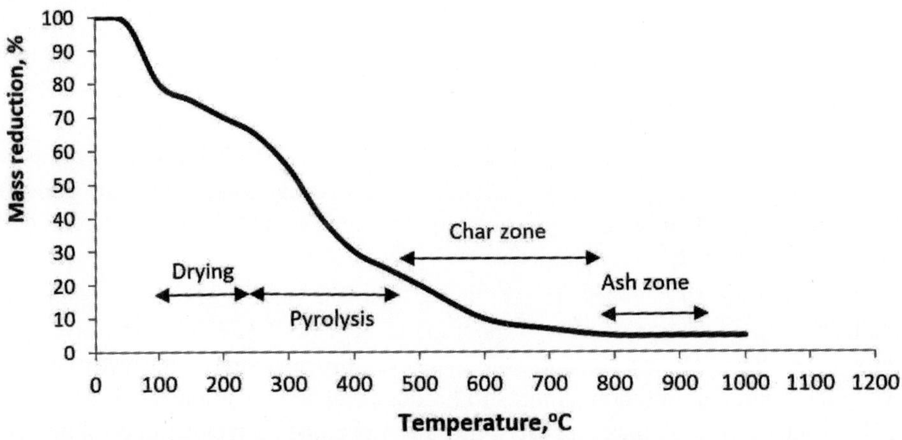

Fig. 14.2 Thermal degradation curve of biomass

It may be seen that the moisture is about 20 to 25 per cent which is removed in the drying zone. Volatile matter constitutes about 50 per cent of the total mass which is removed from the solid biomass in the pyrolysis zone. At about 450°C only char remains, constituting about 25 per cent of the total mass. Char combustion is completed at about 800°C, and the left over is ash constituting about 5 per cent of the mass.

14.8 Calculation of Air-fuel Ratio

Proper air supply is the major factor for efficient combustion and hence calculation of air-fuel ratio is essential for design of combustion equipment. The theoretical oxygen requirement for a biomass fuel essentially containing C and H as the combustible components can be obtained from the basic combustion reactions. According to the chemical equations, 12 g (1 mole) of carbon combines with 1 mole (32 g) of oxygen and

2 g (1 mole) of Hydrogen. Each gram of carbon requires 32/12 g of oxygen and each gram of hydrogen require 16/2 g of oxygen. This is called the 'Stoichiometric' oxygen requirement for combustion. Corresponding quantity of air is called 'Stoichiometric air requirement'. The percentage of oxygen in air is taken as 21 per cent on volume basis and 23 per cent on mass basis. The molar mass of air is taken as 29 kg/mole.

Methodology

When a carbohydrate burns in air, one unit of carbohydrate (corresponding to 1 carbon atom i.e., CH_2O for glucose) can produce only one unit of CO_2. Similarly, water produced will be half the amount of hydrogen as two hydrogen atoms are required for one water molecule. One oxygen atom is required for one water molecule. The total oxygen required will be the sum of these two minus the oxygen present in the biomass. Oxygen required for complete combustion of a biomass (basically carbohydrate) with general chemical formula CH_xO_y can be expressed as below:

$$CH_xO_y + \left(1 + \frac{x}{4} - \frac{y}{2}\right) O_2 \rightarrow CO_2 + \left(\frac{x}{2}\right) H_2O \quad ...(14.12)$$

The oxygen and air requirement for complete combustion of a biomass fuel can be calculated if the elemental composition is known from the ultimate analysis. This composition is often expressed as 'dry ash free'. Here, the moisture and ash fraction are eliminated as they do not give any energy by combustion nor they do require oxygen. Sometimes the elemental composition is given as 'as received', which means the percentages are inclusive of ash content and moisture. This can be converted to 'dry ash free' if we know the moisture content and ash fraction.

Example 14.2: Calculate the stoichiometric air fuel ratio for a biomass fuel with the following ultimate analysis data.

1. 50% carbon, 6% hydrogen and 44% oxygen (dry, ash free).
2. 41.5% carbon, 4.98% hydrogen, 36.52% oxygen, 12% moisture, 5% ash (as received)

Solution (1):

Consider 100 kg of fuel

1 mole of carbon requires 1 mole of oxygen and 1 mole of hydrogen requires half mole of oxygen, for complete combustion.

Constituent	Weight, kg	kg-mole (weight ÷ mol. weight)	Stoichiometric Oxygen requirement, kg-mole
Carbon	50	50 ÷ 12 = 4.17	4.17
Hydrogen	6	6 ÷ 2 = 3	1.50
Oxygen	44	44 ÷ 32 = 1.375	(-)1.375
		Total = 4.17 + 1.50 – 1.375	4.295

Oxygen required for 100 kg of fuel = 4.17 + 1.50 – 1.375 = 4.295 kg-mole

Air required = 4.295 ÷ 0.23= 18.674 kg-mole = 18.674 × mole weight of air = 18.674 × 29 = 541.55 kg

So, 1 kg of fuel requires 5.4155 kg of air for complete combustion.

Air: Fuel ratio is 5.4155:1

Solution (2):

The 'as received' composition should be converted to 'dry ash free' composition first.

Dry Ash free composition × Conversion factor = As received composition

The conversion factor, CF = (1 – Moisture fraction – Ash fraction).

In this case, CF = 1 – 0.12 – 0.05 = 0.83

Dry ash free ultimate analysis:

C = 41.5 ÷ 0.83 = 50%; H = 5.98 ÷ 0.83= 6% and O = 36.52 ÷ 0.83 = 44%

The dry ash free composition is same as above and the A/F ratio calculation is also same.

14.9 Factors to be Considered for Design of Combustion Equipment

Combustion of fuel is done for making available the maximum amount of energy by ensuring that the reactions are complete. Due to difficulties in mixing air with the solid fuels, incomplete combustion causing loss of energy in the unburnt or partially combusted volatile matter may occur. Hence the supply of air in sufficient quantity and introducing it in the proper way to the fuel bed are important aspects. If too much excess air is provided, the heat of combustion will be taken away by the flue gases. The supply of primary air at the fuel bed as well as secondary air in the combustion zone (for burning of volatile matter) in excess of the stoichiometric air requirement is ensured. The following aspects are to be considered in designing combustion equipment:

(i) Stoichiometric air requirement should be calculated based on the composition and excess air (usually 1.2 times) should be supplied.

(ii) The fuel bed should be such that there is free movement of primary air through the fuel particles and maximum contact is given between the fuel and oxygen.

(iii) The secondary air in sufficient quantity should be supplied for ensuring complete combustion of the volatile pyrolysis products.

(iv) There should be sufficient draft provided for ensuring fresh air entry to the fuel bed and combustion zone as well as removal of gaseous combustion products. Sometimes a chimney needs to be provided for getting required natural draft. In forced draft systems a blower is used to force air.

(v) The heat loss from combustion chamber should be prevented so that the fuel particles or the volatile matter going out of fuel bed is not cooled below the ignition temperatures.

(vi) The fuel feed system should be able to carry out feeding of required quantity of fuel for the designated combustion intensity and thermal power rating.

(vii) Sufficient space should be provided in the combustion chamber so as to contain the combustion products for sufficient duration of time essential for heat transfer. Sufficient volume to hold the expanded gases is also required.

(viii) The fuel bed should be such that there is facility for removal of ash from unburnt fuel.

14.10 Combustion Equipment

Properly designed equipment for combustion is required for ensuring efficient combustion and proper heat transfer to the required application. They range from simple firewood stoves (*chuhla*) to large furnaces for various thermal applications. In all these equipment, incomplete burning of fuel is the major factor responsible for low thermal efficiency as well as air pollution due to smoke. Smoke is a suspension of unburnt carbon particles with the combustion products such as CO_2, CO and water vapour along with traces of other gases like SO_2 (if the fuel contains sulphur as in the case of fossil fuels). Some of these unscientific designs are sometimes having a thermal efficiency as low as 10 per cent. Hence, properly designed equipment can save lot of fuel, resulting in drastically reducing the CO_2 emission and air pollution. Biomass furnaces and stoves are the most common combustion equipment for industrial as well as domestic applications. Biomass furnaces are commonly used in rural industries like brick making or pottery along with several other applications. Biomass stoves of various designs are also widely used in developing countries.

14.11 Biomass Furnaces

Furnaces are basically thermal energy plants used for diverse applications including production of hot gases, hot water, steam etc. Biomass fired furnaces are commonly used for a variety of uses where large quantity of thermal energy is required for industrial use or for production of steam. Biomass furnaces basically consist of a 'grate' enabling the biomass fuel to be burned effectively, a 'combustion chamber' and an 'air mixing chamber' or 'heat exchanging unit'. The grate is a supporting structure for the fuel bed generally made of cast iron. The combustion chamber is the space for exothermic reactions and the surface is generally lined with fire bricks to withstand high temperatures and to prevent loss of heat energy. Furnaces for producing hot air have a cold air inlet and an outlet for hot air (Fig. 14.3).

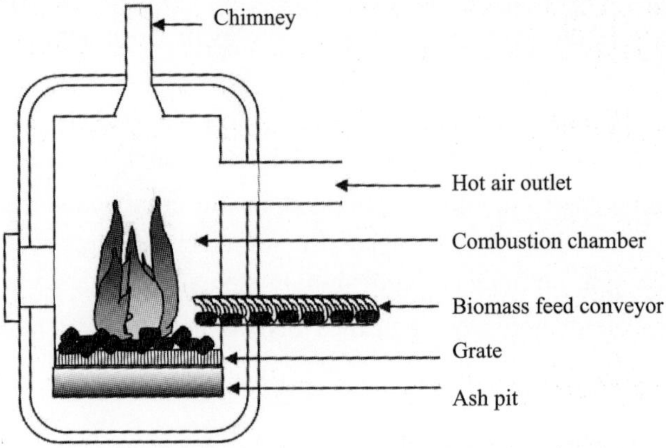

Fig. 14.3 Biomass fired horizontal grate furnace

Most common application of biomass furnace is for drying of grains. Depending on the type of biomass used as fuel, the design can be different. Most biomass furnaces are of fixed grate type. There are different designs for grate and furnaces are sometimes classified based on the type of grate used. They can be horizontal grate, inclined grate or stepped grate. Cyclone furnaces and fluidized bed furnaces are more advanced designs in which pulverised solid fuels or very light weight particles like rice husk or coir pith are used as fuel. In cyclone furnaces, rotary motion is imparted to the flame as well as the fuel particles. This enables good mixing of fuel particles with air and results in high efficiency compared with fixed grate furnaces. Fluidised bed combustion systems are also advanced and highly engineered systems in which the fuel particles are suspended by an upward air flow.

Efficiency of furnaces

In any heating system, heat losses occur through various means. The major heat losses are due to escape of unburnt fuel, incomplete combustion of fuel and heat losses through the walls. If the furnace is used for heating air, the efficiency,

$$\eta = \left(\frac{\text{Heat content of flue gas}}{\text{Energy content of the fuel}}\right) \times 100 \qquad ...(14.13)$$

$$\eta = \frac{QH}{mC} \times 100$$

Where,

Q is the discharge rate of hot flue gases, $Nm^3 \cdot h^{-1}$

H is the enthalpy of flue gas. $MJ \cdot Nm^{-3}$

m is the feed rate of fuel, $kg \cdot h^{-1}$

C is the lower heating value of biomass fuel, $MJ \cdot kg^{-1}$

Note: Normal cubic metre per hour ($Nm^3 \cdot h^{-1}$) is the volumetric flow rate of gas at temperature 0°C and pressure of 1.014 bar.

If the furnace is used for other purposes, then the unrecovered heat lost through flue gases also is a reason for low efficiency.

14.12 Biomass Stoves

Stove or *chulha* is a device designed to facilitate conversion of chemical energy of the fuel into heat energy by exothermic combustion reactions and transfer it to the cooking vessel containing food material to be cooked. Traditional cook stoves have the problem that they produce indoor air pollution due to smoke. This was identified as one of the important reasons for respiratory diseases of rural women in India. Another disadvantage is the fact that due to low efficiency the fuel consumption is high. This has got significance at the micro level that increased use of fuel for cooking a given quantity of food results in unnecessary wastage of fuel which the rural women folk collect with great effort. At macro level, this enhances carbon emission contributing to global warming. Hence, increasing the efficiency of house hold biomass stoves is a great concern in developing countries. The Government of India had implemented the 'National Programme on Improved *Chulhas* (NPIC)' for development and popularisation of improved biomass stoves as early as 1984, which continued until 2002. Under NPIC a variety of efficient stove models were developed and promoted. The details of some of them are given in Table 14.1. Millions of improved stoves have been installed under this programme saving huge quantities of fire wood, thereby preventing millions of tonnes of CO_2 emission.

14.12.1 Improved biomass stoves

Improved biomass stoves are scientifically designed gadgets for optimal fuel utilization, proper heat flow and reduction of indoor air pollution. In addition to avoidance of smoke they aid for drudgery reduction in cooking. They meet all cooking needs of a household achieving better combustion and efficiency but minimizing health and safety hazards. As these are mostly manufactured by rural technicians, they have a favourable impact on rural employment generation also.

Table 14.1 Some improved stove designs and their features

Sl. No.	Name	Organisation	Main Feature
1.	*Parishad Stoves*	Kerala Sasthra Sahithya Parishad	Chimney is provided
2.	*Sukhad Chulha*	Rajasthan Agricultural University	Double pot stove with chimney
3.	*TNAU Chulha*	Tamil Nadu Agricultural University	Single pot and double pot versions without chimney
4.	*CPRI Chulha*	Central Power Research Institute	Portable, metallic non chimney stove

The overall thermal efficiencies of the improved stoves are about 30 to 40 per cent compared to 8 to 12 per cent of traditional stoves. This is achieved by (i) complete combustion of fuel, (ii) better heat transfer and (iii) reduced heat loss.

Qualities of Improved biomass stoves:

The qualities of improved biomass stoves can be listed as follows:

(i) Consume less fuel
(ii) Burn fuel as completely as possible
(iii) Eliminate or carry away smoke from cooking area
(iv) Serve both slow and fast cooking
(v) Reduce heat losses
(vi) Save cooking time
(vii) Minimize drudgery

14.12.2 Types and main components

Biomass stoves may be classified based on consideration such as, portable or fixed, construction materials (clay or metal), number of pot holes, with chimney or without chimney etc.

Main Components:

(i) *Combustion chamber or fire box:* Combustion chamber, also called fire box is the space where fuel is fired and the combustion reactions takes place. It should have sufficient volume to accommodate fuel and combustion products. The cooking vessel should rest over the fire box and should be strong enough to hold the vessel with the food material. Wood is arranged in a criss-cross manner in the fire box so as to ensure air movement through the fuel bed. The entry of primary air is through the fuel feeding door of fire box and is directed to the fuel bed through the grate. There is no separate inlet for secondary air which enters to the flame region and mix with volatile matter. Sufficient air enters through the fuel feed door for ensuring almost complete combustion of the fuel. The height of fire box is in the range 200 to 300 mm. (Fig. 14.4).

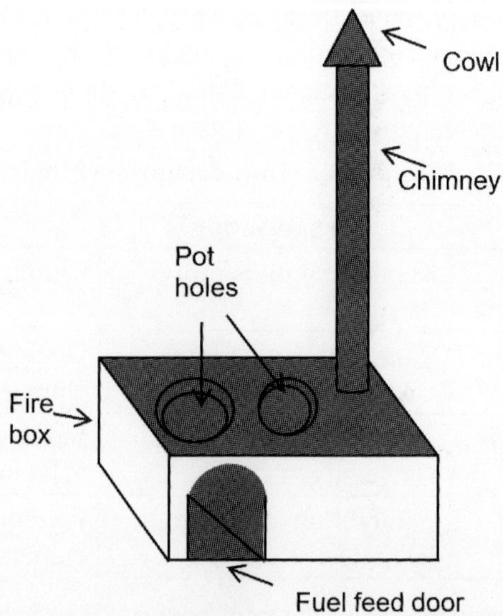

Fig. 14.4 Double pot biomass stove

(ii) *Pot holes and seat:* Diameter of pot hole should be selected based on the diameter of common cooking vessels. The pot seats are provided on the top of fire box and generally have an angle of 45° so as to fit the bottom of round bottomed vessels without gap. In the case of double pot stove, the pot holes should be arranged such that the two pots do not touch each other and the bottom of the pots are at a proper height to ensure maximum heat transfer.

(iii) *Tunnel or flue gas channel:* In multi- pot closed stove, the space under the vessels placed on two or more pot holes are connected to each other by means of tunnels. The space under last pot hole and chimney are also connected by means of a tunnel (Fig. 14.5).

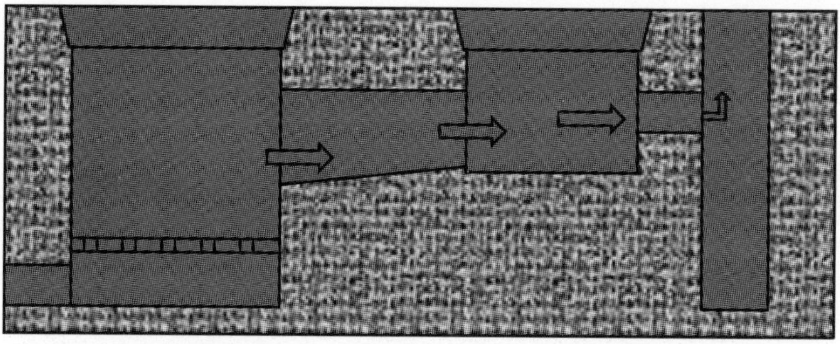

Fig. 14.5 Arrangement of tunnel in a double pot stove

(iv) *Grate:* It is the part of the stove which holds the fuel. It is generally of circular shape and is made as a disc with perforations. It can be made with either metal or clay and placed about 3 to 5cm above the base of the fire box. The fuel rests on the grate and allow entry of primary air through the perforations. The air enters from underneath and pass through the fuel bed arranged on it. The ash formed during combustion falls down through the perforations and do not interfere with air passage through the fuel bed (Fig. 14.6).

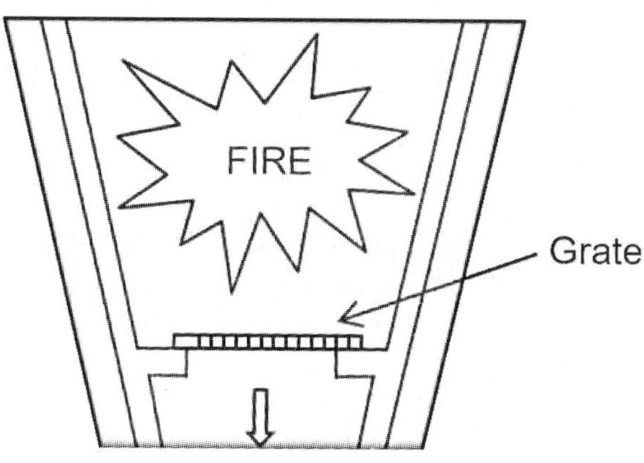

Fig. 14.6 Grate arrangement in a single pot stove

(v) *Baffle:* Hot gases from the fire box have a tendency to move out through the chimney due to the draft. But this will result in wastage of the heat of flue gases. For enhancing efficiency, hot flue gases should not be allowed to move fast and leave through the chimney to the atmosphere. A baffle is an obstruction to orient fire and hot gases towards working vessels generally fixed under the second pot hole.

(vi) *Damper:* As the name indicates, it is used to restrict the entry of air into the firebox so as to control the rate of fuel burning and thereby the intensity of combustion. Usually, a piece of metal sheet is used to close the fuel feed door partly as and when required. The air entry can be regulated by sliding the sheet.

(vii) *Chimney:* Chimneys are required for taking the flue gases out of the cooking area so as to avoid indoor air pollution. It is useful in providing sufficient draft to draw fresh air into the fire box. Usually, a pipe made of asbestos cement, fired clay or metal is fixed to the rear end of a closed stove. It is supported on the wall with clamps and taken out vertically through the roof. In some cases, it is provided outside the kitchen and a bent pipe connects it with the rear end of the stove. The height and diameter of the pipe used for chimney is a deciding factor for the draft. For domestic stoves the chimney diameter of 75 mm is sufficient. Chimney should be cleaned at regular intervals and a provision can be made at the bottom point to remove the ash. Cap provided at the top of the chimney to avoid rain water is called ‘cowl’.

CHAPTER

15

Densification of Biomass for Energy Conversion

The rural economy in most developing countries like India is predominantly based on agricultural production. Huge quantities of agricultural byproducts and wastes are produced seasonally from agriculture. Most of this biomass are either discarded or disposed inefficiently, which may lead to pollution. There was a grave concern over burning of paddy straw in Indian states like Punjab and Haryana which created serious air pollution issues in many areas including the national capital. Direct burning of biomass even for thermal energy purpose is often inefficient. Low energy density is an inherent drawback of crop residues when they are considered for energy production. The energy recovery from the biomass can be improved by reforming it by the development of efficient physical conversion technologies. The production of uniform size pellets or briquettes from biomass like rice husk, saw dust, wood chips or other plant residues not only improve the conversion efficiency but also render it more convenient in transportation, handling and storage. India has a good biomass energy potential in the order of 61,000 MW.

15.1 Briquetting

Briquetting can be defined as the densification or compaction of loose biomass having a low bulk density into a product of higher bulk density. The compacting of biomass to form briquettes is done by the application of heat and pressure. Biomass briquettes can be used either for heating households and in small-scale establishments, or even for industrial power generation by burning or gasification. The advantages of briquettes in comparison to the original biomass can be summarized as below:

- Better convenience of storage and transport as it is compact.
- The feeding and charging system for furnaces are less complicated.
- When used in gasifier, better gasification characteristics can be achieved.
- Improved combustion characteristics can reduce particulate emissions.
- Uniform size and shape of biomass briquettes are advantageous in the design and operation of equipment.
- Biomass can serve as a good substitution for fossil fuels like coal.
- Universal availability of raw materials, especially in less developed countries.
- Locally available raw material lowers overall fuel costs.
- Capable of local employment generation.
- Decrease fuel use and GHG emission.
- Reduce deforestation as more efficient fuel is made available as briquettes.

The raw materials for briquetting depend on the locally available materials which are mostly agricultural and forest residues like saw dust, coffee husks, sugarcane bagasse, tree leaves, charcoal dust, terrestrial and aquatic weed biomass, and various other residues and by-products.

15.2 Principles of Briquetting Process

Most raw materials available from agricultural sector are ligno-cellulosic biomass, which contain lignin in the cellular structure of plant biomass. Lignin gets fluidised when the feed material is simultaneously subjected to high pressure and temperature. Lignin binds individual particles into a compact structure to form the briquette. Under high pressure the materials may bind together by the increased adhesive or cohesive forces, which result in the formation of interlocking bonds between the solid particles. Often, a binding agent is also used to hold together the particles of biomass. The binding agent can also be a biomass material like starch or a partially decomposed fibrous organic material, to provide the fibres necessary to physically hold the biomass particles in the briquette together. Depending on the characteristics of the biomass material, the process conditions like pressure applied and the binder used may vary. The binding methods used for briquetting of different types of feed stock can also be different.

The broad process of briquetting may be described as shown in the flow diagram (Fig. 15.1)

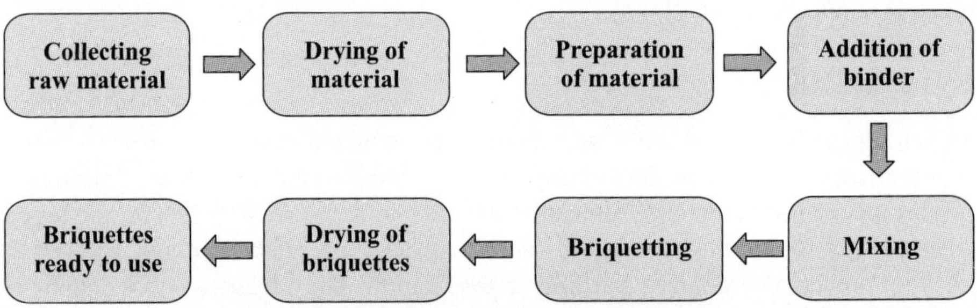

Fig. 15.1 Process flow diagram of briquetting

The following methods are generally employed to achieve densification of biomass:

(i) Direct densification by the use of binding agents

(ii) Binderless briquetting

(iii) Densification by pyrolysis with the use of binder

Depending upon biomass characteristics, three pre-process sequences (Fig. 15.2) are generally employed before the actual briquetting operations which include pre heating, densification and cooling.

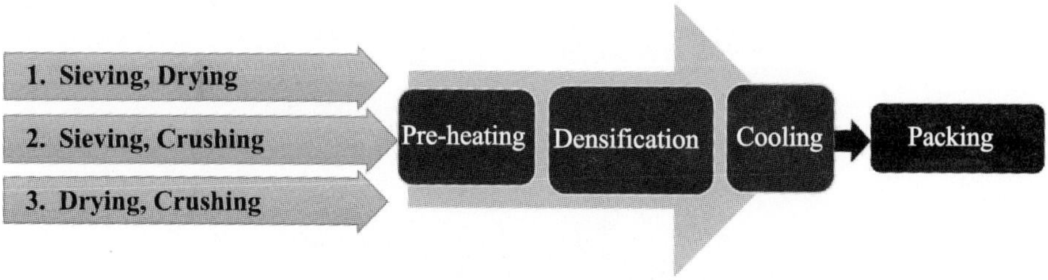

Fig. 15.2 Pre-process sequences in briquetting process

15.3 Briquetting Techniques

According to a classification done by FAO (1990), briquetting could be categorized depending on the types of presses used such as piston presses, screw presses, roller press and manual presses. The process is also classified on the basis of pressure used into high pressure, medium pressure and low-pressure processes. The medium pressure process is usually assisted by a heating device, whereas a binding agent is employed in low pressure process.

15.3.1 Piston press densification

Piston presses mainly make use of two different technologies, (i) hydraulic press and (ii) die and punch technology.

In the hydraulic press process, biomass materials are initially compacted in the vertical direction and subsequently in the horizontal direction. The feed materials up to 22 per cent moisture content is usually processed in this type of press. The process is slow and hence this slowness of operation helps to reduce the wear and tear of the machine parts. Briquettes produced by these types of presses usually have the dimensions 450×160×80 mm with an approximate weight about 5 kg. The power requirement of these presses is in the range 35 to 38 kW for an hourly output capacity of 1600 to 1800 kg.

The 'die and punch' technology is also known as 'ram and die' technology. The mechanical arrangement consists of a reciprocating ram which presses the biomass into the die at a very high pressure. The biomass compressed and extruded from the die forms a compacted and denser product. This machine produces briquettes of a standard size, 60 mm in diameter. The power required by this type of machines is about 25 kW for a working capacity of 600 to 700 kg per hour (Fig. 15.3).

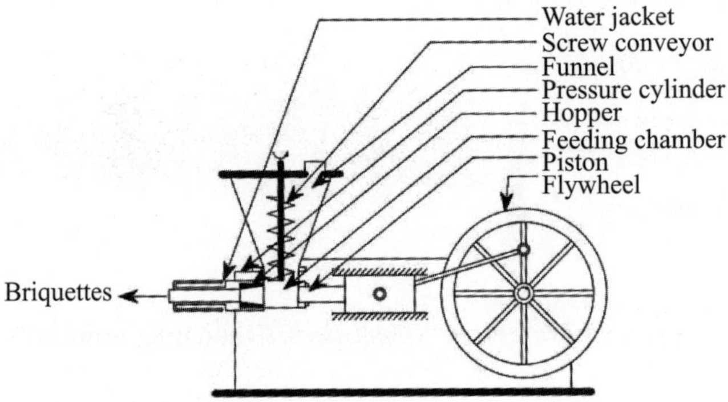

Fig. 15.3 Piston Press briquetting machine

15.3.2 Screw press densification

In the screw press briquetting process, biomass is continuously extruded by means of screws through a tapered die. The application of high pressure and the friction increases temperature of the feed stock. This elevated temperature leads to the release of lignin present in the biomass to act as binder. The briquettes obtained through this method has a hole in the centre and the outer surface is carbonised. This enables better air contact which aids better combustion. The screw press has a compaction ratio in the range 2.5:1 to 6:1. The briquettes formed by this process has a standard size of 60 mm diameter (Fig. 15.4).

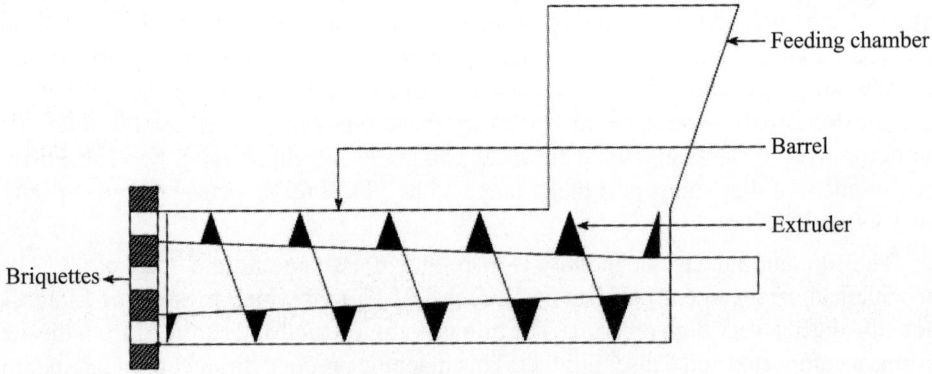

Fig. 15.4 Screw press briquetting machine

15.3.3 Roll press densification

A briquetting roller press consists of two rollers rotating in opposite directions. The feedstock falling in between the two rollers is compacted into pillow-shaped briquettes. Briquetting biomass using roll press requires a binder and they are suitable for carbonized biomass to produce charcoal briquettes (Fig. 15.5).

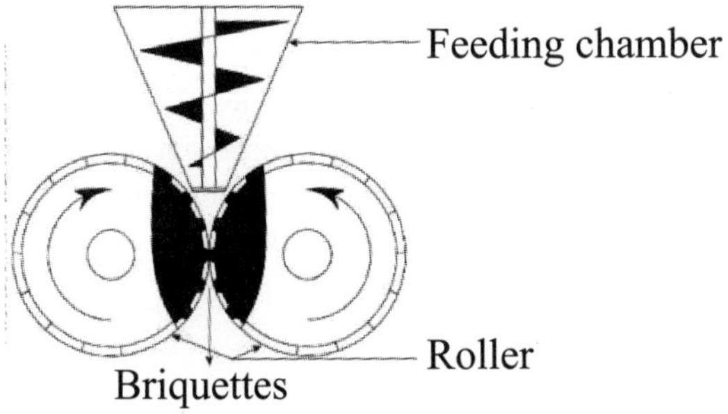

Fig. 15.5 Roll press briquetting machine

15.3.4 Low-pressure process

Low pressure compaction is usually done manually and hence also referred as manual briquetting process. There are different types of manual presses, which are specially designed for biomass briquetting. One example of such a press is the manual clay brick making press, which can also be used for both charcoal and biomass feedstocks. The low-pressure technology is suitable for undried plant raw materials like sugar cane residues. The high moisture feed materials can be shaped under low pressure in simple block presses or extrusion presses. The high-density briquettes thus formed should be sun dried before it can be used. The dried briquettes have little mechanical strength and are liable to crumbling easily. Low capital cost and operating cost along with the non-requirement of skilled operators are the main advantages of the low – pressure briquetting.

15.4 Operating Parameters in Briquetting

The factors affecting briquetting process of biomass are as follows:

(a) **Temperature and pressure:** Temperature at which the process take place influences the compression strength of densified biomass. Higher density of the product is usually observed at high temperature for a given pressure. The maximum strength of biomass pellets is generally obtained at a temperature around 220°C.

(b) **Moisture content:** Moisture content of the raw material is an important parameter as it facilitates heat transfer. The appropriate moisture content for the process is between 8 to 12 per cent, whereas high levels of moisture should be avoided as it may lead to steam formation resulting in an explosion when subjected to pressure.

(c) **Drying:** The drying of briquettes may depend on factors including initial moisture content, particle size, types of densifiers etc.

(d) **Particle size and size reduction:** The compaction process become easier with finer particle size as it provides a larger surface area for better bonding. In general, the size of raw material particles should be reduced by shredding to less than 25 per cent of the densified product. The size reduction is usually done with a hammer mill for most biomass, but materials like wood and straw may require chopping before the process.

15.5 Utilization of Briquettes

- Cooking and water heating in households.
- Heating in some agro-processes such as curing of tobacco, drying in tea factories etc.
- Providing heat in poultry houses.
- Firing ceramics and terracotta in pottery as well as in brick industry.
- Can be used in improved cooking stoves.
- Can be used as fuel for gasifiers.
- Powering steam boilers.
- As domestic fuel for fire places, cooking etc.

15.6 Limitations of Briquetting Process

- Only solid biomass materials are suitable for briquetting.
- It cannot be used as fuel for IC engines.
- Capital investment and maintenance cost for equipment.
- Requirement of covered storage facility for raw material as well as finished product.
- Combustion of biomass briquettes gives a maximum attainable temperature of about 1000°C only, which may not be sufficient for some industrial applications.
- The thermal energy availability per unit volume of briquettes is low in comparison to coal.
- The bonding properties of the briquettes are limited and cannot be increased beyond a certain limit.
- All types of organic wastes are not suitable for briquetting.
- The expertise required for installation and operation.

15.7 Pelletization of Biomass

Pelletization is a similar compacting process except that it uses smaller dies of approximately 30 mm, so that the small sized products are called *pellets*. Pellets are suitable as a fuel for a wide range of applications. As they are suited for automatic feeding, they find application in many industries where there are such requirements. It can also be used for domestic applications. The pelletizer consists of a number of dies bored into a thick steel ring or disc. The feed materials are forced into the dies by means of two or three rollers. This action softens the lignin in the raw material resulting in reshaping of the fibres into a pellet shape. Pellets are in general cylindrically shaped with variable length and a diameter ranging from 5 to 30 mm. The capacity of pelletizers can be in the order of 1000 kg per hour. The high initial capital investment and requirement of high energy input are the main drawbacks of pelletizing machines.

The pelletization process involves three phases as shown below:

Fig. 15.6 Pelletization Process

The pre-treatment process includes a number of preliminary steps such as raw material storage, removal of impurities, size reduction, material transportation and drying followed by mixing and pre conditioning. The size reduction of raw material to a uniform size is an essential pre-requirement. As the material need to pass through the die holes, the particle size should be smaller than the die hole to prevent blocking of the holes. Drying of feed material is an important step and the exact moisture content to produce quality pellets is specific to each type of raw material. As a general recommendation for good product quality, the moisture percentage of the raw material should be between 10 and 20 per cent. Sometimes mixers are used after the raw material has been shredded to proper size and dried to required moisture content. Mixing is done to get a consistently blended raw material. This is required when the raw material has significant variation in moisture content or their characteristics relevant in palletisation like the binding properties or material density. Mixing may also be required when heterogenous materials are used. Additional constituents also may be required after the raw material has been mixed to get sufficient binding quality. These additives may be helpful in increasing the output from the pelletiser or to improve the quality of the pellets produced. The post treatment operations are cooling and storage.

Types of pelletizers: Pelletizer machines are classified according to the shape of the die. There can be 'flat die' pelletisers commonly used in small and medium units and round die pelletisers, which are employed in medium to large-scale industries.

As the terminology indicate, flat die mills use a slotted flat die. The pre-processed raw material in powder form is introduced into the top of the die. When the die rotates, a roller presses the powder through the holes in the die, whereas a cutter on the other side of the die shears the pellet when it is extruded out from the die. The round or ring dies have radially aligned slots throughout. The die is positioned vertically and the powdery raw material is fed into the centre of the die. Rollers are used to extrude the pellets which get sheared by cutters.

CHAPTER 16

Thermo-Chemical Gasification and Pyrolysis of Biomass

The effective conversion of biomass to energy greatly depends on the selection of appropriate conversion technologies. The term *'Thermo-chemical conversion'* implies a process in which a chemical conversion is accomplished by the application of thermal energy. In the case of biomass, the different thermochemical energy conversion processes are combustion, gasification and pyrolysis. While combustion process is the oxidation of biomass to completely convert it then and there to energy, gasification and pyrolysis are processes which may produce gas, liquid or solid (biochar) outputs. In gasification, a combustible gas mixture is produced by a series of thermo-chemical reactions occurring in a partial supply of air.

16.1 Thermo-Chemical Gasification of Biomass

In the gasification process, carbonaceous materials like biomass are converted into a mixture of gases, called 'producer gas'. Biomass gasification is one of the commercially accepted technologies for sustainable and eco-friendly use of bioenergy. Producer gas in the range of 2 to 2.5 Nm^3 can be obtained from one kg of air dry (10 per cent moisture) biomass. The energy conversion efficiency of the biomass gasification process is in the range of 60 to 70 per cent. Producer gas is a mixture consisting of carbon monoxide, nitrogen and hydrogen in addition to carbon dioxide, methane and traces of other gaseous hydrocarbons. The output gas may also have impurities like tar, ash, char particles etc. The incombustible nitrogen in the air remains as such in the product gas. The typical composition of producer gas (by volume) is; N_2: 53 to 55 per cent, CO: 28 to 30 per cent, CO_2: 5 to 8 per cent and H_2: 10 to 12 per cent. The percentage composition of individual components can vary widely depending on the equipment, process parameters and feed stock. The presence of incombustible constituents viz. nitrogen and CO_2 have the effect of lowering the heating value and hence producer gas has a low heating value 5800 kJm^{-3}.

This gas may be used for direct thermal applications or to power gas turbines that are suited to low calorific value fuels. Usage of gasifiers instead of traditional direct combustion devices such as biomass stoves will lead to much savings of fuel. Even the improved biomass stoves have an efficiency in the order of 20 to 25 per cent only, whereas producer gas can be burned in gas burners with an efficiency of about 60 per cent. As the technological advantage of gasifier use for thermal applications have been proved, these systems are used widely. The saving in fuel along with the increased thermal efficiency have favourable effects on GHG mitigation.

16.2 Biomass Feed Stocks Suitable for Gasification and their Relevant Properties

A wide variety of crop residues and agro-industrial by-products as well as solid wastes like sugarcane bagasse can be subjected to gasification. Low moisture content biomass is generally considered suitable for gasification. Thermo-chemical as well as physical characteristics of the feedstock influence the quality of the output gas. Wood, charcoal and plant stalks with low moisture contents are the best choices for gasification. High moisture content of the feed stocks will cause energy loss in drying. Hence feed stocks of moisture content less than 30 per cent is preferable. Agricultural wastes like rice husk, sugar cane trash, sugar cane bagasse, groundnut shell, cotton stalk etc. were all used and are considered as good choices due to their desirable qualities. Pre-treatment of biomass feed stocks (shredding) is performed in some cases to improve conversion efficiency and gasifier performances.

Biomass properties important in gasification: Many biomass properties including physical and chemical properties affect the gasification process and are also relevant in the design of gasifiers. Among physical properties, moisture content, bulk density and particle size are important. Properties found from proximate analysis such as, fixed carbon, volatile matter, moisture content, ash content and gross calorific value are also relevant in design and operation of gasification equipment. Elemental composition of the feed stock (percentage of Carbon, Hydrogen, Oxygen, Nitrogen and Sulphur obtained from ultimate analysis) is also significant in gasification process.

16.3 Gasification Process

Gasification is a complex thermo-chemical process accomplished in a closed reactor vessel. The reactor is usually an air sealed chamber. The reaction chamber within which the thermo-chemical changes occur is called *'gasifier'*. Gasification takes place at high temperatures in the presence of a limited supply of air. But, a wide variety of processes have been developed over the years in which different gasifying agents like air, steam, nitrogen, carbon dioxide, oxygen or a combination of these may be used. The process of gasification is a combination of several overlapping steps which includes drying, pyrolysis, partial oxidation and reduction.

The typical gasification process may be summarized as:

CH_xO_y (biomass) + O_2 + $H_2O \rightarrow CH_4 + CO + CO_2 + H_2 + H_2O$ (unreacted water) + C (unconverted char) + tar + trace species. ..(16.1)

The process involves a number of steps or phases which are occurring in a continuous manner in the gasifier and are shown in Fig. 16.1.

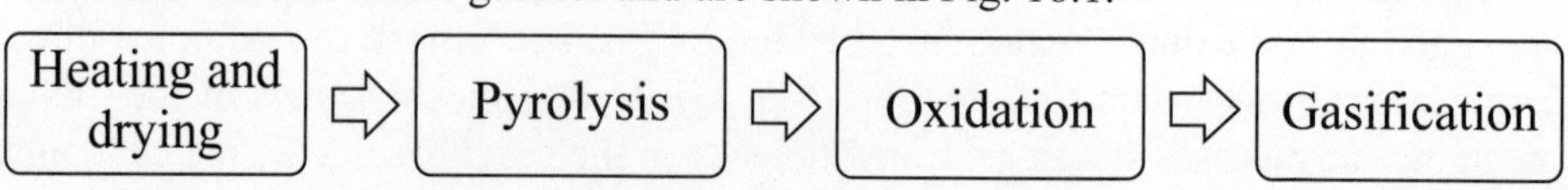

Fig. 16.1 Steps in gasification process

(a) *Drying:* Biomass used for gasification contains moisture in the range of 7 to 30 per cent and the removal of excess moisture from the feedstock is the first step. Drying occurs at temperatures in the range 100 to 125°C and the water vapour is also utilised in the chemical reactions which follow in subsequent steps.

(b) *Pyrolysis:* Pyrolysis refers to the thermal decomposition of biomass in an oxygen deficient environment. Biomass pyrolysis in a gasifier occurs at about 150 to 200°C and involves release of products in three different phases viz. solid, liquid and gas. The ratio of products is influenced by the chemical composition of biomass fuels as well as the operating conditions.

(c) *Oxidation:* Air introduced into the gasifier reaction zone contains oxygen and water vapour. The inert gases and nitrogen are not involved in chemical reactions. The exothermic oxidation reactions are required to raise the temperature of the reaction chamber so as to create a favourable condition for the reduction reactions. The carbon in the biomass is converted to carbon dioxide and hydrogen is converted to water vapor. These reactions take place at the temperature of 700 to 2000°C. Partial oxidation of char may also result in the formation of some CO.

The reactions at the oxidation zone are as follows:

$$C + O_2 \rightarrow CO_2 + \text{Heat } (+393.5 \text{ kJ}) \quad ...(16.2)$$

In the above reaction 12 g of carbon is oxidized with 32 g of oxygen in the air to yield 44 g of carbon dioxide and 393.5 kJ of heat.

Hydrogen in fuel reacts with oxygen in the air, producing steam. The heat of combustion is 286 kJ per mole.

$$H_2 + \frac{1}{2}(O_2) \rightarrow H_2O + \text{Heat } (+286 \text{ kJ}) \quad ...(16.3)$$

(d) *Reduction:* The reactions in which the combustible components of producer gas are evolved are reduction reactions. The principal reactions that take place in the reduction zone are given below:

(i) Boudouard reaction

$$CO_2 + C \rightarrow 2CO - 172.6 \text{ kJ/mole} \quad ...(16.4)$$

(ii) Water-gas reaction

$$C + H_2O \rightarrow CO + H_2 - 131.4 \text{ kJ/mole} \quad ...(16.5)$$

(iii) Water shift reaction

$$CO_2 + H_2 \rightarrow CO + H_2O + 41.2 \text{ kJ/mole} \quad ...(16.6)$$

(iv) Methane reaction

$$C + 2H_2 \rightarrow CH_4 + 75 \text{ kJ/mole} \quad ...(16.7)$$

16.4 Major Process Parameters for Gasification

The gasifier performance and product gas quality as well as the overall energy conversion efficiency is dependent on certain operational parameters. Temperature in different zones can be regarded as the most important parameter. Increased temperatures reduce tar content and there is a slight increase in gas due to cracking of tar into lighter gaseous components. The calorific value is dependent on the proportion of combustible constituents in producer gas. If CO_2 content in the gas increases, it results in lowering of calorific value. Hence the temperature profile should be such that the proportion of combustible constituents viz. CO and H_2 are maximized.

Equivalence ratio (ER) is another important parameter which determines the quantity of oxygen available for reactions. The producer gas composition is greatly influenced by equivalence ratio. Equivalence ratio is the ratio of the amount of oxygen fed into the gasifier to the amount of oxygen actually required for complete combustion of fuel. This means that if excess oxygen is introduced, more CO_2 will be produced which will adversely affect the calorific value of product gas. For effective gasification, the recommended ER is in the range 0.2 to 0.4, which means only 20 to 40 per cent of the oxygen (required for complete oxidation of fuel) is to be supplied.

The residence time of the feed material is another important aspect which decides gas quality and overall performance. Most biomass materials contain oxygen and hence sufficient residence time should be provided. Different gasifying agents like steam, CO_2 and nitrogen can be used which affect the process and the quality of the product gas. Catalysts are used in certain gasification processes and their choice is also relevant in the process. Superficial velocity of gas movement inside the gasifier chamber also affects the process. The pressure inside gasifier can either be equal to atmospheric pressure or can be higher. Higher pressure is preferred for certain applications. For thermal applications and small-scale gasifiers, atmospheric pressure is adequate.

16.5 Biomass Gasifiers

The gasifier is the reactor within which all the chemical reactions take place. Gasifiers are classified into three categories viz. fixed bed, fluidized bed and entrained flow gasifier. As biomass feedstocks differ much in their characteristics, the selection of appropriate type of gasifier is important for smooth operation and steady production of gas. Further, the choice of one type of gasifier over other is dictated by the quality of fuel required. Some applications require less tar or reduced ash content for which a particular type of gasifier will be suitable. In applications where the gas is directly burned for thermal energy, simple and low-cost gasifier designs may be appropriate. Some special types of biomasses like coir pith or rice husk may require specially designed gasifiers.

16.5.1 Fixed –bed gasifiers

In the fixed bed gasifier, the fuel is fed to the top of the gasifier and descents through the four different zones; drying, pyrolysis, oxidation and reduction. Conceptually, the gasification process in these gasifiers takes place in a fixed bed. The fixed bed gasifiers are simple in design, easy to operate with moderate energy conversion efficiencies. Another operational advantage is that they require minimum pre-treatment for feed materials. Hence, they are preferred over other advanced designs for small scale plants used for local applications. The general operation temperature of a fixed bed gasifier is about 1000°C. The product gas consists of 40 to 50% N_2, 15 to 20% H_2, 10 to 15% CO, 10 to 15% CO_2 and 3 to 5% CH_4, with a net calorific value of 4 to 6 $MJNm^{-3}$. The energy content of the product gas can be up to 75% of the biomass energy content. Some energy is lost in the production of CO_2. Other energy losses being attributable to the heat loss due to sensible heat in the output gas, other heat losses from the gasifier (radiation) and the heat content of the ash.

High tar content, ash content and low calorific value of the gas are regarded as the major drawbacks of fixed bed gasifiers. They can be of three types viz. updraft, downdraft and cross draft gasifiers based on the direction of flow of air inside the gasifier.

16.5.2 Updraft gasifier:

As the name indicate, the flow of air is in the upward direction. As the fuel moves down against the air flow, they are also called counter-current gasifiers. The gasifying agent is air which comes from the bottom of the gasifier and the gas generated leaves through a gas outlet located at the top. As shown in Fig. 16.2, the air first interacts with the char at the bottom and combustion reactions occur to produce H_2O and CO_2. The temperature of this zone rises to 1000°C due to the heat evolved in exothermic reactions. This heat helps to dry the incoming biomass from the top and also provide heat necessary for the breaking down of solid biomass in the pyrolysis phase. The hot gases move through the unreacted char, so as to actuate various reduction reactions to produce H_2 and CO.

As the product gas is withdrawn from the low temperature zone, cracking of tar may not happen and results in a high tar content in the product gas. Hence the gas from an updraft gasifier requires a cleaning process to remove tar content before its application as fuel in internal combustion (IC) engines. Production of electricity using IC engine driven generators is an important application of producer gas, especially in agro-industries where residues are gasified for in-situ energy production. As the tar cleaning require additional process and equipment, updraft gasifiers are considered to be not suitable for fuel gas production to be used in IC engines.

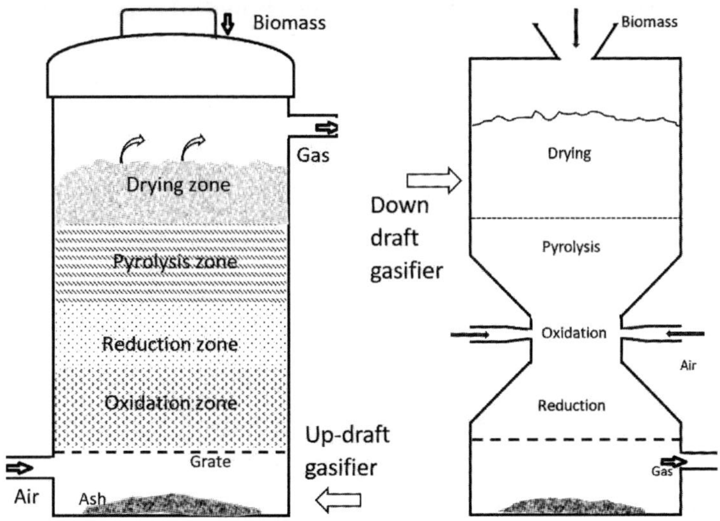

Fig. 16.2 Up-draft and Down draft gasifiers

16.5.3 Down draft gasifier

They are called down draft gasifiers as both the feed materials and air are introduced at the top of the gasifier. They are also called co-current gasifiers as the feed as well as air moves in the same direction. As shown in Fig. 16.2, this gasifier generally has a constriction called 'throat', for the purpose of controlling the movement of feed and also to have an even distribution of air through the fuel bed. The pyrolysis process mostly takes place at the throat area followed by combustion reactions at a temperature range of 1200 to 1400°C. The hot gases thus produced pass further downwards through the hot char bed. The reduction reactions occur at the bottom, where H_2 and CO are produced which leave the gasifier through the gas outlet at a temperature range of 900 to 1000°C. As the hot gas leaves through the bottom, this heat is not properly utilized for drying and pyrolysis. As a significant amount of heat is wasted through the exit of the hot gases, the overall energy efficiency of these gasifiers gets reduced. As the gas moves through the hot region of the gasifier before its exit, the tar gets cracked and hence, tar content in the product gas is less. In the case of updraft gasifier, the gas pass through the fuel bed which acts as a filter for ash and dust. But in downdraft gasifier, the gas directly goes out and hence has high particulate matter content.

16.5.4 Cross draft gasifier

In cross draft or cross flow gasifiers the movement of air and gas are across the gasifier body. The feed material moves downwards while air is introduced from the side. The product gas is withdrawn from the opposite side of the gasifier vessel at the same level where air enters the gasifier chamber. The drying and pyrolysis zones are located at the top of the gasifier vessel. The combustion zone is formed at the region where air is introduced to the reaction chamber. The reduction zone is located in a region adjacent to the gas outlet point. The disadvantages of down draft gasifier are applicable to this type also as the temperature of the gas leaving the system is about 800 to 900°C. The overall energy conversion efficiency for the process is low and the gas has higher tar content (Fig. 16.3).

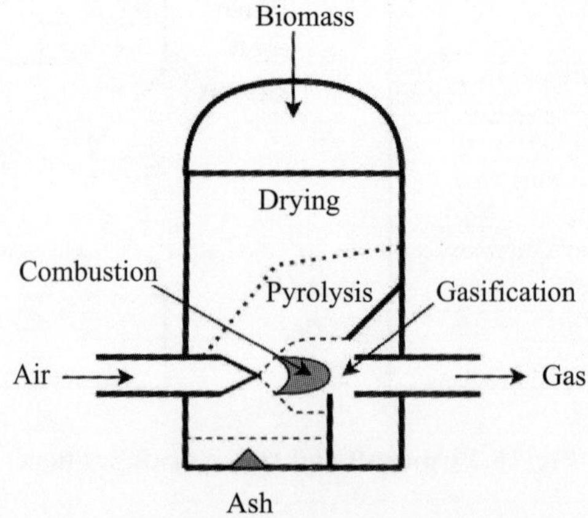

Fig. 16.3 Cross draft gasifier

16.5.5 Fluidised bed gasifiers

Fluidised bed gasifiers are advanced type of gasifiers in which the feed material along with the granular carrier undergoes fluidization in the gasifier chamber. They have been extensively used for gasification of coal for a long time. Due to their high efficiency, industrial scale and commercial biomass gasification systems utilize them if the feed stock is suitable for the process. Compared to the fixed bed gasifiers they have the advantage that a uniform temperature distribution can be achieved throughout the gasification zone. This is achieved by creating a fluidised bed of fine-grained natural or synthetic carrier particles (ceramic or silica) in the gasifier chamber into which the air stream is introduced from bottom. Air is injected at the bottom of the reactor vessel through a distributor comprising of a perforated plate (grate) and nozzles so that the carrier material is suspended in a fluid-like state. Biomass feed is processed to the required particle size and air (or steam) streams from the bottom of the chamber forcing the fuel particles upward through the fluidized bed of carrier particles as there is a high velocity gas stream. Pyrolysis and gasification reactions occur as the feed material moves upward through the hot bed. The gas produced from fluidized bed gasifiers are likely to have high particulate matter along with a medium tar content (about 10 g per Nm^3).

Three types of fluidised bed gasifiers are common: (a) Bubbling fluidised bed (b) Circulating fluidised bed and (c) Dual fluidized bed.

(a) *Bubbling Fluidised Bed:* Bubbling fluidized bed gasifier (Fig. 16.4) has a grate at the bottom of the gasifier chamber through which air is introduced. The biomass feed material is properly processed and is introduced to the fluidized bed of granular carrier, kept above the grate. The air enrry is controlled so as to regulate the temperature inside the reaction chamber within the optimum range. The gasification products are char and gas along with some tar. As some amount of tar may stick to the surface of the carrier particles and subsequently get cracked, the tar content is low for the product gas ($< 1\text{-}3\ gNm^{-3}$) (Fig. 16.4).

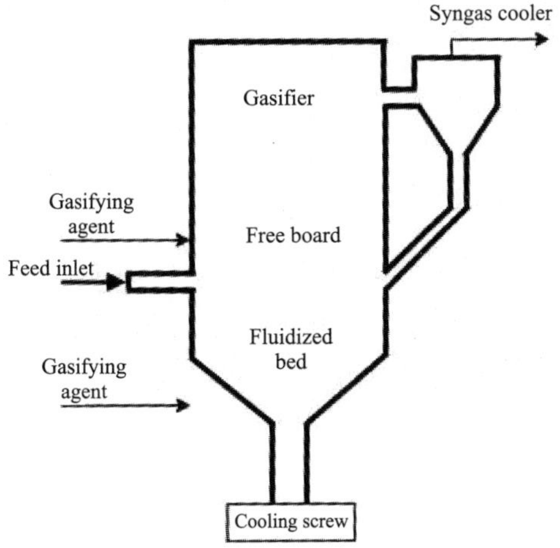

Fig. 16.4 Configuration of bubbling fluidized bed gasifier

(b) *Circulating Fluidised Bed:* In this type of fluidized bed gasifier, the carrier particles are carried over from the gasifier chamber through the gas outlet which in turn is recaptured with a cyclone separator (Fig. 16.5). The carrier material is circulated between the gasifier chamber and the cyclone separator and hence the name. The carrier material moves along with the feed and leaves the reaction chamber to be separated and returned again to the chamber. In the cyclone separator, ash is removed and the carrier particles along with the char is returned back to the gasifier vessel whereas the gas produced is taken out. The operation pressure in the CFB gasifier is higher than the atmospheric pressure and this is an advantage in some cases, where compressed gas is the requirement as in the case of a gas turbine. As the system is more sophisticated the installation and operational expenses are comparatively higher than fixed-bed gasifiers. The tar content of the gas is higher than that of the fixed bed types with a similar particulate content.

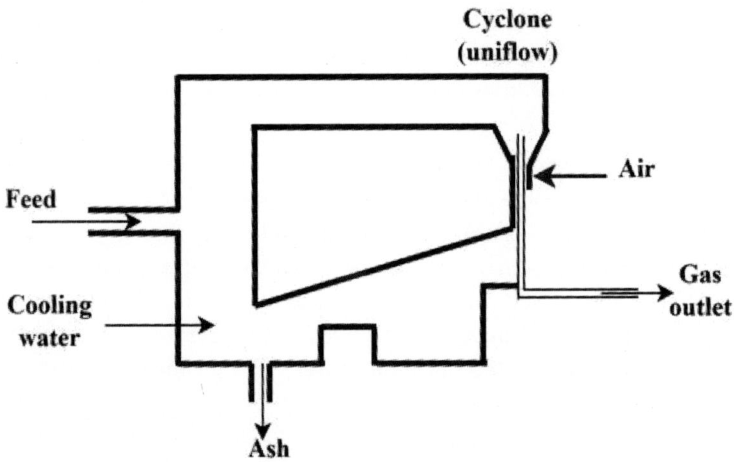

Fig. 16.5 Configuration of circulating fluidized bed gasifier

(c) *Dual fluidised bed gasifier:* The dual-fluidized bed gasifier consists of two separate fluidized beds, one for pyrolysis and the other for gasification reactions (Fig. 16.6). The biomass fed into the base of the pyrolysis reactor is usually fluidized by an upward stream of steam. The second bed acts as a combustion zone where char is burned with air stream creating a fast-fluidized bed. The combustion reactions elevate the temperature of the bed material to around 850 to 900°C. The fluidised bed in the first reactor (pyrolysis) is heated with an inert material like sand using heat from the second reactor. The carrier or bed material acts as a medium for heat transfer between the beds.

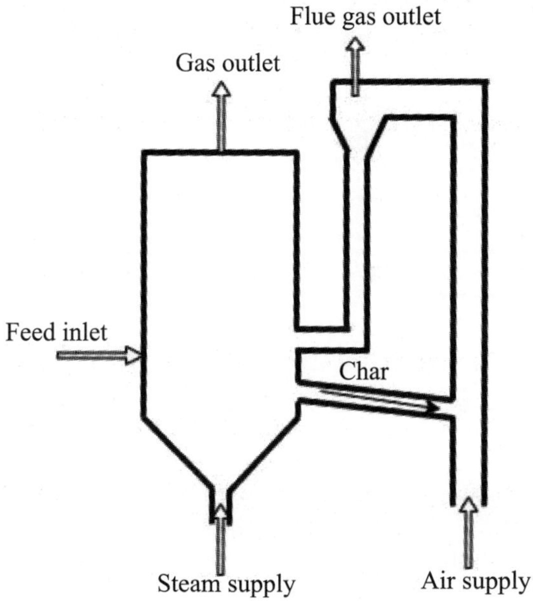

Fig. 16.6 Schematic diagram of dual fluidized bed gasifier

(d) *Entrained bed gasifiers:* Feed materials and air are fed concurrently in entrained bed gasifiers and the residence time is short. The system is operated at a high pressure and temperature. As the operation temperature is higher than the ash melting temperature, a slag is formed. Entrained bed gasifiers are generally employed for commercial coal gasification. These types of gasifiers require finely graded feed materials to ensure proper operation of the gasifier system. The problems with application of entrained bed gasifiers for biomass is not restricted to preparation of finely pulverized and graded material but also owing to the adverse biomass properties. The biomass is cohesive in nature with poor flowability, which may lead to problems associated with continuous feeding and adequate mixing.

16.6 Pyrolysis of Biomass

The word 'pyrolysis' has a Greek origin in which '*Pyro*' means 'heat' and '*lysis*' means 'rupture'. Biomass pyrolysis is defined as the thermal decomposition of biomass materials in the absence of oxygen and is also referred as destructive distillation. Pyrolysis is a term often used in a wider context and hence difficult to be defined precisely when applied to biomass. The process of thermal decomposition of biomass in an atmosphere where significantly less oxygen is present (much lower than required for gasification) is also termed as pyrolysis. In the gasification process, the interest is on production of a fuel gas, whereas the aim of pyrolysis is often production of liquids, either for fuel or more often for other applications (Fig. 16.7).

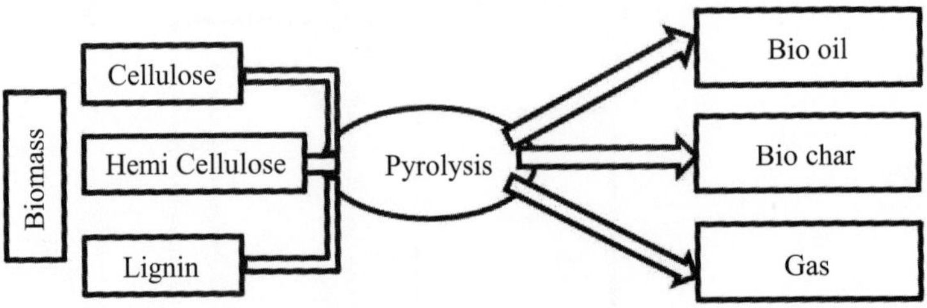

Fig. 16.7 Biomass pyrolysis process

Lignocellulosic biomass can be broken down and converted to bio-oil (a condensable liquid), solid char and a mixture of gaseous products comprising primarily of CO_2, CO, H_2 and CH_4. During the initial stage of biomass pyrolysis, the organic compounds will undergo thermal decomposition in the absence of oxygen. This thermal decomposition will start at a temperature range of 350 to 550°C and will continue up to 700 to 800°C. This reaction at higher temperature releases volatiles from biomass resulting in the formation of char. Heat transfer occurs between the hot volatiles and cooler solids in the unreacted fuel. Tar will be formed as a result of condensation of some of the heavy components in volatiles. There can be primary as well as secondary pyrolysis reactions which will occur simultaneously. The reactions include thermal decomposition, reforming, water gas shift reaction, recombination of free radicals etc. These thermochemical reactions depend on residence time as well as the operation temperature and pressure. As biomass materials vary in their physico-chemical properties, the kinetics of the reactions are very complex. Studies has already indicated that biomass co-pyrolysis with other feedstocks like waste plastic and discarded rubber tires of automobiles can not only enhance physico-chemical properties of pyrolysis products but is also effective in recycling of such wastes.

Biomass feed stocks suitable for pyrolysis: The biomass residues suitable as feed stocks for pyrolysis are rice straw, rice hulls, peanut hulls, oat hulls, wheat straw, sugar cane bagasse and wood. Freshly collected crop residues may have moisture content as high as 60 per cent. This moisture level shall preferably be reduced to below 30% by sun drying or passive drying during storage. Active drying using waste heat exchange or silo drying can reduce the moisture content to even much lower levels.

Ligno-cellulosic biomass contains cellulose, hemicellulose and lignin. The materials with higher hemicellulose content are best suited for pyrolysis as the hemicellulose component can easily be subjected to pyrolysis. Cellulose is the next easier component whereas lignin is not easily subjected to pyrolysis. It has been observed that cellulose pyrolysis around 500°C yields bio-oil as the liquid component and lignin yields bio-char as the solid residue.

16.7 Different Biomass Pyrolysis Processes

The pyrolysis can be classified into three based on the operating conditions viz., slow, fast, and flash pyrolysis. Table 16.1 shows the operating parameters and yield of different types of pyrolysis.

Table 16.1 Typical operating parameters and products for pyrolysis process

Pyrolysis process	Solid residence time (s)	Heating rate ($K.s^{-1}$)	Particle size (mm)	Temp. (K)	Product yield (%)		
					Oil	**Char**	**Gas**
Slow	450 – 550	0.1 – 1	5 – 50	550 – 950	30	35	35
Fast	0.5 – 10	10 – 200	<1	850 – 1250	50	20	30
Flash	< 0.5	>1000	<0.2	1050 – 1300	75	12	13
(Source: Jahirul *et al.*, 2012)							

(a) *Slow pyrolysis:* As the terminology indicates, slow pyrolysis is performed at slow heating rates and the operation temperature in the reactor vessel is also less than 500°C. The solid residence time as well as the vapour residence time are also high (5 to 30 minutes). The product components in the vapour phase further react with each other at lower temperatures resulting in the formation of solid char and liquid products. Slow pyrolysis is not recommended for good quality bio-oil production due to its technological limitations. The prolonged residence time may cause cracking of the primary products which can lower the yield of bio-oil and affect its quality adversely. Another disadvantage of the process is consumption of more energy due to the long residence time and low heat transfer rate.

(b) *Fast pyrolysis:* Fast process is preferred when the objective of pyrolysis is production of liquid fuel. In this popular process, the biomass is heated rapidly by providing a high heating rate in the absence of air. The proportion of different components in the products may vary depending on the feed stock. Typical values of components in products of fast pyrolysis can be 60 to 75 per cent of oily products, 15 to 25 per cent of solids (biochar) and 10 to 20 per cent (weight basis) of gaseous products. Enhanced production of liquids can be achieved by providing a low reaction temperature, high heating rate and short residence time.

The characteristics of fast pyrolysis are:

(i) High heating rate
(ii) Rapid cooling of vapours
(iii) Rapid cooling of aerosols
(iv) High heat transfer rate
(v) Very short vapour residence time
(vi) Precision control of reaction temperature

(c) *Flash pyrolysis:* Rapid devolatilization of biomass is the characteristic feature of flash pyrolysis which takes place in an inert atmosphere. High reaction temperatures in the range of 450 to 1000°C with a high heating rate up to 3000°C per second is adopted in the process in which the gas residence time is also very short. Fast pyrolysis can produce solid, liquid and gaseous products with a bio-oil yield up to 75 per cent.

16.8 Charcoal Production

The carbonisation of wood by heating lead to the formation of charcoal. This is also referred often as wood pyrolysis. Traditionally charcoal is made in kilns with an overall yield of about 20 per cent by weight. When wood is heated, volatiles escape out from wood leaving behind fixed carbon. During the carbonization process, browning of wood appears when the temperature reaches 220° and the wood become deep brown-black colour at about 280°. Low grade charcoal can be made at temperatures starting from 300° which is soft and starts burning at 380°. High grade charcoal is harder and brittle which is obtained at higher temperatures. High grade charcoal has a high ignition temperature in the order of 600 to 700° (Table 16.2).

Table 16.2 Comparison between charcoal and coal

Coal	Charcoal
• Calorific value is 20 to 25 $MJ \cdot kg^{-1}$	• Calorific value is 30 to 32 $MJ \cdot kg^{-1}$
• Non-renewable fuel	• Renewable fuel
• It is formed naturally from decayed plant and animal matter beneath the earth crust by the action of pressure and heat for millions of years	• It is formed by the carbonization of wood or organic matter in the absence of oxygen.
• Mineral	• Organic in nature
• The process of formation takes a very long time in the order of millions of years	• It can be produced in a very short time
• Centralised distribution from coal mines	• Decentralised production is possible

Traditional charcoal kiln: In traditional earth kiln, wood is arranged radially on a platform in a circular fashion about 1 to 1.5 m high. The wood is stacked in a pyramid shape and then covered with mud so as to form a closed chamber. A chimney is sometimes fixed at the kiln to facilitate draught. After covering the entire lot with vegetation and soil, firing is done from the top. There can be regional variation in the construction of traditional charcoal kilns. Charcoal is a good solid fuel with better burning characteristics compared to wood. It does not produce smoke and can replace coal. Traditional charcoal making results in wastage of energy through the escape of unutilized volatiles and is not an environment friendly technology. Modified versions of charcoal making equipment with provision for gas collection and utilization can be more efficient and environment friendly (Fig. 16.8).

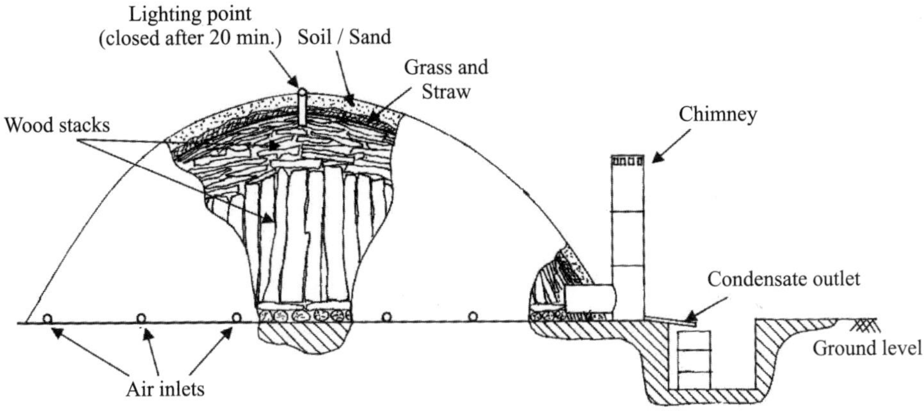

Fig. 16.8 Traditional charcoal kiln

16.9 Environmental Aspects of Charcoal Making

Charcoal production in tropical regions of the world especially in less developed countries has a drastic effect on ecology and environment. In the light of environmental concerns, governments, public and non-government organizations are discouraging the felling of trees for charcoal production. The most commonly noticed adverse impact is deforestation as forest woodlands are cleared unethically in many parts of the world. This ultimately results in complete destruction of forest ecosystem due to degradation of forests leading to deforestation. The main reason for elevated forest degradation is the fact that much of the charcoal in tropical countries is made in traditional earth and pit kilns with a low wood-to-charcoal conversion rate of about 20 per cent. Emissions of GHGs due to charcoal production in tropical ecosystems in 2009 was estimated to be above 70 million tonnes of carbon dioxide and 1.3 million tonnes of methane. Hence, proper policy interventions as well as technology interventions are required to address the issues of environment degradation and GHG emission due to charcoal making. It may not be possible to prevent charcoal making in many countries due to socio economic reasons.

CHAPTER 17

Biodiesel

Liquid fuels are dominating the transport sector and among the liquid fuels, diesel fuel plays a prominent role. This increased preference is due to the fact that diesel engines have 25 to 30 per cent better fuel economy owing to their higher thermal efficiency compared to gasoline (petrol) fuelled spark-ignition engines. Moreover, diesel fuel is much safer to store and contains more usable energy than petrol. Due to this reason, diesel engine powered cars, trucks, generators and heavy machinery lead the race. Even though diesel engines have higher thermal efficiencies, the possibilities of incomplete combustion are there leading to higher smoke production. The NO_x emission also is higher in diesel engines. Even though there are environmental concerns associated with the use of diesel engines, the transport system in many countries depend mostly on diesel engine powered automobiles.

17.1 Need for Alternative Fuel for Diesel Engines

The global demand for petroleum products was about 100 million barrels per day in 2019, in which the leading product was diesel fuel (28.4 million barrel per day). Taking into account the number and size of diesel engines in use they are creating much concern especially in thickly populated cities. In New Delhi, diesel engines are not allowed in public transport system. Similar restrictions are being imposed on diesel engines owing to the health hazards associated with vehicular pollution. In addition to the environmental concerns, the global fossil fuel reserve is limited and if the current consumption prevails, it is estimated that the global oil reserve will finish in about 50 years. Alcohol has severe limitations for its use in diesel engines. This clearly explains the need for an alternative renewable fuel to substitute diesel fuel. Taking into account the wide usage of diesel fuelled engines, developing a fuel that needs a new engine type or major engine modification may not be economically feasible. It is better to have a renewable and environment friendly fuel suitable for diesel engine which does not necessitate major engine modifications.

17.2 Vegetable Oil as Fuel for Diesel Engine

When diesel engine was exhibited in 1900 by Rudolf Diesel the German inventor at Paris, he used peanut oil as fuel. But the modern diesel engines have been refined during the last century to a great extent so as to suit the properties of petroleum derived diesel fuel. Today's diesel engines are hence much different from the original engine developed by Rudolf Diesel, which use the petroleum derived fuel. The petroleum derived diesel fuel is a mixture of hydrocarbons. The ideal diesel fuel for modern diesel engines may be regarded as a hydrocarbon having a carbon chain length in the range 12 to 18.

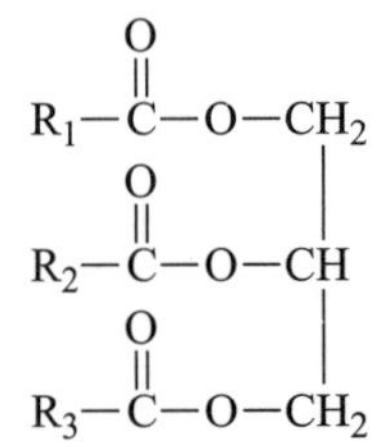

Fig. 17.1 Chemical structure of triglycerides
(R_1, R_2 and R_3 are hydrocarbon chains of fatty acids)

As vegetable oils are renewable and carbon neutral, the major environmental concerns in the use of fossil-based diesel fuel are eliminated to a greater extent. The fuel properties of few commonly available vegetable oils are given in Table 17.1. The problem associated with the use of vegetable oil as a substitute to diesel engine fuel is its poor atomization quality due to high viscosity, incomplete combustion due to lower volatility and problems associated due to carbon deposit on the fuel injector and valve seats. Most vegetable oils have a low cetane number. These adverse properties are mainly due to their complex molecular structures (Fig. 17.1). They are comprised of triglyceride esters having a mixture of saturated, unsaturated or cyclic hydrocarbons with oxygenated components like carboxylic acids, ketones, aldehydes and alcohols. Hence, to use vegetable oil as fuel in diesel engines, either the engine has to be modified to suit the fuel properties of vegetable oil or the vegetable oil has to be modified in a way to conform to the ideal diesel fuel properties (similar to petroleum derived diesel).

If we want to use vegetable oils as fuel the following arrangements may be required:

(i) The fuel is injected into the diesel (CI) engine cylinder at very high pressures so as to break it into fine droplets. As the viscosity of vegetable oils are higher than that of diesel fuel, a desirable level of atomization cannot be achieved. But for most vegetable oils, the viscosity gets reduced at elevated temperatures. Hence heated fuel lines may be provided to increase the temperature of oil so as to reduce the viscosity.

(ii) Additional filters may be required in the fuel lines to remove larger particles in the fuel stream, as vegetable oils are likely to get solidified and form solid lumps during cold conditions. The problem can be aggravated especially while starting in the morning.

(iii) To reduce the effect of deposits inside the engine cylinder and easy starting, it is better to start and stop the engine using diesel fuel. This necessitates starting the engine with diesel fuel and switchover to vegetable oil after about 10 to 15 minutes when the engine gets heated up. Similarly switching over to diesel 10 to 15 minutes before stopping is also recommended so as to minimize the chances of deposits formed with vegetable oil. This requires two fuel tanks and a fuel switchover system.

The possible ways to modify the vegetable oil so as to avoid or minimise the engine modifications include subjecting vegetable oils to pyrolysis, micro-emulsification and transesterification. Pyrolysis or thermal cracking of vegetable oils results in a diesel like fuel but is a complicated process, which includes dehydration, decarboxylation, hydrolysis, polymerization, aromatization, dehydrogenation, etc. Micro-emulsification is a better approach than thermal cracking. It is effective in reducing the viscosity and emissions of vegetable oils, but the problems associated with carbon deposition and incomplete combustion prevails.

Transesterification of vegetable oil is considered as a better fuel modification approach. The products of transesterification are a mixture of simple esters called as 'Biodiesel' and perform similar to diesel fuel. Though the process of transesterification was known much earlier, the increased interest to use this technology for production of alternative diesel fuel was seen during the current century.

Table 17.1 Fuel properties of few vegetable oils

Vegetable oil	Kin. Viscosity ($mm^2 \cdot s^{-1}$, at 40°C)	Density ($g \cdot cm^{-3}$, at 21°C)	Cetane number	Flash point (°C)	Cloud point (°C)	Pour point (°C)
Soybean	33.1	0.914	38.1	254	-3.9	-12.2
Rapeseed	37.3	0.912	37.5	246	-3.9	-31.7
Sunflower	34.4	0.916	36.7	274	7.2	-15.0
Corn	35.1	0.91	37.5	277	-1.1	-40.0
Safflower	31.6	0.914	36.7	246	-3.9	-31.7
Cottonseed	33.7	0.915	33.7	234	1.7	-15.0
Peanut	40	0.903	34.6	271	12.8	-6.7

(Source: Canakci and Sanli, 2008)

17.3 Biodiesel

Biodiesel became a popular diesel fuel alternative as the present-day diesel engines could be run with this vegetable oil derivative without much modifications. The chemical process required to modify vegetable oils was very simple and did not require sophisticated technology. Many people even produced biodiesel at domestic levels and used in their automobiles. The popularity of biodiesel increased as the production process was very simple, reliable, cost effective and could produce high quality fuel using locally available renewable feed stock. The chemical ingredients were also low cost and easily available.

Biodiesel is the common name given to diesel engine compatible fuel produced from vegetable oils or animal fats. Chemically they are mono-alkyl esters of fatty acids having long chains. These esters are produced when triglycerides in natural oil and fats are subjected to transesterification process using alcohols such as CH_3OH or C_2H_5OH in presence of a catalyst such as NaOH and KOH. The alcohol breaks the triglyceride chain so that three simple esters are produced as shown in Fig. 17.2. R_1,

R_2 and R_3 represent the hydrocarbon chains of the triglyceride. R' is the hydrocarbon part of the alcohol. The chemical composition of the biodiesel produced depends on the chemical structure and composition of the triglycerides in the oil (or fat) used as feedstock. The carbon chain length of these mono-alkyl esters varies between 8 to 24. Even the same feedstock collected at two different seasons may not have the same composition as the fatty acid composition may change with respect to climatic conditions, soil conditions and agricultural practices. As the change in feed stock composition affects the composition of biodiesel as well, there can be differences in its physical and chemical properties.

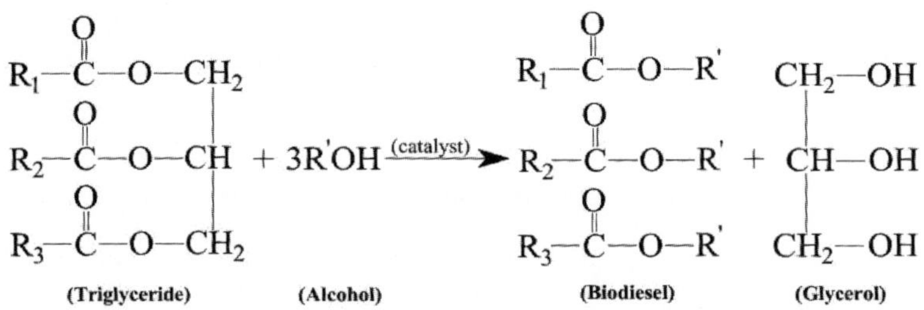

Fig. 17.2 Chemical reaction involved in the transesterification process of triglycerides

Stoichiometrically one mole of triglyceride needs three moles of alcohol to detach glycerol as seen from the chemical structure of the triglyceride. It should be noted that the reaction is reversible and to push the equilibrium towards the product side (biodiesel + glycerol), excess amount of alcohol (quantity varies depending on the feedstock material used) is used. Studies have shown that methanol is a very effective alcohol and NaOH is an effective catalyst for the transesterification process.

17.4 Biodiesel Production Process

Biodiesel can be prepared either using a batch, semi-continuous or continuous system. Batch and semi-continuous systems are used for preparing small quantities especially for lab-scale production. Continuous system is adopted for large commercial scale production. According to the feedstock used for biodiesel production there will be some changes in the procedure to be adopted and quantity of alcohol and catalyst used, but the process remains the same. The general operations required for biodiesel production is described in Fig. 17.3. Proper safety precautions have to be taken while preparing biodiesel as both the alcohol and catalyst used in the preparation are harmful to human body. The alcohol has to be mixed with the catalyst, added to the feedstock maintained at the optimum temperature range and continuously stirred for 20 to 30 minutes. The mixture is allowed to settle for 12 hours so that the glycerol and soap (formed if any) will be settled at the bottom. The biodiesel remaining at the top layer has to be transferred to a different vessel and washed (agitated with equal volume of water) to remove dissolved impurities. After each wash, the mixture is allowed to settle and the biodiesel removed for the next wash. The pH of the wash water has to be checked after separating biodiesel and washing has to be continued till the pH of the wash water is almost equal to the initial pH. Because of washing, the biodiesel is likely to contain moisture, which needs to be removed by heating or by using other dehydration techniques.

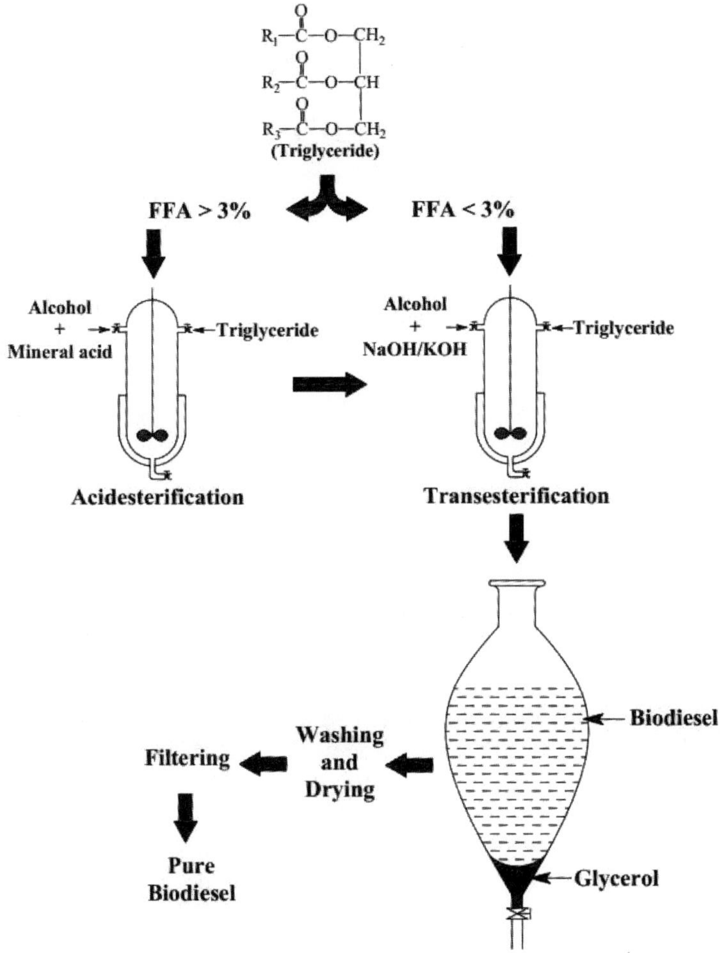

Fig. 17.3 Process flow diagram of biodiesel production process

17.5 Parameters that Influence Biodiesel Production

Globally about 350 vegetable oils are considered as potential biodiesel feedstocks. From most of these materials we can get up to 98 per cent biodiesel yield under optimum conditions. The quantity and quality of the biodiesel produced are dependant of the following parameters that influence the transesterification process:

(a) *Percentage of FFA in the feedstock:* The percentage of 'Free Fatty Acids (FFA)' in the feedstock selected is very crucial in deciding what process has to be adopted in biodiesel production. The FFA reacts with the catalyst used (NaOH or KOH) and can form soap. To get better biodiesel yield through transesterification process, it is advised to use feedstocks having FFA content below 3 per cent. The acid esterification process can reduce the FFA content to the required levels for the transesterification process. Feedstocks having FFA content above 3 per cent have to be subjected to acid esterification (pre-esterification process) with mineral acid as the catalyst (H_2SO_4 is the most common), followed by the usual transesterification process.

(b) *Moisture content in the reagents used:* Water content in the feed stock is one of the key factors that has to be monitored similar to FFA. The ingredients used for the production has to be dry, or else it will increase the quantity of soap produced and decrease biodiesel yield. Soap and glycerol are formed when triglycerides react with aqueous NaOH or KOH. The formation of soap cannot be completely avoided during the process, but can be reduced by using dry ingredients.

(c) *Temperature:* Temperature maintained during the reaction process is one of the major factors that decides the quantity and quality of biodiesel yield. The optimum temperature range for the reaction depends on the feedstock selected, e.g., for sunflower oil 60 to 70°C, palm oil 50 to 65°C, coconut oil 50 to 60°C, etc. As the temperature increases, the time required for reaction decreases. Increasing the temperature beyond the optimum range will increase the side reactions and will result in the thermal decomposition of the alcohol esters of fatty acids produced. Higher temperature results in increased vaporisation of the alcohol used, reducing the availability of alcohol for the reaction. Combined effect of these issues decreases the quantity and quality of biodiesel produced. It is generally advised not to maintain the temperature above 70°C and below 40°C for most feed stocks.

(d) *Ratio of triglyceride to alcohol:* An excess amount of alcohol is needed to shift the reaction equilibrium towards the product and the triglyceride to alcohol ratio often affects the quantity of biodiesel produced. The ratio depends on the catalyst and process used. While using the alkali catalyst (NaOH and KOH), it is advised to have a triglyceride to alcohol ratio of 6:1. Adding alcohol beyond this requirement will not influence the reaction, but will increase the cost of production.

(e) *Ratio of triglyceride to catalyst:* Biodiesel production is strongly affected by the catalyst used (like NaOH for transesterification and H_2SO_4 for acid esterification) and its concentration. Increasing the catalyst quantity beyond the optimum level is not economical and may increase the viscosity of the biodiesel produced, creating difficulty in separation.

(f) *Stirring time as well as speed:* Continuous stirring is very vital to increase the contact between the ingredients and there by speed up the reaction process. Continuous stirring for a period of 20 to 30 minutes with an rpm ranging between 400 to 600 is recommended for faster reaction and also helpful in preventing the reversible reaction.

(g) *Reaction time:* Transesterification is a slow process and sufficient time has to be given to complete the reaction. Studies have shown that up to 30 minutes, the rate of biodiesel production increases and there after it may decrease. There can be slight variation with respect to different feedstocks. In general, it is advised to provide a reaction time between 30 to 60 minutes for higher biodiesel yield.

(h) *Purification process:* The extent of washing and drying determines the quality of biodiesel and its shelf life. The biodiesel has to be washed with sufficient quantity of water such that all the dissolved impurities such as soap, glycerol etc. are removed. It is better to wash the biodiesel with water in the ratio 1:1 at least 3 to 5 times. After washing, the moisture content in the biodiesel has to be removed and stored dry.

17.6 Fuel Properties of Biodiesel

The fuel properties of biodiesel changes with the feedstock used and the method adopted for its preparation. A comparison of important fuel properties of few biodiesels with petroleum diesel fuel with respect to its usage in CI engines is given in Table 17.2.

Most international standards prescribe certain limits or ranges for important parameters relevant for biodiesel to be used as a fuel. There are about 11 essential parameters as given below:

(a) *Flash Point:* It is the temperature at which sufficient vapour is produced by the fuel so that it can ignite. The flash point of biodiesels is much higher than that of diesel fuel, implying that biodiesel is less flammable than diesel and are safer for storage.

(b) *Water and Sediment Content:* Presence of water content reduces the heat value of fuel and can cause corrosion to engine parts. Though biodiesel and diesel are insoluble in water, biodiesels have more affinity to water than diesel. On storage, biodiesel may produce more sediments compared to petroleum diesel.

Table 17.2 Properties of biodiesel in comparison to diesel fuel

Fuel	Kin. viscosity ($mm^2 \cdot s^{-1}$, at 40°C)	Density ($g.cm^{-3}$, at 21°C)	Cetane number	Flash point (°C)	Cloud point (°C)	Pour point (°C)
Diesel fuel	2.0 – 4.5	0.820 – 0.860	51	55	−18	−25
Soybean biodiesel	4.08	0.884	50.9	131	−0.5	−4
Rapeseed biodiesel	4.83	0.882	52.9	155	−4	−10.8
Sunflower biodiesel	4.6	0.88	49	183	1	−7
Tallow biodiesel	5	0.877	58.8	150	12	9
Yellow grease biodiesel	5.16	0.873	62.6	–	9	12

(Source: Canakci and Sanli, 2008)

(c) *Kinematic Viscosity:* It is the resistance to flow for a fluid and generally depends on the molecular weight of the chemical components of the fluid. This is a very important property for a diesel engine fuel as it determines the degree of atomization of the fuel when injected into the engine cylinder. The viscosities of biodiesels are generally at par with the petroleum diesel and do not create much problem with engine operation.

(d) *Sulphated Ash Content:* This parameter is concerned with the operating life of the engine parts as it is a measure of the inorganic contaminants. Improper processing of biodiesels can lead to the presence of catalyst residues, soaps etc. The ash produced from these components in the engine cylinder may act as abrasive particles aggravating wear of components thereby affecting engine life.

(e) *Sulphur Content:* The quantity of Sulphur in biodiesels have to be controlled for maintaining the emission standards with regard to particulate matter. Animal fat-based biodiesels have very low levels of Sulphur.

(f) *Cetane Number:* The fuel needs to be self-ignited inside the engine cylinder in a compression ignition engine. The self-ignition ability of the fuel is given by 'cetane number' and hence, is a very important property for a diesel engine fuel. Generally, biodiesels have a higher cetane number than the regular petroleum-diesel.

(g) *Carbon Residue:* Carbon residues deposited inside the engines as a result of fuel combustion can cause serious damages to the engine and thereby affect the engine life. Biodiesels are oxygenated fuels and burn better than the diesel fuel leaving much lesser residues than petroleum-diesel.

(h) *Acid Number:* This indicates the quantity of acids present in the biodiesel. Keeping the acid content within the prescribed limits is highly important for the life of the engines as it can corrode the engine parts. Even though the quantity of acids in biodiesels vary with the feedstock used for its production, the acid values of common biodiesels are below the prescribed limits.

(i) *Free and Total Glycerin:* This parameter explains the effectiveness of the conversion and purification process adopted, there by indicating the quality of the biodiesel produced. The amount of free glycerin is a measure of the ineffectiveness of the washing/purification process. The amount of total glycerin is an indication of unconverted glycerides in the biodiesel prepared.

(j) *Phosphorus, Calcium and Magnesium Content:* These are minor components associated with phospholipids and gums. The presence of these elements above the prescribed limit will enhance emulsification or sedimentation in fuels, which can cause problems in the operation of engines.

(k) *Oxidative Stability:* Biodiesels are susceptible to oxidation, which is considered as one of the major drawbacks concerning its shelf life. Oxidation of biodiesel induces polymerization and formation of gums and sediments affecting the effective operation of the engine.

The permissible limits of 11 essential properties of biodiesel as per different international standards is given in Table 17.3.

Table 17.3 Different international standards for important biodiesel parameters

Property	**Unit**	**ASTM D6751**	**EN 14214**	**Australian Standard**	**Brazilian Standard**
Flash point (Min.)	°C	130	120	120	100
Water and sediment (Max.)	% Volume	0.050	0.050	0.050	0.020
Kinematic viscosity, 40°C	$mm^2 \cdot s^{-1}$	1.9 – 6.0	3.5 – 5.0	3.5 – 5.0	3.5 – 5.0
Sulphated ash (Max.)	% Mass	0.020	0.020	0.020	0.020
Sulphur (Max.)	% Mass	0.0015	0.0010	0.0050	0.0010
Cetane number (Min.)	–	47	51	51	45
Carbon residue (Max.)	% Mass	0.050	0.030	0.030	0.050
Acid number (Max.)	mg $KOH \cdot g^{-1}$	0.80	0.50	0.80	0.80
Free glycerine (Max.)	% Mass	0.020	0.020	0.020	0.020
Total glycerine (Max.)	% Mass	0.240	0.250	0.250	0.380
Phosphorous (Max.)	% Mass	0.001	0.001	0.001	0.001
Oxidative stability, 110°C	Minute	NR	4	6	6

(Source: Patel and Shah, 2015)

17.7 Engine Performance on Biodiesel Use

Biodiesels can be used in engines in pure as well as blended with diesel fuel. The generalised engine performance when biodiesels are used as fuel can be summarised as below:

(i) Biodiesel can be used as fuel in diesel engines without any major engine modifications. The only modification required is the change of rubber components which come in contact with fuel.

(ii) Biodiesel is a renewable carbon neutral fuel and thus they are environment friendly. The thermal efficiencies are higher. They have lesser emissions of hydrocarbons, CO, CO_2 and SO_x in exhaust.

(iii) Biodiesel fuelled engines have slightly higher NO_x emission than diesel fuel. This problem can also can be minimized by properly adjusting the fuel injection timing.

(iv) Biodiesel being an oxygenated fuel (having oxygen molecule in its structure) have lesser calorific value than diesel fuel. Due to this reason, biodiesel produces 3 to 5 per cent lesser engine power and torque and have higher specific fuel consumption than diesel fuel.

(v) Biodiesel have a lubricant property and hence cause lesser engine wear and tear compared to petroleum diesel. This is likely to increases the operating life of the engine.

(vi) There are many concerns about deposits and clogging while using poor quality biodiesels. But good quality biodiesels do not cause clogging or deposits in engines.

(vii) Due to higher cloud and pour points than petroleum-diesel, using biodiesel in cold weather conditions may cause problems such as clogging of the filters and injectors. Generally, biodiesels operate well up to about 5°C. At lower temperatures, blending with diesel or addition of antifreeze agents can be adopted.

17.8 Major Environmental Aspects of Biodiesel Production and Use

Biodiesel has attracted the global attention as an alternative to petroleum-diesel as it is more environment friendly along with favourable economics. The possibility of conversion of used cooking oil to biodiesel is a promising option in waste management. The major aspects are briefly described below:

(i) Biodiesel is a carbon neutral, renewable, biodegradable and nontoxic fuel. It can play a great role in reducing greenhouse gas emission and global warming as it is an eco-friendly, clean burning fuel. It is estimated that using biodiesel instead of petroleum diesel can result in reduction of 56 to 86 per cent of greenhouse gas emissions, which means a reduction of 75.5 million metric tonnes of carbon pollution. To reduce an equivalent amount of carbon, it is required to plant 1.9 billion trees. In addition to this, it will bring about a reduction of 47 per cent particulate matter, 67 per cent of hydrocarbon emissions and considerable amount of smog. Another benefit is that biodiesels are less harmful to both terrestrial and aquatic ecosystems in case of an accidental spill.

(ii) Biodiesel usage is beneficial to human health as it can reduce the emission of cancer causing polycyclic aromatic hydrocarbons (PAH) and nitrated polycyclic aromatic hydrocarbons (nitro-PAH) produced by diesel fuel.

(iii) Even blending biodiesel with diesel is beneficial to the environment. It was found that diesel having 20 per cent biodiesel can reduce emissions of hydrocarbons by 30%, CO by 20% and particulate matter by 15%.

(iv) When comparing biodiesel production with petroleum diesel production, it can reduce 79 per cent of waste water and 96 per cent of hazardous waste production. Biodiesel production have a positive energy balance and for every unit of energy input for production it yields 5.5 units of energy in the output product.

(v) The energy inputs for biodiesel production process that have major environmental impacts are electricity and alcohol (methanol which is the most effective one), as major portion of the electricity and methanol is produced from non-renewable resources. This can be further improved by producing methanol from biomass by different (gasification or pyrolysis) routes and different sources.

(vi) The life cycle assessment with regard to the production and use of biodiesel as an alternative fuel raises concerns over land use change and may affect food production system, if agricultural lands are converted for oil crop cultivation for biodiesel production. Hence the use of used cooking oils as the raw material for biodiesel production needs to be encouraged.

17.9 Commercial Biodiesel Production

Commercially biodiesel can be produced using either batch reactors (for small-scale plants) or Continuous reactors (large-scale refineries). Though many catalysts, both alkaline (NaOH, KOH, carbonates, sodium methoxide and potassium methoxide) and acidic (H_2SO_4, HCl, H_3PO_4 and sulfonic acid) have been successful in biodiesel production, industrial process mainly rely on the homogeneous mixture of methanol and alkaline catalyst. This is due to the fact that alkaline catalysed process has high reaction rate, cost effectiveness and give higher yield of biodiesel. Other advantages include requirement of lower reaction temperature and lesser alcohol to oil ratio.

17.9.1 Small-scale Biodiesel plants

Small-scale biodiesel production units have social, economic and environmental advantages as they contribute to renewable energy production, rural development and waste management by utilizing locally available raw materials. The batch reaction units used for small scale biodiesel production either employ a single stage reaction process or two stage reaction process. The reactors operating on single stage process have a reactor with a washing and drying unit. The process is applied for producing biodiesel from feedstocks having less than 3 per cent FFA. The premixed alcohol-catalyst solution is mixed with vegetable oil maintained between 60 to 65°C in the reactor and stirred for one hour for the reaction to complete. After this the mixture is allowed to settle (time required for settling depends on the type of feedstock used). During this period the settleable glycerol and soap present in the biodiesel produced is settled at the bottom and separated out. The biodiesel thus obtained is then transferred to washing and drying unit. Here the biodiesel is washed using excess amount of water to remove the dissolved soap and glycerol. After each wash the water with dissolved soap and glycerol get settled at the bottom, which is separated out. Then the biodiesel is dried to remove the water content present. Biodiesel production using this system is time consuming and is generally adopted for very small scale commercial or lab-scale production. The schematic diagram describing the production process is given in Fig. 17.4.

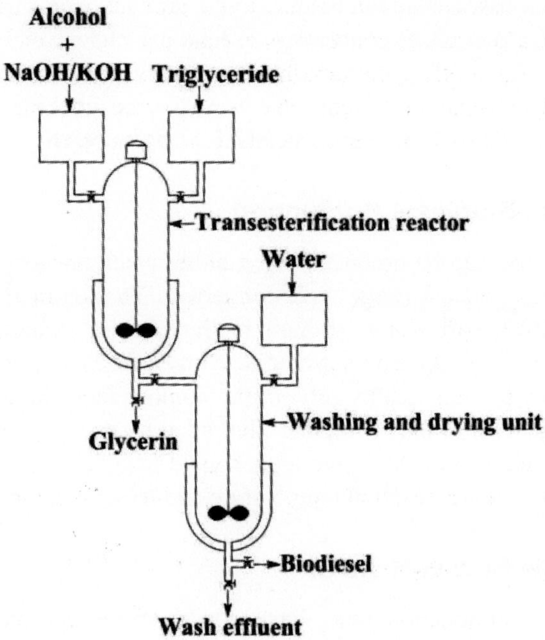

Fig. 17.4 Process flow diagram for small-scale single stage biodiesel reactor

Two-stage reaction process is adopted to increase the degree of biodiesel conversion by reducing the inhibiting effects during the reaction process. The process requires lower alcohol to oil stoichiometric ratio, requires lesser production time and can produce biodiesel yields as high as 98.5 per cent. The process flow explaining the two-stage process is detailed in Fig. 17.5. In the first stage nearly 80 to 90 per cent by weight of required alcohol and the catalyst are added prior and transferred to Reactor-1. As the product is transferred to Reactor-2, the glycerol produced in Reactor-1 is removed. In Reactor-2 the rection is completed by adding the remaining alcohol and catalyst. The crude biodiesel produced is then transferred to the washing and drying unit, where it is washed to remove the dissolved soap, glycerol and other soluble impurities, dried and transferred to the storage tank. The molar ratio of oil: alcohol used in the process ranges from 1:4 to 1:20 depending on the feedstock used. The time required for completing the reaction process is 30 minutes. The two stage reactors can also be used to produce biodiesel from high FFA feedstocks. While using high FFA feedstocks, Reactor-1 will be used for acid esterification and Reactor-2 will be used for transesterification. Thus, two stage reactors are more flexible in handling various feedstocks for biodiesel production.

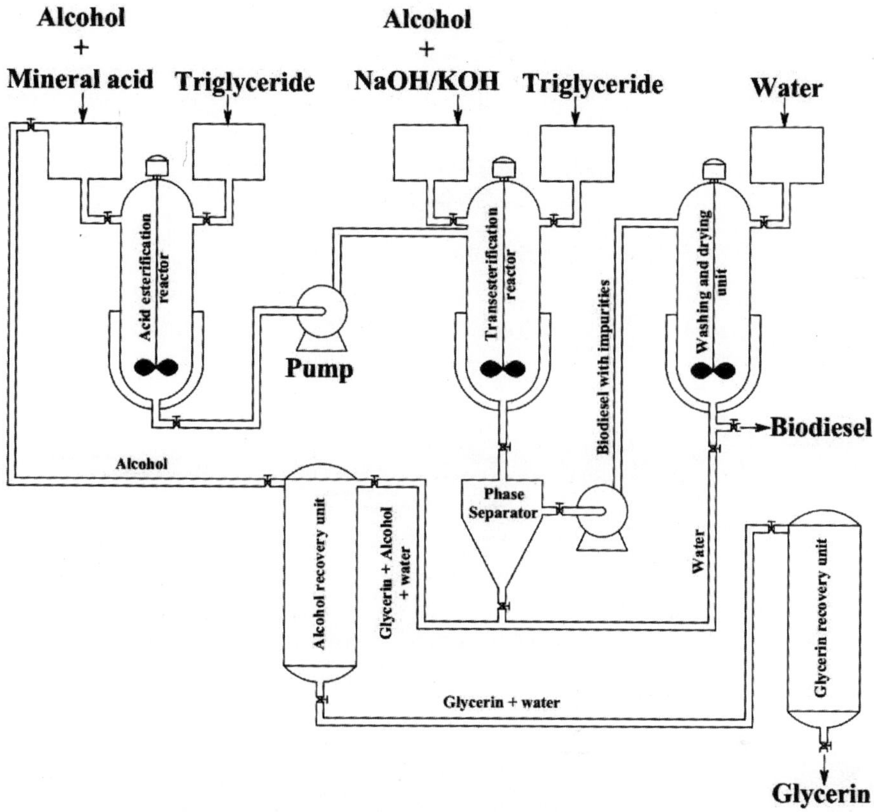

Fig. 17.5 Process flow diagram for small-scale two stage biodiesel reactor

17.9.2 Large scale biodiesel producing refineries

Biodiesel refineries have many constraints when compared to the small-scale facilities for production. The plant has to be designed such that the available resources are used in an optimal way, meet the safety and environmental standards set by the government, reduce wastage by recycling the excess chemicals used during the process and ensure continuous supply of biodiesel satisfying the national and international standards. To achieve this, commercial refineries adopt continuous stirred tank reactors or plug flow reactors with 2 to 6 reactors arranged in a cascade pattern to complete the process. The size of the subsequent reactors goes on decreasing as glycerol rich phase is removed from each reactor. The general process flow diagram adopted by industrial biodiesel refineries is given in Fig. 17.6.

Depending on the type of feedstock used, the residence time in each reactor varies between 6 to 12 minutes, the temperature is maintained at 60°C. NaOH is the widely used catalyst and its quantity varies between 0.25 to 1.0 per cent of the weight of oil depending on the type of feedstock. The purity of biodiesel produced by commercial refineries varies between 97.5 to 99.2 per cent.

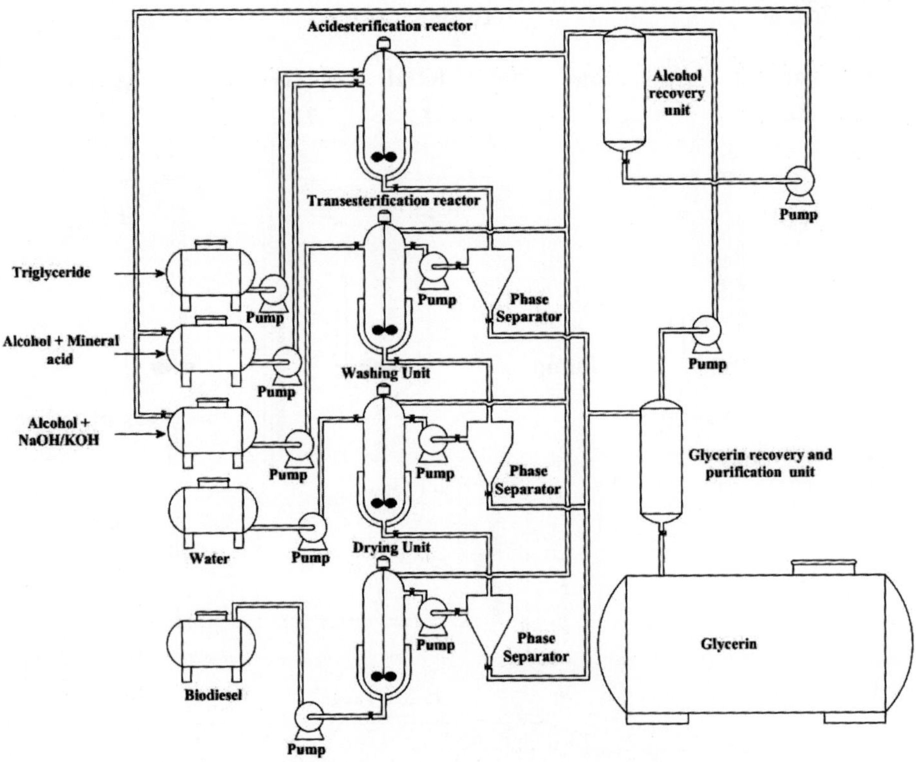

Fig. 17.6 Process flow diagram of large-scale biodiesel refinery

17.10 Global Status of Biodiesel Adoption

Many countries of the world are producing biodiesel commercially. Indonesia is leading in global biodiesel production followed by USA, Brazil, Germany and Argentina. India holds the 26th rank in global production. Renewable Energy group in USA is the leading producer of biodiesel and USA is the leading consumer of biodiesel followed by Brazil, Thailand, China and Argentina. India is the 13th largest consumer of biodiesel in the world. The cost of biodiesel is lowest in Indonesia. According to the 2018 Biodiesel Market Report, there will be an increasing trend in both biodiesel production and consumption till 2023. It should also be noted that encouraging changes in bio fuel policies put forth by many countries have led to this increasing trend. The feedstock used for biodiesel production by different countries are different and it is based on the choice of oil producing crop which is suitable and prominent in the region. For example, Argentina mainly uses rapeseed oil, Brazil uses Soybean, Peru uses palm and jatropha oils, whereas Thailand and Indonesia use palm oil.

17.11 Adverse Effects of Increasing Biodiesel Production

Taking into consideration the priority to reduce the emission of GHGs caused by diesel fuel, both developed and developing countries are framing laws and implementing projects to increase use of renewable energy. Biodiesel is a renewable fuel with lot of social and environmental benefits. At the same time, it is also true that increasing biodiesel production can also have some adverse effects, which are briefly discussed below:

(a) *Reduction in agricultural land:* Increasing biodiesel production means increasing the oil producing crop area. As more area is allotted for oil production, it will result in the reduction of land available for crop production. The global population is increasing and so also is the demand for agricultural produce. It is estimated that the food shortages will reach a new height by 2050 and the developing countries of the world will be affected most. There will be increased pressure on agricultural sector to produce more from the limited land with the available resources, which will be further aggravated if agricultural land is converted for the production of biodiesel feedstock.

(b) *Deforestation:* Rapid expansion of oil crop area will also lead to extensive deforestation. This was seen in Indonesia where a huge percentage of forest land was converted for oil palm plantations. In Malaysia also, new oil palm plantation expansions consumed a very high fraction of existing forest land along with agricultural land. There is a global concern over the conversion of the existing Amazon Forest lands to energy farms.

(c) *Social issues:* The biodiesel industry requires more labour than the petroleum industry. Since the feedstock production areas are usually concentrated in the rural sector, it creates more jobs and will curb the rural-urban labour migration. But as the plantations extend in area, it may lead to displacement of local communities and may also compromise labour rights. There is a concern over ongoing expansions in Amazon region as it is likely to force out indigenous communities. A similar situation was experienced by the rural local settlements of Malaysia as a result of expansion of oil palm plantations. The wide spread cultivation of Jatropha in the so-called waste lands may lead to denial of the rights of people on local community lands, which were used by villagers for grazing cattle as well as for their firewood needs.

(d) *Emissions caused by Biodiesel:* Biodiesels produces comparatively lesser emissions than diesel fuel, more over they are carbon neutral and cause much lesser harm to the environment. But the emission of Carbonyl compounds viz. formaldehyde, acetaldehyde, acrolein and acetone are much higher than that of petroleum diesel. This is due to the presence of oxygen molecule in the biodiesel molecular structure.

(e) *GHG emissions:* Biodiesel is considered environment friendly as it only emits carbon which the crop absorbed during its growing period. Hence, it is highly regarded as a fuel alternative to diesel fuel by the global community which can reduce GHG emissions. Taking into consideration the present scenario of operations to increase biodiesel production which produce indirect land-use change (ILUC), the GHG reduction brought by biodiesel may be small in some situations. It is roughly estimated that nearly 80% for the new land area used for producing feedstock is deforested land. This implies that naturally existing area which can bring about lager volumes of carbon sequestration is destroyed for comparatively smaller carbon reduction.

(f) *Water resource:* Water is unavoidable in all stages of biodiesel production, starting from feedstock production to the final purified form of biodiesel. Nearly 1 to 3 litres of clean water is needed for washing one litre of biodiesel. Clean potable water is becoming scarce and is one of the most demanded commodities of the growing population. Making inexpensive and clean water available to its citizens is becoming a major challenge to any country. This challenge will be intensified as the biodiesel production is increased. In addition, as the feedstock required for biodiesel preparation will not be used for human consumption, there is more chance for excessive use of chemical inputs to get higher production. Mitigating the pollution caused to the water bodies due to leaching of chemical fertilizers, pesticides and other contaminating sediments will become a serious problem in the future, affecting both aquatic and terrestrial ecology.

(g) *Effluent from biodiesel refineries:* Biodiesel refineries let out huge quantities of effluents containing soap, glycerol and unrecoverable catalyst used. In addition, the glycerol is one of the major by-products of biodiesel preparation process. Though it is of high demand by other industries, as the biodiesel refineries scale up their production it will be surplus and may result in disposal issues.

(h) *Health issues of workers in biodiesel industry:* Although biodiesel is considered as much safer and cleaner fuel to use, it involves the use of highly hazardous chemicals at different stages of its production. Thus, exposure to these chemicals beyond prescribed limits can cause serious health problems and even death for the workers involved.

CHAPTER 18

Energy from Microalgae

Microalgae are microscopic algae belonging to the group of organisms called 'phytoplankton'. They are capable of photosynthesis and plays a major role in absorbing atmospheric CO_2 and contributing greatly to oxygen supply. Energy production from microalgae is considered highly attractive as a source of renewable fuel from an environmental standpoint because they trap atmospheric carbon dioxide and can be grown in waste or salty water. They survive in fresh water and marine ecosystems and is being commercially used in various sectors such as pharmaceuticals, cosmetics, aquaculture and nutraceuticals in addition to their use for energy. The use of microalgae as food is reported since 1939, where it was used as soldier's food during the second world war. Commercially micro algae form a raw material for a variety of products. They are useful in (i) increasing the nutritional value of food and animal feed, (ii) as ingredients in cosmetic products, (iii) production of biofuels, (iv) capturing the atmospheric CO_2 and (v) waste water treatment.

Its use in the energy sector is mainly focussed on the oil stored in it. In-fact they have higher oil content per dry mass than many other presently available biomass sources and have the advantage that they require lesser area for production. The possibility of growing them in waste water makes it economically attractive due to reduction in production cost. The major bottleneck in their commercial production is their high sensitivity to temperature. Maintenance of optimum temperature conditions can result in the consumption of large quantities of electrical energy in colder climatic conditions. The other constraints in its commercial production can be attributed to the problems associated with the isolation of effective strains, management of the production system, cost-effective way of harvesting and oil extraction and inadequacy of commercially feasible technologies for using the residual biomass after oil extraction. In an effort to overcome these bottle necks several investigations were taken up by various scientific institutions during the last several decades, especially in producing genetically engineered algae.

This chapter briefly discuss their cultivation, factors affecting the growth and development, methods of harvesting and the possible ways of producing energy from microalgae.

18.1 Cultivation of Microalgae

Mass production of algae can be broadly divided into two, (i) the 'Clean system' using fresh water fortified with nutrients and (ii) the 'Waste system' in which sewage or other waste waters are utilized. The second system is often attributed with the dual advantages of waste treatment and energy production. In the early efforts to cultivate microalgae, *Chlorella* and *Scenedesmus* were the prominent strains. In clean systems, algal strains *Chlorella*, *Scenedesmus* and *Spirulina* are usually cultivated. In waste water systems, strains like *Micratinum, Oocystis, Oscillatoria, Clamydomonas, Euglina* are considered.

The growth of microalgae under batch cultivation undergoes the following stages:

(i) Adaptation stage
(ii) Exponential growth stage
(iii) Linear growth stage
(iv) Diminishing growth stage
(v) Stationery stage
(vi) Mortality stage

Depending on the type of microalgae, the conditions required for its effective growth may vary. Generally microalgae are grown under either phototrophic, heterotrophic, mixotrophic or photoheterotrophic conditions. The Phototrophic cultivating condition is the most commonly used system. Here, the microalgae use light (preferably sunlight) as the source of energy and inorganic carbon (CO_2) is utilised to produce energy through photosynthesis. Some species have the ability to use organic carbon even under dark conditions. In other words, they use organic carbon both for energy and as carbon source. This system is referred as heterotrophic and can result in a higher biomass production. If the microalgae use both organic compounds and inorganic carbon for its growth and development, the system is said to be in mixotrophic condition. These types of microalgae can grow either in phototrophic or heterotrophic conditions. Microalgae that grow using light as the source of energy and organic compounds for carbon source are said to be photoheterotrophic. This system is rarely used in microalgae cultivation for energy purpose. Microalgae growing in heterotrophic systems can produce more oil but can get contaminated easily and likely to have higher production costs. Phototrophic system, generally adopted in open pond cultivation system is easy to scale up and have the benefit of using CO_2 from the flue gas of factories. Though phototrophic systems produce lesser oil than the heterotrophic systems, it is preferred by largescale growers over heterotrophic systems. This is due to its easiness in operation and better stability compared to others systems.

Microalgae can be effectively cultivated both in open pond and photobioreactors. Each system has its own pros and cons and the basic requirements for effective production of these photosynthetic organisms are availability of light and nutrients to get high land use efficiency. Any cultivating system should (i) be capable of providing sufficient sunlight, (ii) be effective in transferring nutrients, (iii) enable simplified operations with minimum contamination and (iv) have low initial and operating cost. The different systems used for the production are briefly described below:

18.1.1 Cultivation in open pond and water reservoirs

It is the simplest and the cheapest way for large scale microalgal cultivation with low investment as well as minimum maintenance and operation costs. The additional benefits are low energy demand and easiness for scaling up. But these systems often face with the problem of low cell concentration, making the harvesting a tedious and technology demanding operation. The lower cell concentration is often the result of runoff rain water entering the pond. This excess water inflow cause change of salinity and pH, increase turbidity and induce leakages and erosion of banks. In addition to this, wash of protozoa and bacteria into the pond can result in the increased levels of toxic contaminations which further cause reduction in population density. Ponds (both natural and artificial) equipped with agitators are found to be beneficial in increasing the population density of micro algae, as majority of them remain suspended with lesser sedimentation. Presently Hutt Lagoon, Australia is the largest natural water body where micro algae are being cultivated commercially for production of β-carotene (pigment, used in food supplements) (Fig. 18.1).

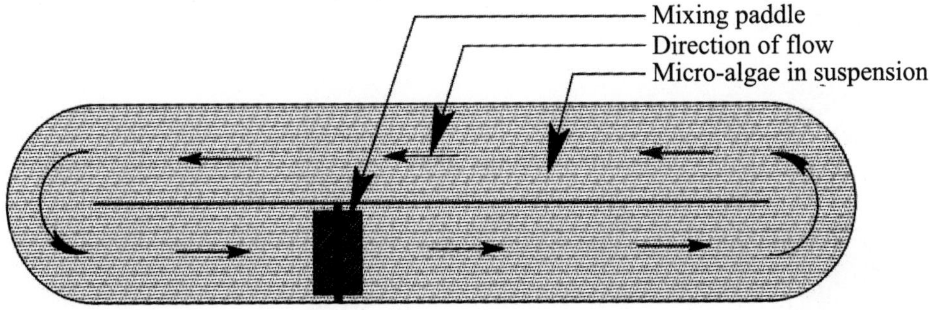

Fig. 18.1 Schematic diagram of open pond for microalgae cultivation

Photo bioreactors are closed systems and are compact, making it easy to control the various parameters as required. Here the better nutrient management and increased metabolic efficiency can yield higher biomass production compared to open pond systems. Apart from experimental bioreactors, practical designs enabling scale up to commercial units have been seldom developed. Other problems like higher initial and operating cost are also limiting factors for their adoption. The different types of photobioreactors include vertical column photobioreactor, flat-plate photobioreactor and horizontal tubular photobioreactor.

18.1.2 Vertical Column Photobioreactors (VCP)

VCPs are made of transparent tubes fixed vertically with an air pumping system at the bottom. The air pumped from the bottom helps in mixing and enable proper mass transfer of CO_2 along with removal of O_2 produced during photosynthesis. The mixing caused by the pumped air eliminates the need for a mechanical agitator. There are two types of VCPs, airlift and bubble column photobioreactor (Fig. 18.2).

Bubble column photobioreactor is cylindrical in shape and requires low initial cost. It does not have any moving part and have comparatively better heat and mass transfer ability with efficient release of O_2. The cylinder height should be greater than two times the diameter.

The Airlift photobioreactor have a design similar to the traditional bioreactor with two interconnecting zones, gas riser zone and downflow zone. The gas riser zone receives the air from the air pump giving an upward draft to the liquid. In the downflow zone, the medium flows down toward the bottom and circulates in both the zones. Based on the mode of circulation, airlift photobioreactor can be further classified into two types, internal loop and external loop airlift photobioreactors.

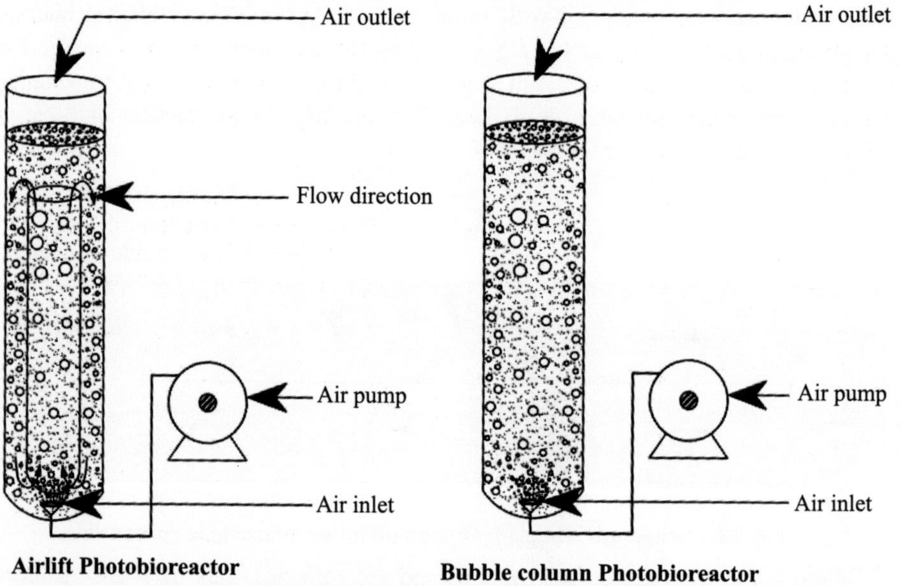

Fig. 18.2 Vertical Column Photobioreactor

18.1.3 Flat-Plate photobioreactor

Flat-plate photobioreactors have rectangular cross section with light penetrating the unit through the sides. The mixing of reactor mass is done by the upward draft of the air pumped from the bottom. They are associated with higher space requirements, light energy requirements and cleaning problems. There is a lower production per unit volume of this type of photobioreactor and there exists difficulty in scaling-up due the increased hydrostatic pressure required with increased volume (Fig. 18.3).

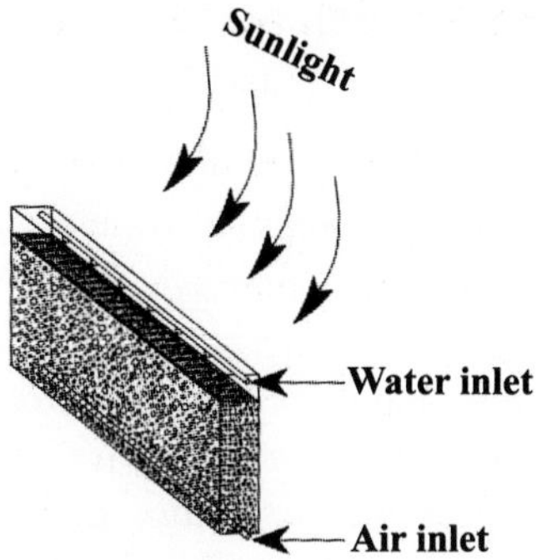

Fig. 18.3 Schematic sketch of flat-plate photobioreactor

18.1.4 Horizontal Tubular Photobioreactor

This type of photobioreactor is widely used for commercial microalgae production under closed environments. They are made of small diameter transparent pipes. The diameter of the pipes is kept small so as to increase the light penetration. Mixing and circulation of the liquid is achieved as a result of the air pumped. As a result of long horizontally laid tubes, the air-residence time in the photobioreactor is high. This helps the system by providing more dissolved CO_2 but gas transfer depends on the flow characteristics and the air supply technique adopted. Lager surface area to culture volume ratio, easiness in controlling temperature and contaminants and easy scaling-up facility are the notable advantages. But the horizontal tubes may partly block the penetrating light and the increased oxygen concentration may inhibit photosynthesis in these systems. Apart from these major issues, cleaning difficulty is another problem associated with this type of photobioreactor (Fig. 18.4).

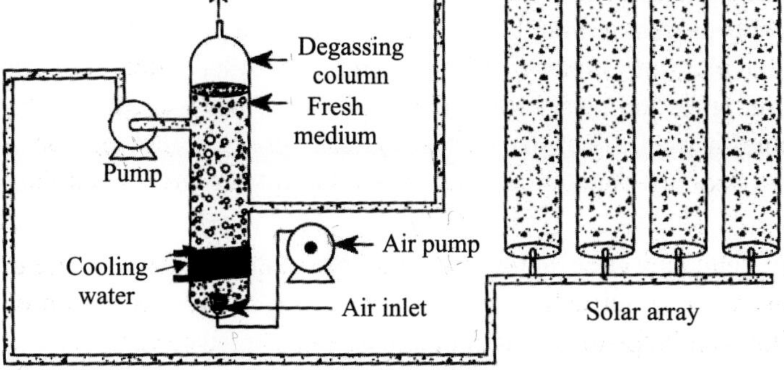

Fig. 18.4 Schematic sketch of Horizontal Tubular Photobioreactor

18.2 Factors Affecting the Growth and Biofuel Production from Microalgae

Energy and carbon availability is crucial for growth of microalgae whereas nutrients have an important role. The importance and relevance of various factors is as below:

(a) *Supply of light:* Duration, intensity and type of light is an essential and critical factor for the growth and development of microalgae. Ideally, there should be a uniform distribution of light penetrating throughout the medium to attain better growth. This is not practically possible due to the shading effect caused by the microalgae cells. The formation of biofilm further blocks the penetration of light into the medium. These problems can be reduced to a greater extent by continuously mixing the emulsion containing the microalgae.

(b) *Carbon source:* Carbon is one of the most important constituents required for the growth and development of microalgae. The various sources of carbon that can be used include carbon dioxide, methanol, acetate, glucose etc. But carbon dioxide or bicarbonates are the most widely used source as other organic carbon sources are costly. In addition to the economic aspect, use of CO_2 helps in addressing the global warming effect caused by CO_2 emission.

(c) *Nitrogen source:* Microalgae usually accumulate lipids when cultivated under stress. Stress conditions are usually caused by deficiency of essential nutrients like nitrogen and variations in pH along with other unfavourable conditions. Limiting the supply of nitrogen is an effective method as microalgae decompose proteins into lipids when cultivated under limited nitrogen supply. But limiting the supply of nitrogen for longer duration will affect the growth of the microalgae and in turn will affect the overall oil/lipid production. Thus, limiting the nitrogen supply has to be done strategically such that both the growth and oil/lipid production is not affected.

(d) *Temperature:* In commercial cultivation, temperature plays an important role in attaining the targeted yield. Maintaining a temperature between 15-30^{0}C is essential for the optimal growth and development of microalgae. Temperatures below 15^{0}C will slow-down the growth while higher temperature above the allowable range will result in denaturation of essential proteins/enzymes and affect the cellular physiology. This will result in the reduction of microalgal biomass production.

(e) *pH:* The optimum pH requirement for the cultivation of microalgae is in the range between 7-9. Maintaining the pH in the optimum range is very essential for the solubility and availability of carbon dioxide and carbonates which is essential for the proper functioning of the intracellular enzymes. If the pH is decreased below 7 into the acidic range, it will affect the nutrient uptake, produce intracellular metabolites and can induce toxicity. pH above 9 will affect the uptake of CO_2 and the process of cell division.

(f) *Mixing:* This is a very important parameter and it determines the distribution of supplements such as nutrients and gas to the cells, avoids stagnation of the cells, prevents settling and helps to maintain the microalgae in suspension.

18.3 Microalgae Harvesting Methods

Harvesting of microalgae is a difficult and costly process due to lower mass fraction and presence of excess organic matter. The common harvesting techniques are centrifugation, flocculation, filtration or screening, gravity sedimentation, flotation and electrophoresis. Selecting the most suitable harvesting method depends on density, size and value of the desired product of microalgae. Generally, the harvesting operation is completed in two steps, bulk harvesting and thickening. During the bulk harvesting step, the microalgal biomass is separated from the bulk suspension. Usually flocculation, flotation or gravity sedimentation methods are adopted for this purpose. The total solid content of the harvested mass after bulk harvesting ranges between 2 and 7 per cent. After bulk harvesting, thickening process is done to increase the concentration of the harvested slurry. Generally, filtration and centrifugation are the commonly used methods during the thickening process of the bulk harvested micro algae. The thickening process is a highly energy demanding process. The commonly used harvesting techniques for harvesting micro algae are briefly described below:

(a) *Centrifugation:* This method is adopted to separate microalgae by spinning the slurry around an axis. The microalgae are separated from the liquid by the property of differential densities. Even though it is a costly process, we can achieve a harvesting efficiency between 80 and 90 per cent.

(b) *Flocculation:* In this method microalgae are aggregated to larger particles using different agents and made to settle. The different flocculation processes include, auto-flocculation, chemical coagulation and combined flocculation. In auto-flocculation the microalgae settle automatically when the pH of the solution is increased due to the precipitation of carbonate salts with algal cells. Induced flocculation by addition of coagulants is called chemical coagulation. Based on the nature of the coagulants added, it can either be inorganic flocculation or organic or polyelectrolyte flocculation. The commonly used inorganic coagulants are iron-based or aluminium-based coagulants. The inorganic flocculation method has many drawbacks. The requirement of large quantity of flocculant to initiate the process and the production of large quantity of sludge is one among the disadvantages. The process is highly sensitive to pH, making the process complicated. The effectiveness of the coagulant is dependent on the species of microalgae and the harvested mass is contaminated with either aluminium or iron salts which in turn has to be removed by adopting further purification processes. Using organic flocculants such as chitosan do not contaminate the harvested microalgal biomass. The method of judiciously combining organic and inorganic flocculants is referred as combined flocculation process.

(c) *Electrolytic process:* This method of harvesting is completed in three steps. First step is the generation of coagulants through the electrolytic oxidation of the sacrificial electrode. It is followed by breaking of the liquid-microalgae emulsion by destabilizing the particulate suspension. Final step is the formation of aggregate flocs of microalgae. Using this harvesting method, a harvesting efficiency of 80 – 95% can be achieved.

(d) *Gravity sedimentation:* This method of harvesting fully depends on the density of microalgal particles and can be effective only with microalgae having higher densities. To increase the sedimentation, lamella separators and sedimentation tanks are often used along with flocculation.

(e) *Filtration and screening:* This is the process of separating the microalgae from the emulsion using the principle of difference in sizes. The emulsion containing the microalgae is passed through sieves such that the microalgae is filtered out from the emulsion. Micro-strainer and vibrating screen filters are commonly made use of in this method of harvesting. This method of harvesting is simple, easy to operate, requires comparatively lesser investment and have a capability of harvesting 70–89% of microalgae from the emulsion.

(f) *Flotation:* This is a gravity separation process, more effective than the sedimentation method and can be effectively used on microalgae having a size less than 500μm in diameter. The emulsion containing microalgae is pumped with air such that the air or gas bubbles formed get attached to the solid particles and move them to the liquid surface, from where it is easily removed. Based on the size of the bubbles used for the separation, the process can be classified into dissolved air flotation, dispersed flotation and electrolytic flotation. In dissolved air flotation, air is pumped at a pressure such that it produces bubbles having a diameter in the range of 10 – 100 μm. Dissolved air flotation method can be more effective than settling method if flocculation method is applied on the emulsion as a pre-treatment process. The dispersed air flotation method uses bubble diameter in the range of 700 – 1500 μm. Here the bubbles are formed using a high-speed mechanical agitator along with an air injection system.

(g) *Electrophoresis techniques:* In this method an electric field is used to drive the microalgae out of the solution. The hydrogen generated as the result of the electrolysis of water adheres to the microalgae, which in turn brings it to the surface. This method is considered to be more environment friendly, versatile, energy efficient, safe and cost effective.

(h) *Comparison of different harvesting methods:* Among the different harvesting methods available, centrifugation and chemical precipitation are the most economically feasible methods, of which centrifugation method is having a marginal advantage over chemical precipitation. Higher flow rate requirements are associated with mass microalgae cultivation systems. Higher flow rate keeps the microalgae moving upwards than in the downward direction, thus favouring flotation harvesting method over sedimentation method. The gravity sedimentation method can be effectively adopted only for microalgae like Spirulina. Filtration is an effective method but micro strainers often have problems in handling emulsions. They often get clogged with increased bacterial and algae slime on the microfabric necessitating regular cleaning. Electrolytic method does not require chemicals, but high-power requirements and fouling of the cathodes are the problems encountered.

18.4 Different Options for Biofuel Production from Microalgae

As microalgae are known to have high lipid content, they can be used to produce biodiesel. As it has the ability to accumulate starch and cellulose, it can also be effectively used to produce bioethanol and can be anaerobically fermented to produce biogas. While considering microalgae for the production of biofuels, the two major challenges that pose hurdles in its wide acceptance are cost involved in production and the difficulties in effective harvesting both in terms of efficiency and cost. The different choices of biofuel production from microalgae are depicted in Fig. 18.5.

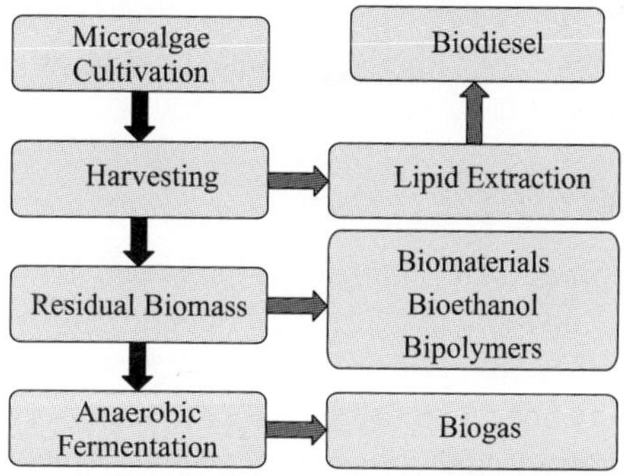

Fig. 18.5 Products obtained by processing microalgae

18.5 Biodiesel from Microalgae

Commercial production of biodiesel from microalgae is attracting interests of many scientists. Microalgae such as *Chlamydomonas reinhardtii* accumulate up to 25 per cent lipids once the hydrogen production is stopped. Once the microalgae are harvested, they can be dried to extract the oil and then the oil is converted to biodiesel. The possible ways of extracting oil from micro algae are by mechanical press, solvent extraction or supercritical fluid extraction method. Out of these methods, solvent extraction using the solvents, hexane, methanol, chloroform or mixed solvent are found to be more effective.

Drying of the microalgae is an essential prerequisite before extracting the oil. This is done to avoid moisture interacting with the solvents used for extracting the oil. Drying can be done either using mechanical driers or under diffused sunlight. After drying the microalgae and prior to oil extraction, it is subjected to a process called disruption of cells. This is done to increase the extraction of oil and is usually done by expeller press or by bead beating method. The next step is the extraction and purification of the oil from the micro algae. This is an intensive process both in terms of energy and cost. The purified oil is then subjected to transesterification process to get biodiesel.

18.6 Bioethanol from Microalgae

As micro algae can produce starch and cellulose without competing with the food crops with regard to land and water, they are regarded as a potential feedstock for the production of bioethanol. If the hydrolysis process to break down starch and cellulose (saccharification) can be simplified to make the overall process cost-effective and energy efficient, industrial scale production of bioethanol from microalgae can be attractive.

As lignin is absent in microalgae, the saccharification process is easier, thereby making significant cost reduction in the overall process. The harvested microalgae biomass has to be disrupted, for which acids (concentrated or dilute) are utilised. Then the starch is extracted using water or other organic solvents. Studies have shown that hydrolysis of these starch yielded sugars consisting of xylose, glucose and arabinose. Studies have also been successful in genetically modifying microalgae to boost carbohydrate accumulation. This paves way to techniques for genetically modifying microalgae for possible direct production of ethanol in larger scales.

18.7 Biogas from Microalgae

Of the various energy production alternatives proposed for microalgae, anaerobic fermentation has been found attractive as the microalgae can as such be utilized. This technology has been used for the past 60 years or so. As all the macromolecules of microalgae can be utilized for the anaerobic digestion process, the residual biomass after oil extraction and bioethanol production, otherwise discarded as waste, can be used for biogas generation to produce more renewable energy. Proper design of bioreactors and optimum operational conditions can promote increased biogas yield. Pre-treatment, co-digestion and strain-improvement are other favourable aspects which can boost biogas yield.

Attached film expanded bed reactors have been developed and used for anaerobic fermentation (bio-methanation) of microalgae. These reactors make use of a media for cell attachment and the media bed expansion is achieved by using a recycle pump. A schematic diagram is shown in Fig. 18.6. The algal culture is centrifuged to get a total solid (TS) content compatible for the bioreactor. The HRT is usually in the order of 2 to 10 days. The average CH_4 content of bio-methanation is about 60 per cent.

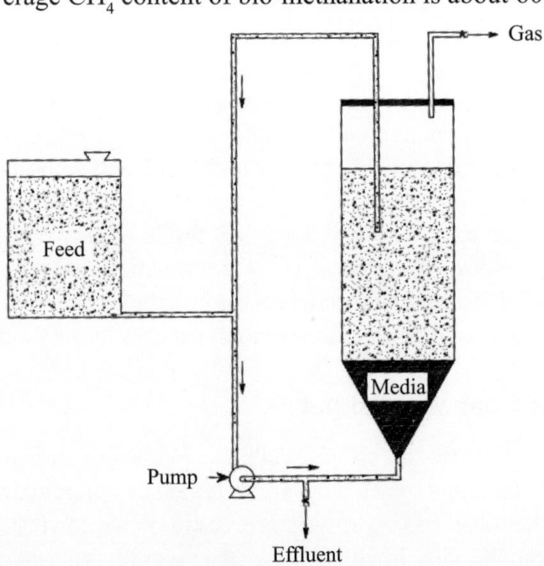

Fig. 18.6 Experimental Attached Film Expanded Bed reactor for microalgal bio-methanation

18.8 Overall Prospects of Energy Production from Microalgae

Cultivation of microalgae for energy is a promising concept as they do not interfere with food production systems, unlike other energy crops. High biomass production rate is a major factor which offer significant opportunity for microalgae as a source of renewable fuels. Efficient and cost-effective harvesting is still a bottleneck in the production of microalgae. If technological interventions can address these issues, there is much scope for these systems. Production of lipids and their conversion to good quality fuels are already practiced and other technologies like thermo-chemical liquefaction, catalytic conversion and hydrogenation are also attempted. Anaerobic digestion for methane production is also a possible technology. The fuel production from microalgae can emerge as an environment friendly alternative in addressing the emission issues related to fossil fuels. The capacity of microalgal systems to convert atmospheric CO_2 to good quality renewable fuel is the most striking aspect when we assess the option on an environmental scale.

CHAPTER

19

Hydropower

The earliest inventions in generating power from the kinetic energy of water were consummated during the period of Han Dynasty in China between 202 BC and 9 AD. Trip hammers driven by a vertically set water wheel were used to beat and hull grain, shape metals in smithy and break ores (Fig. 19.1). Water power was also used in early paper-making process. Accessibility to water power had long been influential in kick-starting economic development. In 1771, Richard Arkwright set up Cromford Mill (Fig. 19.2) in England's Derwent valley to spin cotton utilizing hydropower, which is regarded as one of the world's first factory systems.

Fig. 19.1 Water-wheel powered trip hammer in smithy

(Author: Ken Bgnall, Sheffield-Abbeydale Industrial Hamlet, Available at https://www.geograph.org.uk/photo/2357009 via Geograph, Under lincense CC-BY-SA 2.0)

Some of the key developments in hydropower technology took place during the first half of the 19th-century. Such as, in 1827, a French engineer Benoit Fourneyron developed a turbine capable of generating nearly 6 horsepower, which happened to be the earliest version of the Fourneyron reaction turbine as shown in Fig. 19.3.

Fig. 19.2 Cromford Mill's water-wheel

(Author: Ashley Dace, Cromford Baryte Crusher, Available at https://www.geograph.org.uk/photo/2297009 via Geograph, CC-BY-SA/2.0)

Later in 1849, a British–American engineer James Francis invented the first modern water turbine and named after its inventor, the *'Francis turbine'*. Even today, it is the world's most widely used water turbine in hydropower plants for electricity generation. In the late 1870s, American engineer Lester Allan Pelton invented an impulse water turbine, the *'Pelton wheel'*, patented in 1880.

In the 20th century, Austrian professor Viktor Kaplan (in 1913) developed a propeller-type turbine with variable pitch runner blades known as *the Kaplan turbine.*

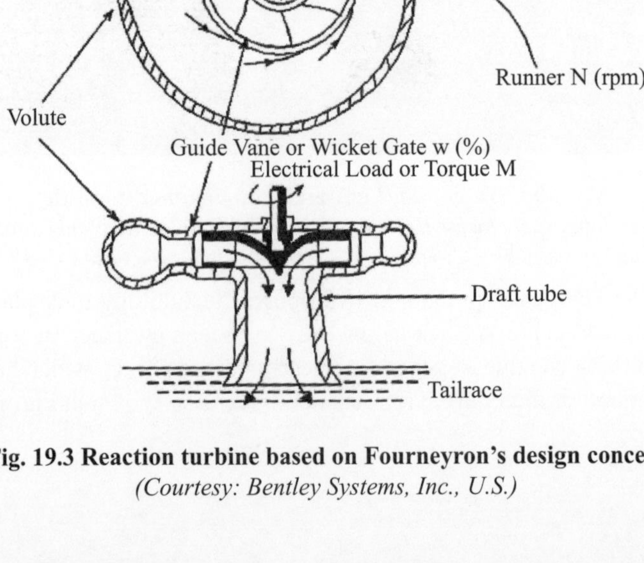

Fig. 19.3 Reaction turbine based on Fourneyron's design concept

(Courtesy: Bentley Systems, Inc., U.S.)

19.1 Concept of Hydel Power Generation

Hydroelectric (Hydel) power was the first and simplest electricity generation technology. Even today, hydropower is one of the foremost conventional energy sources for electricity generation. Other traditional sources are fossil fuels (coal, petroleum and natural gas) and nuclear energy. Hydroelectric resource is considered as a renewable source for generation of electric power, whereas the other two sources are regarded as non- renewable. Hydel power is renewable owing to the recurring nature of the hydrologic cycle. It is also regarded as an environmentally benign source as it produces neither atmospheric nor thermal pollution. It is considered as an ideal energy source in the regions with high annual rainfall with a hilly or mountainous topography and are most ideal in potential locations that are in close vicinity to the main load centres. Some remotely located hydropower sites with large potential can also be justified in consideration of the high cost required for high-voltage transmission lines from far off generating stations. Small hydro-sites may also be economical, principally if they combine storage of water during light loads with electricity production during peaks. The other methods of hydropower generation harness the kinetic energy of the water stream flowing through a waterway directly without impounding in the dam.

The hydropower resources are directly reliant on the hydrologic cycle on the planet. Hence, it is important to have basic comprehension on the hydrology in order to understand the concept of hydropower. The hydrologic cycle has three major phases:

Phase-1: Solar energy heats up the surface water of water bodies like oceans, rivers and lakes resulting in evaporation of water.

Phase-2: Water vapor condenses into clouds and precipitates in various forms such as rain and snow.

Phase-3: Part of the precipitation gets collected in streams and rivers, which eventually run into oceans and lakes, where it evaporates and repeat the cycle all over again.

The quantity of water that reach rivers and streams (either as surface flow or sub surface flow) in a particular geographic area (catchment area) determines the potential for producing hydropower. Climatic and weather factors such as long term and short term (seasonal) variation in the amount and pattern affects the potential of hydel power. Long-term changes in precipitation patterns due to climate change effects can have significant effects on hydropower production.

The major disadvantages of Hydropower are that they are expensive and result in submergence of agricultural or forest land.

19.2 Basic Working Principle of Hydroelectric Power Generation

Hydroelectric power is the electric power produced by means of water turbine-run generators that convert the potential energy of falling water or kinetic energy in a fast-flowing water stream to electrical energy. The resultant rotational motion and torque produced at the turbine shaft by the water stream is applied to an AC generator using suitable mechanical transmission (Fig. 19.4).

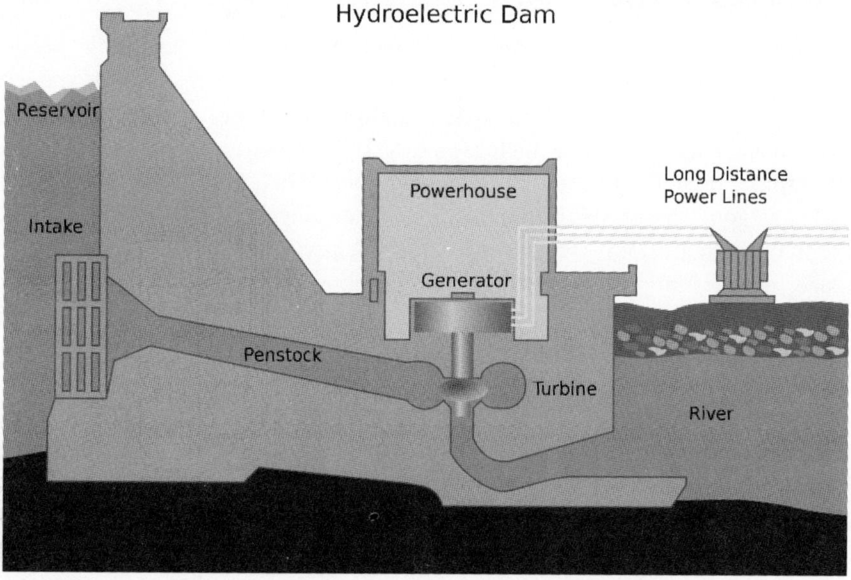

Fig. 19.4 Schematics of hydroelectric power plant

(Author: Tennessee Valley, Schematic diagram of Hydroelectric power plant, Available at https://commons.wikimedia.org/wiki/File:Hydroelectric_dam.svg, GNU Free Documentation License CC-BY-SA/3.0 via Wikimedia Commons)

In the generation of hydroelectric power, water is impounded at a higher elevation and led downwards through large conduits (*penstocks*) to a lower elevation. The difference between these two elevations is known as the *hydraulic head*. The high hydraulic head drive the water stream at a high velocity by converting its potential energy to kinetic energy. At the outlet of penstock, the fast-moving water jet causes turbines to spin. The water turbines in turn drive the generators, which convert their mechanical energy into electricity. The AC electricity produced is then fed to the transformers, to modify the AC voltage suitable for long-distance transmission. The construction that houses the turbines and generators, into which the penstocks deliver the high velocity water stream, is called *the powerhouse.*

The magnitude of power attainable from a water reservoir is directly proportional to the working hydraulic head. Hence, hydroelectric power plants are typically installed at the dams that impound rivers, thereby raising the water level behind the dam and forming a maximum hydraulic head as possible. It is also apparent that a powerhouse working under a high hydraulic head requires a smaller volumetric flow of water than a low-head one to generate the same amount of power. In most of the hydel projects, the powerhouse is built on one side of the dam utilizing part of the dam as a spillway. In certain situations where the river flows in a narrow and steep valley, the powerhouse can be installed within the dam itself.

With the knowledge of magnitude of available hydraulic head and mass flow rate of water, the *Theoretical Potential Hydropower (TPH)* is calculated as:

$$\text{TPH, Watts} = Q \times H \times g \qquad ...(19.1)$$

Where,

Q = mass flow rate of water, kg s^{-1}

H = effective hydraulic head, m

g = acceleration due to gravity, m s^{-2}

Also, the overall efficiency of the turbine-generator can be found out using following formula:

$$\text{Turbogenerator efficiency } (\eta_{turbogen}), \% = \frac{\text{TPH}}{\text{Actual power output}} \times 100 \qquad ...(19.2)$$

19.3 Classification of Hydropower Plants

The hydropower plants can be classified using four different criteria, as follows:

(a) On the basis of nature of hydropower resource

(i) *Fast moving river flow without impoundment: Kinetic energy*

It involves construction of a diversion, sometimes called *run-of-river* that channels a portion of river flow through a canal or penstock to run the turbine-generator.

(ii) *Impoundment reservoir (dam): Potential energy:* The most common type of hydroelectric power plant is an impoundment facility. Large hydel power systems usually use this system where a dam or reservoir is built to store water at a hydraulic head so that the turbines and generators can be powered using the water released from the reservoir through penstock to produce electricity.

(iii) *Pumped storage type:* Another type of hydropower called pumped storage works like a *battery* or a *buffer*, storing the surplus electricity generated by other power sources like solar, wind, and nuclear for later use. It stores energy by pumping water uphill to a reservoir at higher elevation from a second reservoir at a lower elevation. During the leaner demand period, the pumped storage facility stores energy by pumping water from the lower reservoir to the upper. At the peak time of electricity demand the water is released back to the lower reservoir which turns the turbine to generate electricity.

(b) On the basis of available hydraulic head (h)

(i) High-head : $h > 180$ m

(ii) Medium-head : $30 \text{ m} < h < 180$ m

(iii) Low-head : $h < 30$ m

(c) On the basis of plant capacity

(i) Pico-hydropower: up to 5 kW, usually for remote communities

(ii) Micro hydropower: 5 kW to 100 kW, for a small community or rural industry

(iii) Mini hydropower: 101 kW to 2 MW, often off-grid

(iv) Small scale hydro power: 2 to 25 MW, usually grid-connected

(v) Medium scale hydro power: 26 to 100 MW, grid-connected

(vi) Large scale hydro power: above 100 MW, connected to large electricity network

(d) On the basis of duty load

(i) Base-load type – operational round the year

(ii) Peak-load type – operational during the peak period only (e.g., pumped storage)

19.4 Classification of Hydro-turbines

As discussed earlier, hydro-turbine is a mechanical device that converts the energy possessed by water into mechanical energy (rotary motion) by means of hydraulic pressure or kinetic reaction. This mechanical torque developed at the turbine's shaft drives an electric generator by direct or indirect coupling (most often a gearbox). Turbine comprises of a wheel called *runner*, with a number of blades or buckets fixed on its periphery. Hydro-turbines can either be *reaction* or *impulse* type. The turbine's type specifies the mode in which the water-flow makes the turbine runner to spin.

19.4.1 Reaction turbines

This kind of hydro-turbine functions with its runners completely flooded, developing the torque by virtue of water pressure reaction against the runner blades. Reaction turbines are further categorized as Francis (mixed flow) or axial flow types. Axial flow turbines are offered with both fixed blades (Propeller) and variable pitch blades (Kaplan). Both axial flow (Propeller and Kaplan) and Francis turbines can be mounted either horizontally or vertically. Moreover, the propeller-type turbines can be mounted in slanting manner.

(a) *Francis turbine:* A Francis turbine (Fig. 19.5) is the one having a runner with fixed buckets (or vanes), usually nine or more, into which the water enters the turbine in a radial manner with respect to the shaft and exits in an axial direction. The turbine assembly consists of the runner, a water supply volute to convey the water radially to the runner, *wicket gates* to regulate and distribute the flow of water equally to the runner and a *draft tube* to discharge the water out of the turbine.

It has been installed at numerous hydropower plants throughout the world due to its versatility to the wide range of available hydraulic heads. It is employed where head ranges between 15 to 750 m and for output capacities between 0.25 to 800 MW per turbine. A Francis turbine can be operated over a range of flow variations, approximately 40 to 105 per cent of the rated discharge.

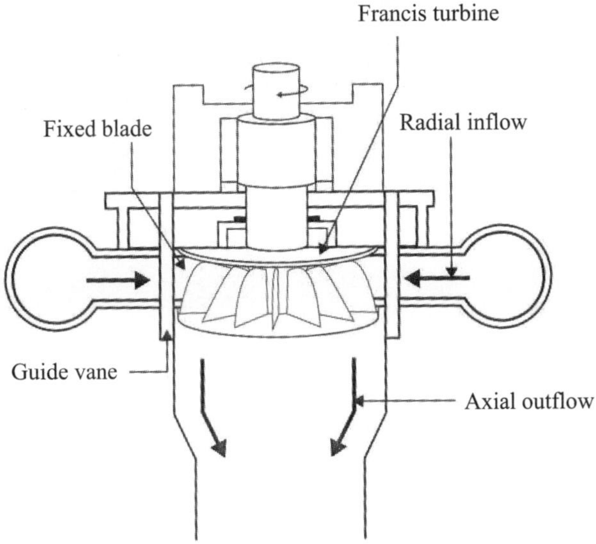

Fig. 19.5 Francis Turbine

(Reprinted from Chapter-8: Hydropower, Power Generation Technologies, 3rd Ed., Paul Breeze, 2019, Francis Turbine, 173-201, https://www.sciencedirect.com/science/article/pii/B9780081026311000080, with permission from Elsevier – License No. 5230750843511)

The allowable head variation for safe operation is in the range of 65 to 125 per cent of design head. With the aid of modern design analysis tools such as CFD (Computational Fluid Dynamics) peak efficiencies in the range of 90 to 94 per cent could be attained for these turbines.

(b) *Axial flow turbine:* Axial flow turbines are those in which water flows through the runner along the axis of turbine rotation. These types of turbines have successfully been applied for net heads up to 75 m with an output power up to 200 MW. However, they are extensively used in applications below 35 m head. *Tubular* turbines (Fig. 19.6) are used below 30 m head and up to 8 MW capacity. *Bulb-type* turbines (Fig. 19.7) can be used to about 25 metres head and up to about 100 MW capacity. In small hydropower plants, bulb-type turbines can be used for low heads, provided that the runner diameter is larger than 1 m.

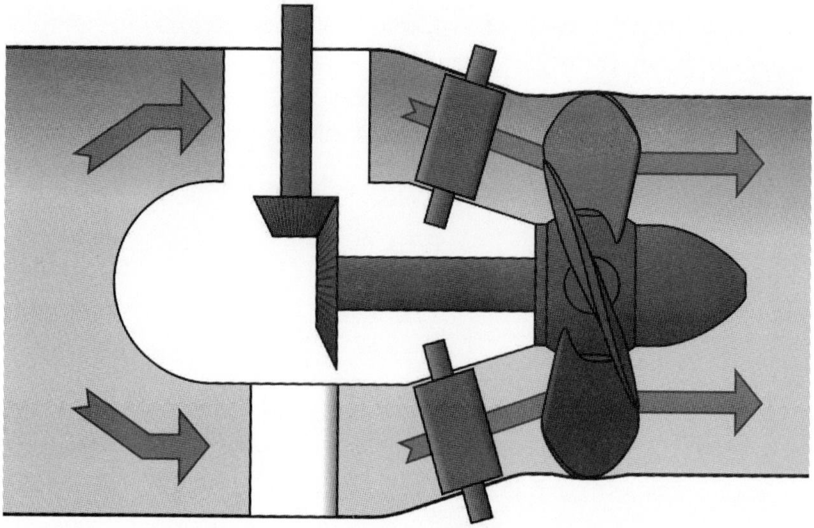

Fig. 19.6 Tubular turbine

(Author: Jahobr, Sketch of a tubular gear turbine, Available at https://commons.wikimedia.org/wiki/File:TubularGearTurbineSketch.svg, GNU Free Documentation License CC0 1.0 Universal Public Domain Dedication via Wikimedia Commons)

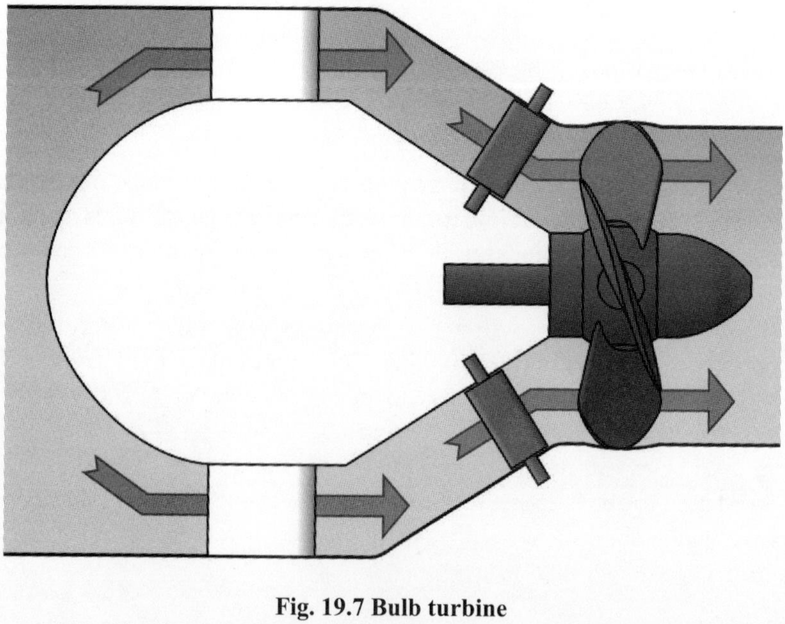

Fig. 19.7 Bulb turbine

(Author: Jahobr, Sketch of a bulb type turbine, Available at https://commons.wikimedia.org/wiki/File:BulbTurbineSketch.svg, GNU Free Documentation License CC0 1.0 Universal Public Domain Dedication via Wikimedia Commons)

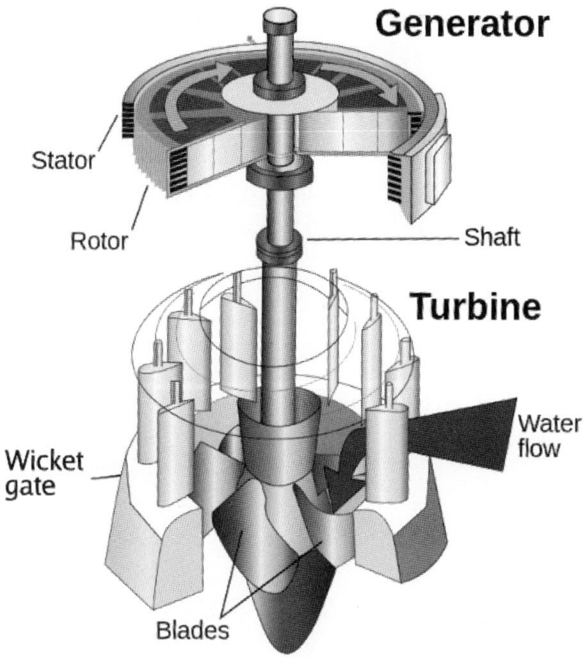

Fig. 19.8 Kaplan turbine with variable pitch runner blades

(Author: U.S. Army Corps of Engineers, Hydraulic turbine and electrical generator, Available at https://commons.wikimedia.org/w/index.php?curid=18581340, in Public Domain via Wikimedia Commons)

Kaplan or propeller-type turbines (Fig. 19.8) have a runner with a number of blades and water passes through the runner in an axial direction with respect to the shaft. There may be three, four, five or six blades with variable or fixed pitch angle. The pitch angle of a turbine blade describes the angle of the blade to the plane of rotation. The major components of this type of turbine include a water supply case, wicket gates, a runner and a draft tube. The draft tube is a diverging pipe which safely disposes the high-velocity low-pressure water jet emerging from the reaction turbine. It allows to dissipate the kinetic energy of the emerging water by gradually increasing the cross-sectional area of the outlet. Provision of draft tube enhances overall efficiency of the turbine, reduces erosion of outlet components as well as eliminates chances of backflow of water from tail race. The tail race is nothing but a channel which carries away the emergent water from the turbine to the downstream side. Axial flow turbine is called 'Kaplan turbines' if the blades are movable, whereas those with fixed blades are termed 'propeller turbine'. There are also '*semi-Kaplan*' axial flow turbines which have movable blades but with fixed guide vanes. They do not have wicket gates as in the former type.

19.4.2 Impulse turbines

Impulse turbines operate with their runner in air and convert the pressure energy of water into kinetic energy of a jet that impacts onto the runner buckets to develop mechanical torque. Basically, an impulse turbine is the arrangement of one or more high pressure jets discharging into an aerated space thereby impacting on the runner buckets. The typical impulse turbine can often achieve efficiencies around 90 per cent and above.

(a) *Pelton turbine:* A single nozzle Pelton turbine (Fig. 19.9) exhibits a nearly flat efficiency curve as shown in Fig. 19.10 and could be run down to loads of 20 per cent of rated capacity with an acceptable efficiency. For the multi-jet Pelton turbines, the range is even wider, since the number of effective jets can be controlled.

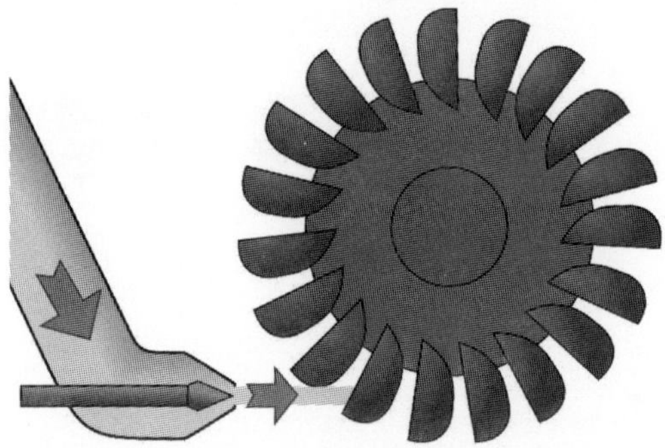

Fig. 19.9 Pelton wheel

(Author: Jahobr, Sketch of a pelton turbine, Available at https://commons.wikimedia.org/wiki/File:PeltonSketch.svg, GNU Free Documentation License CC0 1.0 Universal Public Domain Dedication via Wikimedia Commons)

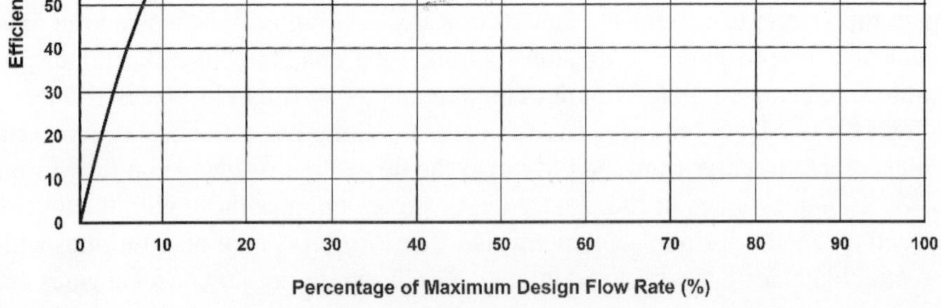

Fig. 19.10 Efficiency curve of a Pelton turbine

The turbine speed is controlled by hydraulically regulated needle nozzles in each jet. Moreover, the deflectors are provided to divert the jets away from the runner in case of emergency shutdown. This mechanism offers a surge protection feature without a pressure relief valve in the penstock. A large amount of stress is avoided, since the hydraulic load can swiftly be removed from the runner without changing the flow rate.

For the same available hydraulic head and flow conditions, increment in the number of jets requires a smaller size runner, resulting in a higher operating speed. Therefore, whether the axis mount is vertical or horizontal, multi-jet turbines tend to be less costly for equivalent output, as the cost of the runner embodies up to 20 per cent of the cost of the entire turbine assembly.

The maintenance cost for an impulse turbine is less than that for a reaction turbine with similar capacity, since they are practically free from cavitation problems. However, an excessive amount of silt or sand in the water would cause more wear on the impulse runner than on a reaction runner. On the other hand, draft tubes are not necessary for impulse turbines, as the runner is always positioned well-above extreme tailrace water level to perform the operation at atmospheric pressure.

(b) *Turgo impulse turbine:* A *Turgo impulse turbine* is a modified turbine based on the Pelton wheel concept. Eric Crewdson originally designed and patented it in 1920. Today, it is commercially being manufactured exclusively by Gilkes Energy in England. The Turgo runner (Fig. 19.11) is a cast wheel with its geometry resembling to a fan blade sealed on the periphery. The water flow is introduced on one side, which moves across the blades and discharges on the other side. The specific speed of turgo runner is considerably higher than the Pelton turbine. Hence, for the same runner diameter as a Pelton runner, the rotational speed of a Turgo runner would be higher. The Turgo runner has a wide operating hydraulic head range between 100 to 300 m. These are available up to 10 MW output capacities per turbine with an overall efficiency of about 80 to 87 per cent.

Fig. 19.11 Turgo impulse turbine

(Author: Jahobr, Sketch of a turgo turbine, Available at https://commons.wikimedia.org/wiki/ File:Turgo_Turbine_Sketch.svg, GNU Free Documentation License CC0 1.0 Universal Public Domain Dedication via Wikimedia Commons)

(c) *Cross-flow turbine:* A cross-flow turbine (Fig. 19.12) is an impulse type turbine with restricted air admission. Similar to the Pelton turbine, it has a flat efficiency curve over a broader array of flow and head (h = 2 to 180 m) situations. It involves a guide vane at the ingress, which leads the flow to a restricted portion of the runner subject to the amount of flow. This process fairly resembles to the multi-jet impulse turbine.

Compared to the other impulse turbines, the cross-flow turbine exhibits a lesser peak efficiency limiting to 80 per cent. Also, the maximum diameter of the runner is limited to 1.2 m, restricting the output per unit. However, the multi-unit systems can be installed to achieve higher overall power plant capacities.

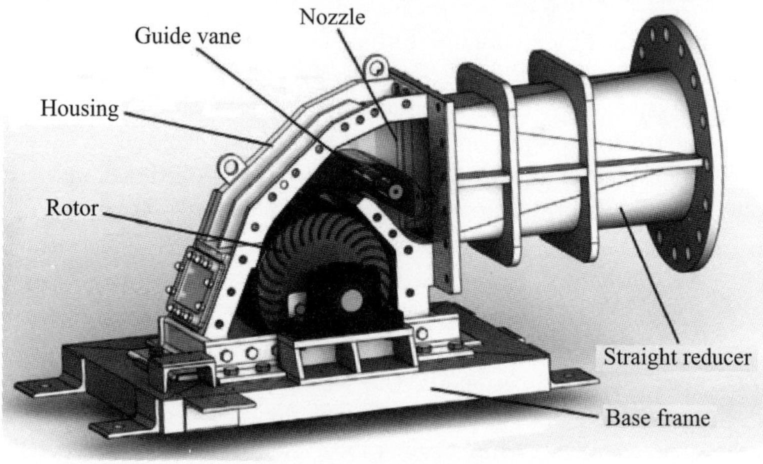

Fig. 19.12 Cross-flow turbine

(Reprinted from Renewable Energy, A novel design and performance optimization methodology for hydraulic Cross-Flow turbines using successive numerical simulations, 1402-1421, https://www.sciencedirect.com/science/article/abs/pii/S0960148121000975?via%3Dihub, with permission from Elsevier – License No. 5236370503879)

Cross-flow turbines often have conical draft tubes creating a slight vacuum in the turbine casing as discussed earlier. This feature helps in preventing the tailwater re-entry in the cross-flow turbines, which is the case for a typical impulse turbine. Air enters into the casing over a regulating air inlet valve to control the pressure. These turbines are free from cavitation, but similar to other impulse turbines, are vulnerable to wear with excessive silt or sand loads in the water. These turbines require less maintenance compared to the other types and its runners are self-cleaning.

(d) *Deriaz turbine:* In practice, a Francis turbine running below 40 per cent of the rated water-flow is susceptible to vibrations and/or power surges. Deriaz turbine (Fig. 19.13) was developed to alleviate these problems. These turbines have mixed-flow runner with variable pitch blades (as in Kaplan runners) guide and gate mechanism, operable with medium heads up to 200 m. The only one installation of Deriaz runner reversible turbine in India is at Kadana pumped storage scheme in the state of Gujarat.

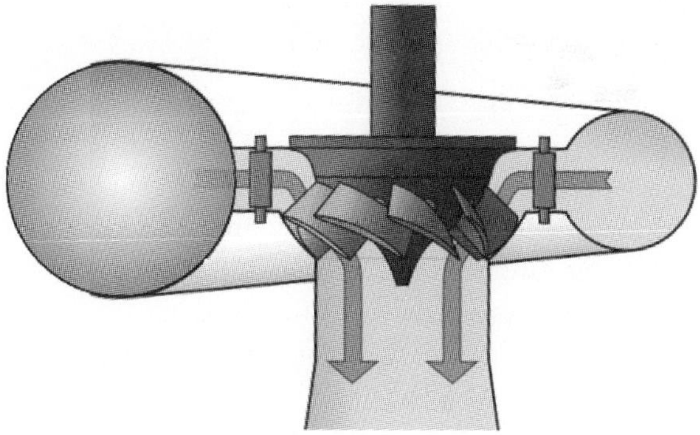

Fig. 19.13 Deriaz turbine

(Author: Jahobr, Sketch of a Deriaz turbine, Available at https://commons.wikimedia.org/wiki/File:Deriaz_Turbine_Sketch.svg, GNU Free Documentation License CC0 1.0 Universal Public Domain Dedication via Wikimedia Commons)

19.4.3 Fish-friendly turbines

Environmental concerns about hydropower includes safety of aquatic life. Some of the fish-friendly turbines for low head applications include water wheels, Archimedes screws and vortex turbines. For high head applications, the Alden turbine and the minimum gap runner turbine may be made use of.

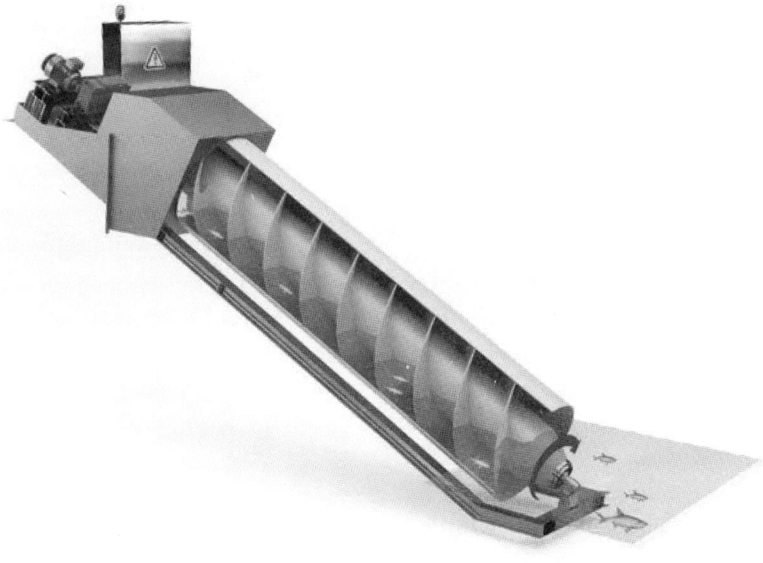

Fig. 19.14 Fish-friendly Archimedean screw turbine

(Courtesy: M/s. FishFlow Innovations, Netherlands)

19.5 Selection of Site

The selling price of electricity, plant factor and interest on capital are factors to be considered in the installation of a hydro power plant. The best location for a hydroelectric power plant is usually found along the path of a river at the river canyon or at the place where the river narrows. This is advantageous for collection of water for power production. The catchment area and the hydrologic characteristics of the area are also to be considered. For storing sufficient quantity of water for the required period, sufficient volume of basin is required above the dam. However, the following factors may be considered in the selection of site for hydel power plant:

(i) Potential based on water availability
(ii) Water storage facility
(iii) Hydraulic head available
(iv) Possibility and rate of sedimentation.
(v) Pollution issues
(vi) Proximity to load centres.

19.6 Turbine Selection Criteria

The major factor for proper operation of the power plant is the selection of turbines. Turbines are generally selected based on the following characteristics:

(i) Available head
(ii) Specific speed
(iii) Rotational speed
(iv) Turbine efficiency
(v) Cavitation
(vi) Arrangement of turbine shaft
(vii) Operation at partial load

The turbine selection should be based on the hydraulic head and for different head ranges different turbines are suitable. Heads are divided into five categories viz. very low (3 to 10 m), low (10 to 60 m), medium (60 to 150 m), high (150 to 350 m) and very high head (more than 350 m). Kaplan turbines are suitable for very low and low head applications, whereas Francis turbines may be used for medium heads. Francis and Pelton turbines may be useful for high heads and Pelton turbines are appropriate for very high heads.

The term specific speed is often used in classifying turbines and is high for turbines which works under high heads with low flow rate. Specific speed is defined as the rotational speed (revolutions per minute) at which a hydraulic turbine would operate at its optimum efficiency producing unit power (one kilowatt) under unit head (one meter). The equation for specific speed derived from non-dimensional considerations is therefore:

$$\text{Specific speed, } N_s = \frac{N \times \sqrt{P}}{H^{1.25}} \times 100 \qquad ...(19.3)$$

Specific speed is low for a turbine which works under low hydraulic head and high flow rate. When the specific speed of the selected turbine is not appropriate, the resultant reduction in rotational speed of turbine will also reduce the capacity of the plant. Francis turbines work with medium-specific speed and can generate high rotational speed. There can be a wide range of specific speeds suitable for a particular head. The specific speed of the turbine should match the rotational speed requirement of the generator. The specific speed of the turbine as well as the frequency and number of pair of poles in the electric generator decides the rotational speed of the turbine.

Selection of a high specific speed turbine operation for a given head may be more economical as a smaller turbine and generator can be used, as there is a saving in the equipment cost. But the requirement of costly installation may offset this economy as in the case of reaction turbines, as they need to be placed lower. The basic criteria for selection of a hydro-turbine based on specific speed as per Indian standards IS:12837 is given in Table 19.1.

Table 19.1 Turbine selection criteria based on specific speed

Turbine type	Head variation, % of rated head	Load variation, % of rated output	Specific speed	Peak efficiency, %
Pelton	120 to 80	50 to 100	15 to 65	90
Francis	125 to 65	40 to 100	60 to 400	93
Deriaz	125 to 65	50 to 100	200 to 400	92
Kaplan	125 to 65	40 to 100	300 to 800	92
Propeller	110 to 90	90 to 100	300 to 800	92
Bulb	125 to 65	40 to 100	600 to 1200	92

(Source: Indian standards IS:12837)

The efficiency of the turbine selected should be as high as possible in consideration of varied operation conditions. Different turbines in the order of their efficiencies are, Pelton $<$ Kaplan $<$ Francis.

Cavitation of turbines is a complex phenomenon that involves the formation and interaction of bubbles inside a liquid medium which results in negative effects on hydro turbines. Cavitation mainly occurs in the case of reaction type turbines and it cause excessive vibration, noise and surface erosion leading to performance deterioration. So, while selecting the turbine, cavitation factor should be determined for that particular turbine so as to make sure that it is safe.

It is recommended that the horizontal shaft arrangement is best suitable for large size impulse turbines such as Pelton turbine. In the case of large size reaction turbines such as the Kaplan turbine, vertical shaft arrangement is recommended. The efficiency of the turbine is maximum when it is running at the rated load. Efficiency reduces at part-load or overload conditions and the selection shall be in consideration of these operating conditions as well.

CHAPTER

20

Ocean Thermal Energy Conversion (OTEC)

At present, our civilization consumes nearly 17.7 TW of power taken from all the energy resources (fossil and renewable). Considering solar insolation received at Earth's outer atmosphere as 1367 $W \cdot m^{-2}$ and the cross-sectional area of the Earth as 1.277×10^{8} m (as R = 6378 km), our planet receives 174,000 TW of solar power continuously. Looking at the numbers, our planet receives solar power almost 10,000 times of current global demand.

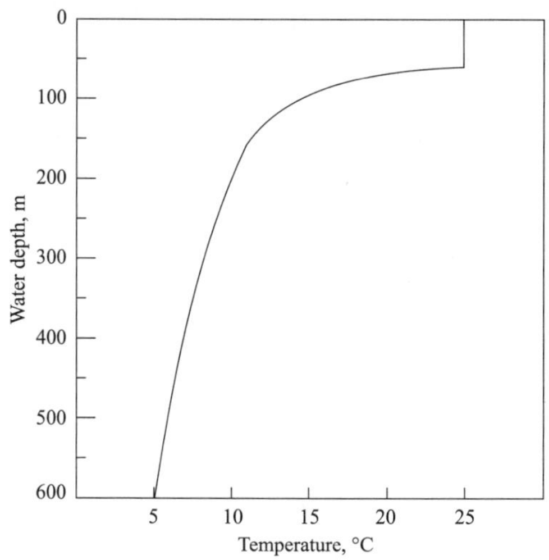

Fig. 20.1 Typical temperature gradient profile of tropical oceans

In order to harness even a small fraction of this energy sustainably, huge collecting areas and enormous storage capacities are necessary, which could be met by the tropical oceans, as they cover 71 per cent of the Earth's surface.

20.1 Temperature Gradient in Oceans and Scope of Energy Conversion

Tropical oceans are the largest solar thermal collectors absorbing solar radiation which heats up the upper layers to around 25°C (Fig. 20.1). This warmed up superficial water from the equatorial belt moves towards the poles melting both the Arctic and the Antarctic frost. The consequential cold-water currents flow back towards the equator at great depths effecting the most gigantic thermosyphon cycle on our planet. Considering the Gulf Stream, which has the flow rate of 9.5×10^{10} $m^{3} \cdot h^{-1}$ of water at a temperature around 293K, exhibits a huge heat engine. Theoretically it handles a heat flow of $\Delta T \times Cp \times Q$, where ΔT is the temperature difference, Cp is the heat capacity of water (about 4.186 $MJ \cdot m^{-3} K^{-1}$) and Q is the flow rate. This amounts to 7.95×10^{12} $MJ \cdot h^{-1}$ or approximately 2200 TW. Though it has some potential as a commercially viable supplemental energy source, only a tiny portion of this vast energy can actually be harnessed with the present *Ocean Thermal Energy Conversion* (OTEC) systems.

Conventional OTEC systems work based on the temperature gradient between the surface and deep waters in the ocean. This temperature difference is about 4K per km depth. According to thermodynamic principles, the efficiency of a heat engine operating between temperatures, T_1 and T_2 is given by:

$$\eta = \frac{T_1 - T_2}{T_1} \quad ...(20.1)$$

The mean surface temperature of ocean water in tropics is about 300K and the practical temperature difference is only about 5K. As a result, the practically achievable efficiency is only about 2 per cent or less. In order to have a reasonable power production, we need large pipe lines and a pumping system to achieve large intake of water, making the capital cost prohibitive. However, there has been some development in the recent past, to make the system more practical.

OTEC is a process or technology for producing energy by harnessing the temperature differences (thermal gradients) between ocean surface waters and deep ocean waters. According to U.S. Energy Information Administration, a temperature gradient of at least 25°C is necessary to power a turbine to produce electricity. The basic concept is illustrated in Fig. 20.2. Warm surface water is pumped through an evaporator containing a working fluid. The vaporized fluid drives a turbine/generator. The vaporized fluid is turned back to a liquid phase in a condenser circulated with cold ocean water pumped from deep ocean. OTEC systems using seawater as the working fluid can use the condensed water to produce desalinated water.

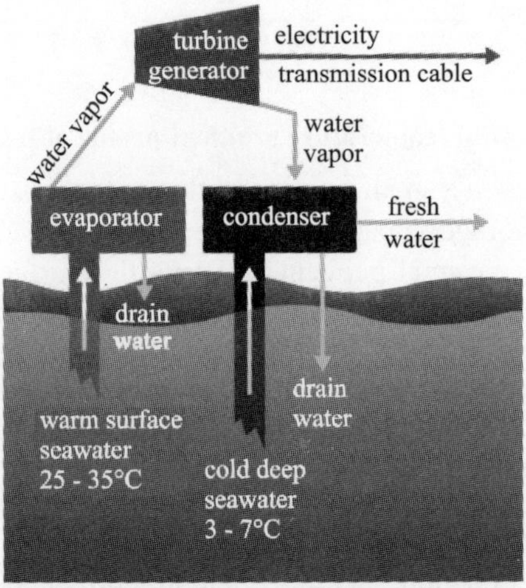

Fig. 20.2 Schematics of OTEC process
(Courtesy: U.S. Energy Information Administration)

20.2 Types of Ocean Thermal Energy Conversion (OTEC) Systems

The two practical configurations projected for energy conversion are Hydraulic turbine (HT) and Vapor turbine (VT) OTECs.

20.2.1 Hydraulic turbine (HT) – OTEC system

This is one of the early configurations for OTEC during 1970s, exploiting the temperature gradient between the superficial and abyssal (deeper) waters to generate the hydraulic head that can drive a conventional water turbine. This approach eliminates the requirement of heat exchangers, reducing the system complexity.

The HT–OTEC configuration suggested by Beck (1976), as illustrated in Fig. 20.3a, comprises of a hemispherical dome in which a long conduit delivers cold water, whereas a small one carries warm water. The dome is evacuated so that, only the low-pressure water vapor inhabits the space above the liquid surface. In this case, other gases dissolved in the water may also be present along with the water vapor which should be scavenged out.

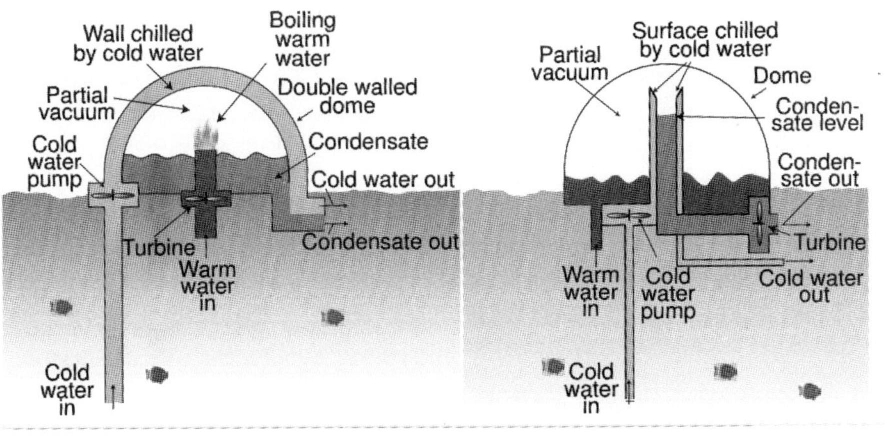

(a) **(b)**

Fig. 20.3 Hydraulic turbine OTEC system

In practice, at a temperature of 15°C, the interior pressure of the container is nearly 15 kPa (0.017 atm.). The upper-oceanic warm water at 25°C should boil at this condition. Since the dome being cooled by low-temperature water pumped-up from the deeper layers, the resultant vapor would condense on the inner surface. The condensate flows back into the ocean, creating a continuous warm water flow inside the dome. This inward warm water movement drives a turbine coupled to the generator producing useful electricity. Since the magnitude of hydraulic head at the warm water entrance is low, larger turbines are required for sustainable energy production.

In 1975, Zener and Fetkovich proposed a modified configuration (Fig. 20.3b.) to partially overcome the obstacle of low hydraulic head. In a modified HT–OTEC system, one of the two concentric cylinders at the dome's core is sealed by a funnel-like surface. The low-pressure vapor formed inside the dome condenses inside this internally cooled funnel and the condensate then runs into the central conduit, at the base of which turbine is mounted. Under the action of enhanced hydraulic head, the condensate drives the turbine more efficiently.

20.2.2 Vapor turbine (VT) – OTEC system

The concept of OTECs based on vapor-driven turbines were developed in the 1980s. These systems have various configurations depending on the type of cycle involved in energy conversion process as given below:

(a) Open cycle – with no distilled water recovery
(b) Open cycle – with distilled water recovery
(c) Closed cycle
(d) Hybrid cycle

The open cycle (Fig. 20.4a) eliminates the necessity of heat exchangers, unless the distilled water recovery is intended (Fig. 20.4b). The low-pressure steam generated in the open cycle system necessitates oversized turbines, which may not be feasible in most of the cases. A closed (Fig. 20.4c) or hybrid cycle (Fig. 20.4d) employs a secondary thermal exchange fluid like ammonia (NH_4) which eliminates the problem of low-pressure vapor. A heat exchanger transmits thermal energy from warm oceanic water to the liquid ammonia, generating relatively high-pressure vapor which then drives the turbine. The secondary thermal fluid with higher coefficient of volumetric thermal expansion ($\beta_{ammonia} = 210\times10^{-6}\ K^{-1}$) than the water ($\beta_{water} = 245\times10^{-5}\ K^{-1}$ at 293 K) enhances the work efficiency of the system.

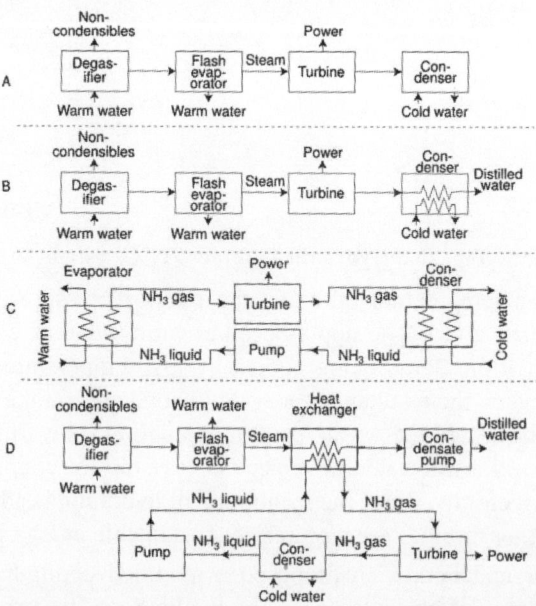

Fig. 20.4 VT-OTEC configurations – (A) Open cycle without distilled water recovery, (B) Open cycle with distilled water recovery, (C) Closed cycle, and (D) Hybrid cycle

20.3 Present Status of OTEC

OTEC power plant designed, developed and installed by Makai Ocean Engineering, USA is the world's largest operational facility of its kind. The power plant has a 100-kW power generating capacity, which is enough to power 120 households in Hawaii. This advanced OTEC facility is situated in Kailua-Kona at the Natural Energy Laboratory of Hawaii Authority (NELHA). In 2015, it was connected to the grid and has potential of supplying base-load power, i.e., it can continuously generate electricity 24 hours throughout the year.

Makai's OTEC plant is a closed cycle vapor turbine system that uses the ammonia as thermal exchange fluid to drive the turbine generator (100 kW) with two advanced heat exchangers. The turbine generator is placed on top of the 12 m tall OTEC tower. Deep seawater can be acquired using a 1000 mm diameter intake pipeline, capable of supply up to 10 m^3 min^{-1} of cold seawater. The amount of warm surface water pumped into the facility each minute is approximately equal to the amount of deep-sea cold water pumped into the facility each minute. At the end of the cycle, mixed volume of warm and cold water is reverted back deep into the sea.

The two heat exchangers supplied by Lockheed Martin Corporation, Washington D.C. have a 2 MW thermal capacity each, designed to handle the seawater flow of 0.25 $m^3 \cdot s^{-1}$. They are the first heat exchangers manufactured using friction stir welding technique which helps in resisting oceanic corrosion.

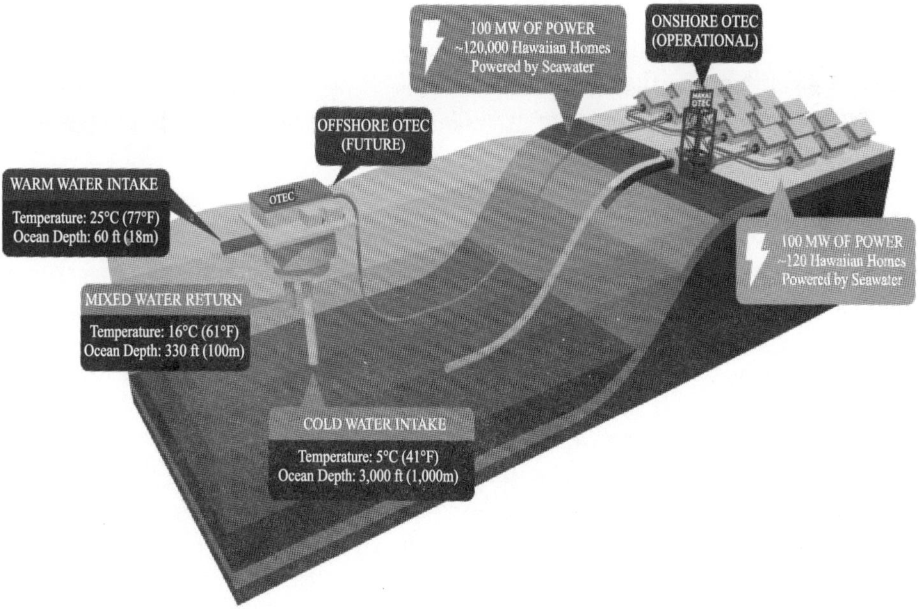

Fig. 20.5 Schematics of the closed-cycle 100 kW VT–OTEC power plant
(Courtesy: Makai Ocean Engineering, U.S.A.)

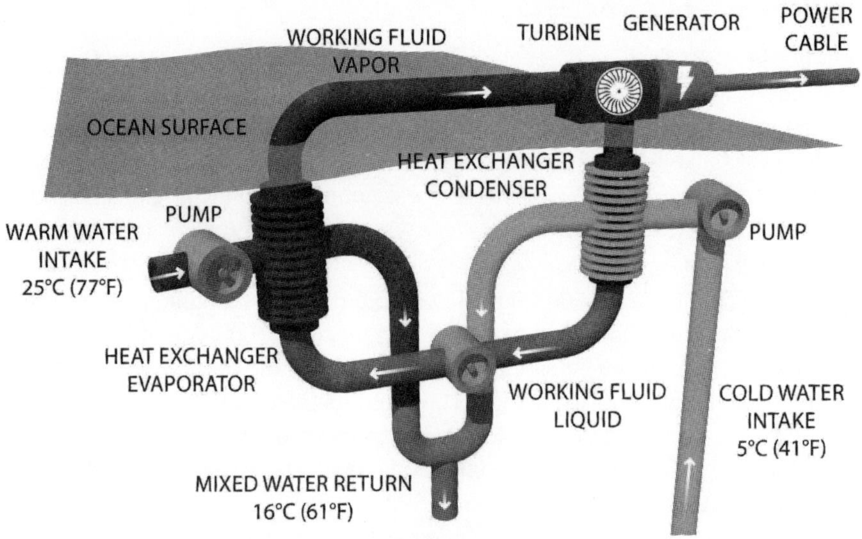

Fig. 20.6 OTEC power plant working mechanism
(Courtesy: Makai Ocean Engineering, U.S.A.)

To generate electric power, large amounts of deep cold saltwater and warm surface seawater are pumped to run a power cycle. Using warm surface saltwater and a heat exchanger, liquid ammonia is vaporized; which in turn drives the turbine generator. A condenser cools the ammonia vapor and transforms it back into a liquid condition while the steam transfers its energy to the turbine. To restart the power cycle, the liquid ammonia is returned to the evaporator of the heat exchanger. The overall efficiency of energy conversion is 2.65 per cent.

20.4 Future Prospects of OTEC

For the future ventures, Makai Ocean Engineering intend to employ the observations from the existing plant to build the subsequent higher power plants, extending output capacities between 1 MW and 100 MW. Their joint venture in association with Lockheed Martin Corporation propose to create 100 MW OTEC plants for tropical island populations such as Hawaii and Guam. Simultaneously, Lockheed Martin Corporation – U.S.A. is also working with Reignwood Group (a Chinese investment company) to design and develop a 10 MW OTEC pilot power plant off the coast of southern China.

Similarly, as a part of the Okinawa-Hawaii clean energy agreement reached in 2010, the Ocean Energy Research Center (OERC) – NELHA is being considered by an international consortium comprising of Japanese and US partners.

CHAPTER 21

Wave Energy

Nearly one third of the earth's surface is covered by oceans and the major quantity of solar energy incident on earth is intercepted by oceans. A part of this energy is used for the surface winds in the ocean and a small portion of this energy is utilised by ocean currents. A considerable quantity of the energy is converted to winds by natural energy transfer near the ocean surface. They blow across oceans creating wind driven waves.

21.1 Energy from Ocean Waves

Ocean wave energy conversion has been identified as a possible technology for renewable energy production. One attraction is the availability of a good energy flux up to 70-kW per metre length of the wave depending on the energy parameters. It is also possible to predict the conditions of waves over a reasonable period of time and is somewhat consistent and regular. Hence many attempts have been made to utilize the energy of waves as they could be developed into a feasible commercial power production venture (Fig. 21.1).

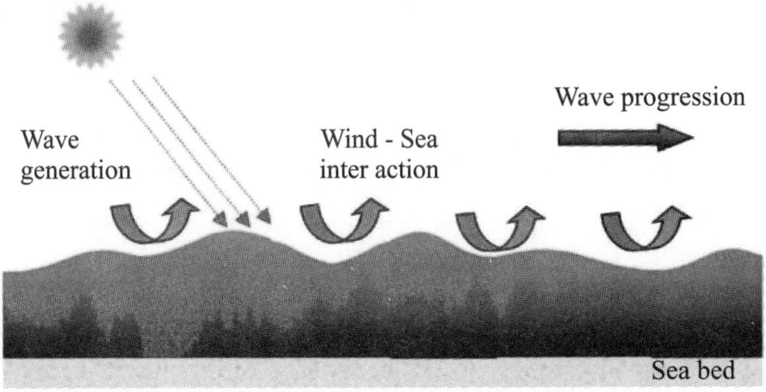

Fig. 21.1 Wave generation and progression

Extraction of energy from waves should be more efficient than from wind as waves contain concentrated energy. Even though the energy of the waves in Indian costs is lower than that in northern and extreme southern latitudes, the estimated potential is about 40,000 MW. Even though ocean waves are random in nature, a wave atlas has been published by the National Institute of Oceanography in Goa, India. Scientific wave data can be obtained from 'wave rider buoys'. The average wave power availability on Indian costs is in the range 5 to10 kW.

21.2 Nature of Wave Energy and its Estimation

Waves travel in the direction of the wind and they possess kinetic energy and potential energy. The movement of the water particles when subjected by wave motion generally follow a circular path. When the depth of the seabed, D is more than half of the wavelength then the water particle moves in a circular path. The radius of this circle decreases with depth and when the wave reaches shallow points near the cost, the motion path becomes elliptical (Fig. 21.2). As the amplitude of the water particles are reduced in elliptical motion, the energy also is reduced.

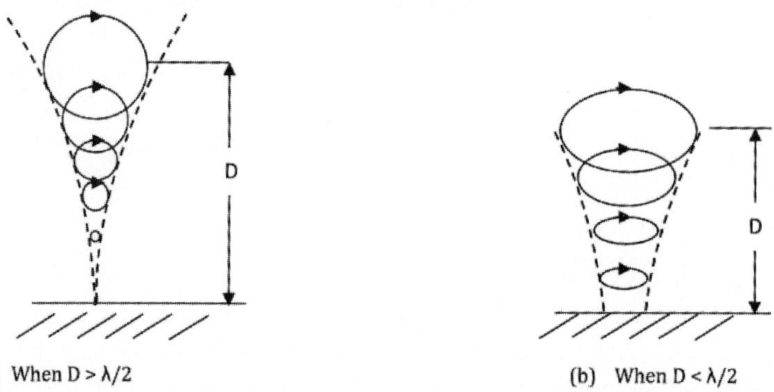

Fig. 21.2 Movement of water particles due to wave action

When the water particle is elevated from its original position its potential energy also is escalated. The total energy is the sum of the kinetic energy and potential energy. The surface waves are composed of continuous sine waves of irregular wavelength and direction. Even though the wave form progresses with a velocity c, the water particles do not have any net forward motion. They have only a rotary motion, in a circle of radius equal to the amplitude (Fig. 21.3).

The following parameters are relevant in estimating the energy of a wave:

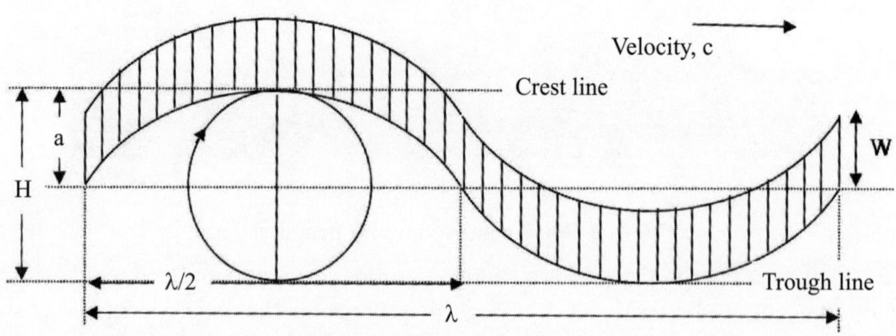

Fig. 21.3 Parameters of wave

(i) Wave Height (*H*) – Trough to crest height measured in metres

(ii) Wavelength (λ) – Distance between two consecutive crest-to-crest or trough-to-trough, m

(iii) Amplitude (a) – Displacement of water particle on the crest or trough extremes from the mean sea level ($H/2$), m

(iv) Period (T) – The time period during which the wave form repeats or time in between two consecutive crests or two troughs, s

(v) Frequency (f) – No. of crests or troughs passing through a given point in unit time ($1/T$)

(vi) Velocity of propagation or phase velocity (c_p) – Distance travelled by the crest or trough (*i.e., the phase*) in unit time, m s^{-1}

(vii) Density of water (ρ) – Mass of water per unit volume at given temperature, kg m^{-3}

(viii) Acceleration due to gravity (g) = 9.81 m s^{-1}

The period (T) is the reciprocal of frequency (f) and the product of frequency and wavelength (λ) give the velocity of the wave. The angular repetency (k) of a particular waveform is given by the number of radians per unit wavelength i.e., $k = 2\pi/\lambda$.

In terms of angular motion, wavelength and period are given as:

$$\lambda = \frac{2\pi g}{\omega^2} \quad ...(21.1)$$

$$T = \frac{2\pi}{\omega} \quad ...(21.2)$$

Therefore, the relationship between wavelength, period and phase velocity can be expressed using the following equation:

$$c_p = \frac{\lambda}{T} = \frac{2\pi g}{\omega^2} \cdot \frac{\omega}{2\pi} = \frac{g}{\omega} \quad ...(21.3)$$

Substituting, in the above equation, we get:

$$c_p = \frac{g}{2\pi} \cdot T \quad ...(21.4)$$

Also, it alternatively gives an expression for wavelength in terms of phase velocity and period as:

$$\lambda = c_p \cdot T = \frac{g}{2\pi} \cdot T^2 \quad ...(21.5)$$

In practical conditions, two sinusoidal waves interfering with varying wavelengths, but having the similar amplitude and the direction of propagation, create a *group* or *packet* pattern. The wave group or packet travels with a velocity (called group velocity, c_g) different from the phase velocity (c_p), owing to frequency dispersion. The group velocity in fact is the energy carrying velocity, with which the wave energy is transferred horizontally in a narrow-band wave field.

In case of shallow water, the difference between phase and group velocities becomes infinitesimally small, as the frequency dispersion is negligible.

However, the frequency dispersion is profoundly existing in deep water, which generally reduces the group velocity equal to half the phase velocity i.e., $c_g = \frac{c_p}{2} = \frac{gT}{4\pi}$.

21.3 Energy of the Wave

Assuming a pure sinusoidal wave propagating in space-time, with the wavelength (λ, m) and wave-front width (*W*, m), the Potential Energy (*PE*) possessed by the wave per unit width of wave-front can be given by,

$$PE = \frac{1}{4} \cdot \rho g \cdot a^2 \lambda W \qquad ...(21.6)$$

Additionally, there exists an equal magnitude of Kinetic Energy (*KE*) associated with the mass of water in motion. Therefore, the total energy carried by the wave per unit width of wave-front is:

$$E = PE + KE = 2PE = \frac{1}{2} \cdot \rho g \cdot a^2 \lambda W \quad \text{in joules} \qquad ...(21.7)$$

The energy density, E_d of the wave is the energy contained per unit wave area ($\lambda \times W$),

$$E_d = \frac{E}{\lambda W} = \frac{1}{2} \cdot \rho g \cdot a^2 \ \text{Jm}^{-2} \qquad ...(21.8)$$

The wave power density, P_d refers to the rate of energy flow per unit width of wave-front obtained by the product of energy density and the group velocity of wave:

$$P_d = E_d \cdot c_g = \frac{1}{2} \cdot c_g \rho g \cdot a^2 \ \text{Wm}^{-1} \qquad ...(21.9)$$

By substituting $c_g = gT/4\pi$ and $a = H/2$, the above equation results:

$$P_d = \frac{1}{2}\left(\frac{gT}{4\pi}\right)\rho g \cdot \left(\frac{H}{2}\right)^2 \qquad ...(21.10)$$

$$\therefore \quad P_d = \frac{\rho g^2 T}{32\pi} H^2 \qquad ...(21.11)$$

21.4 Bottlenecks in Wave Energy Conversion

Even though ocean waves offer a renewable source, to develop the theory into an efficient and feasible working model is troublesome due to the following factors:

(i) The ocean wave patterns are highly irregular.

(ii) Construction of structures for installation of wave machines is highly risky and troublesome due to the fact that very strong waves to the extent of 50 times the normal power may occur during hurricanes striking in a 50-year probability. The construction becomes costly and at the same time the device may become less efficient at normal wave conditions.

(iii) The frequency of waves is generally very low as the wave periods are usually in the range 5 to 10 seconds. These devices cannot be directly coupled to generators to produce electricity as they require higher frequencies.

(iv) If the energy of high amplitude waves is to be tapped, the devices should be installed in deep seas, which is again troublesome. The anchoring of the system in deep sea beds is difficult and the transmission of electricity produced also is complicated.

(v) Sea water is corrosive and the materials of construction need to be corrosion resistant for a long duration.

21.5 Wave Energy Converters (WEC)

The possibility for power extraction from waves was known since 1799, when the first patent on use of ocean wave power was granted to Monsieur Girard of Paris. Since then, thousands of patents have been taken on many wave-machine concepts. In 1974, Stephen H. Salter developed a wave converter which was popularly known as 'Salter's duck' and was later officially referred as 'Edinburgh duck' as the inventor belonged to the University of Edinburgh. Thereafter, many concepts were developed by many researchers.

21.5.1 Types of wave energy converters

Many devices are based on the 'Oscillating water column', which works on the principle of air pressurisation by the movement of a water column inside a chamber. Overtopping devices have a reservoir into which the sea water overtops by the wave action, thereby producing an elevation difference between the water levels in the reservoir and sea level. The power production is done similar to hydroelectric generation. Wave absorbing devices can harness the energy of waves directly. They can be point absorbers like simple buoys which can take the energy of waves from all directions. Multi point absorbers have a number of floating members. Another class of wave energy absorbers are called terminators which comprise of flaps or rollers rotating in an axis perpendicular to the direction of wave propagation. Wave attenuators make use of hinged cylindrical members or catamarans, whose relative motion is utilised by hydraulic or pneumatic pumps. A large number of concepts were evolved during the last two centuries, and many of them belong to the types discussed above.

(a) *Oscillating floating buoy:* This is one among the simplest concepts in which a floating buoy is used to impart up and down motion of the counterweight through pulleys. This reciprocating motion is transmitted to a piston through a suitable arrangement. The buoy is anchored to prevent drifting off from position (Fig. 21.4).

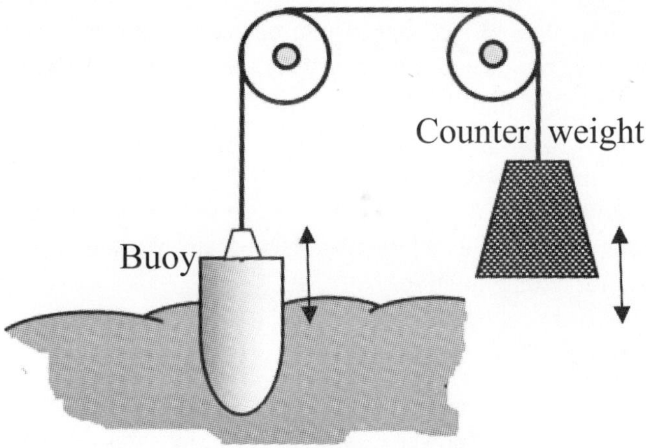

Fig. 21.4 Oscillating floating buoy

(b) *Oscillating air column devices:* These devices consist of a submerged hollow cylindrical structure open below the sea water level. When the wave crest pushes the water upwards, the air inside the chamber is compressed. When the trough reaches the structure, the air is decompressed. Thus, these oscillatory motions of the wave pressurise and depressurise the air inside the chamber (Fig. 21.5). This creates an oscillating air current flowing in both directions. A special type of air turbine called 'Wells' turbine is used to produce mechanical power to turn the generator. Wells turbine has blades having aerofoil sections symmetrical about the plane of rotation. Hence the rotation of the shaft is in the same direction, irrespective of the direction of air flow. A similar device to drive a water turbine also is possible as shown in Fig. 21.6.

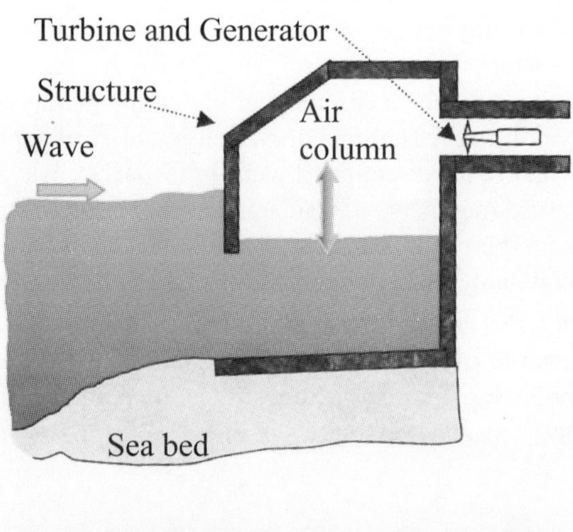

Fig. 21.5 Oscillating air column with air turbine

(c) *Hydraulic and pneumatic piston devices:* A hydraulic piston type wave energy converter has two pistons of varying area and work on the principle of hydraulic pressures. When a crest is formed, the piston with larger area rises up pushing water to a chamber from where hydraulic pressure push a piston of smaller area. Pressurised water can drive a water turbine.

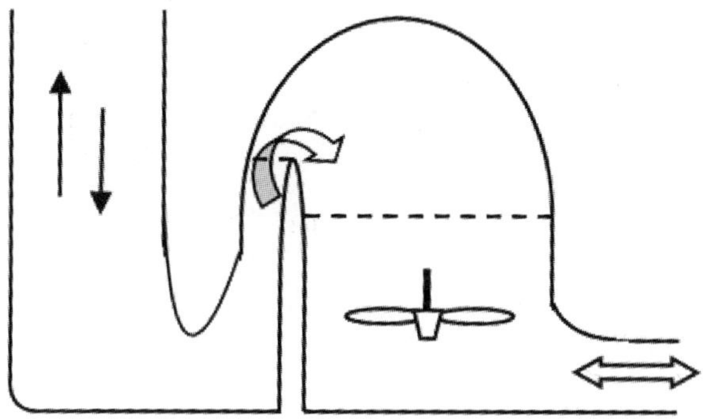

Fig. 21.6 Oscillating air column with water turbine

In a similar pneumatic piston concept, an oscillating float works an air pump to produce compressed air which in turn can energise an air turbine.

(d) *Overtopping devices:* These devices can hold the water pushed in by the waves into their temporary storage and subsequently let it out to run low head turbines.

(e) *Dolphin buoy type:* In this type of device, the buoy is mounted such that the rotary motion of the buoy is converted to reciprocating motion of a connecting rod. A floating roller is used as the buoy which rotates about a shaft connected to a floating generator through gears. This float assembly is connected to another generator unit through a supporting shaft, which oscillates due to the swinging movement of the buoy assembly. Thus, there is two stage generation of electricity (Fig. 21.7).

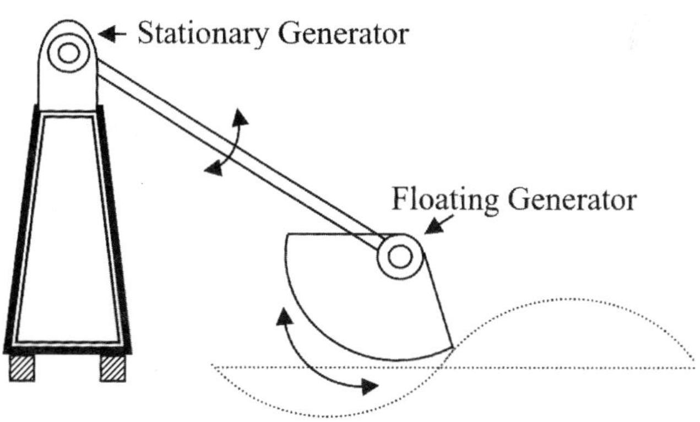

Fig. 21.7 Dolphin buoy type device

(f) *Three-raft converter:* This device consists of three raft like floating bodies hinged together like a chain to enable swinging of individual members by wave action.

Power can be extracted using air pumps or hydraulic pumps. Piston rods are connected to the side members and the middle member has the cylinder (Fig. 21.8).

(g) *Nodding duck wave machine:* This is one among the earliest concepts initially developed by Salter. A number of cam-shaped floating members are connected through hinges. When the crest and trough approach alternatively, the cam shaped float oscillates and transmit power to a generator rotor through a transmission system.

(h) *Surging devices:* These devices make use of the horizontal velocity of the water particles to deflect a flap or flexible member kept against the wave direction. The swinging action of the flap is transferred to a piston through suitable drive mechanism.

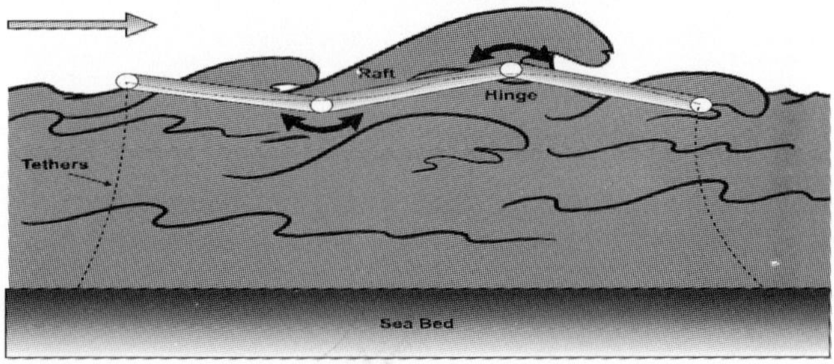

Fig. 21.8 Three-raft converter

21.6 Entrepreneurship in Wave Power Development

Major research efforts in the development of wave energy were initiated in 1970s consequent to the oil crisis. Salter's duck was one among the pioneering works in the University of Edinburgh and was followed by many entrepreneurs and institutions, particularly in Europe. One such development is the 'Archimedes wave swing' device developed by Teamwork Technology in Netherlands in the 1990s. The Scottish firm, 'Pelamis Wave Power' tested their successful device at the European Marine Energy Centre in Scotland during the period 2004 to 2007.

Further, the Pelamis wave energy farm, probably the first commercial wave energy project, having a capacity of 2.25 MW located in Portugal started operation in 2008. They built many machines during 2006 to 2014, but the firm went into liquidation. The Energetech Oscillating Water column was developed by Energetech Ltd., Australia. 'Wave Dragon' was an overtopping type system built by a group under the leadership of Wave Dragon ApS in Denmark. Another Denmark firm 'Ramboll' developed a Point Absorber type converter. 'Mighty Wale', yet another oscillating water column type device developed by Japan Marine Science and Technology Centre is a 120-kW prototype. Ocean Power Technology in USA developed the 'OPT Wave Energy Converter'. Wavegan Ltd. in UK in association with Queen's University in Belfast developed 'LIMPET', the 500-kW oscillating water column device. In India, pilot scale studies were taken up by Indian Institute of Technology, Madras, who installed a prototype oscillating water column type device with a capacity of 150 kW in 1991 at Vizhinjam near Thiruvananthapuram in Kerala state.

CHAPTER 22

Tidal Energy

The periodic rising and falling of the surface of the ocean and connected water bodies (such as gulfs and bays) that occur due to the variation in the gravitational forces exerted on earth by the sun and moon are called tides. These differential gravitational pulls on different parts of the earth create corresponding motions or currents in the world's oceans, usually happening twice in a day. The strong attraction to the part of the ocean which face the celestial body creates a bulge in the ocean in that part, causing a temporary increase in sea level. When the sea level is raised, water from other areas of the ocean is forced to move towards the bulging zone. This phenomenon creates tidal waves which moves towards the shorelines. A tidal generator converts the energy of moving water due to tides into electricity. As tides are created by gravitational interaction of earth with the moon and sun and the earth's rotation, the energy that can be extracted is practically inexhaustible and hence renewable. While almost all other renewable energy sources are derived from the light and heat from sun, tidal energy is an exception.

22.1 Tides and their Nature

Tides, in practice, are long-period waves that flow through the oceans in conjunction towards the sun and moon's gravitational influence. Tidal waves start in the oceans and propagate to the shorelines, where they exhibit a phenomenon of regular rise and fall of the sea level. High-tide occurs when the wave's highest point (crest) reaches a certain location; where as low-tide occurs when the wave's lowest point (trough) reaches a specific location. The tidal range is defined as the height difference between high and low tides (Fig. 22.1). The moon is responsible for around 70 per cent of the tide-producing force, while the sun accounts for 30 per cent.

Tidal waves in the open sea are only distinguished by height variations of a few metres. The mainland, however, has a braking effect on the tidal wave and creates backwaters along the beach, allowing the water level variations up to 10 m. Due to resonances or funneling effect, the tidal range in some coastal regions, such as bays and river estuaries, can reach to 20 m. Although tides occur on a regular basis, no two tides in the same cycle are identical. The tides are exceptionally high (referred to as spring tides) during full moon and no moon days in any month, as the gravitational pulls of the sun and moon operate in an additive manner on these days. In a year, tides that occur at vernal and autumnal equinoxes are even higher because of the sun's and earth's relative positions. Whereas, the tides occur during the 1st and 3rd quarter of a lunar month are the least ones and referred as *neap tides* (Fig. 22.2). A normal mean tidal range is about 1/3rd that of a spring tide range in a given cycle.

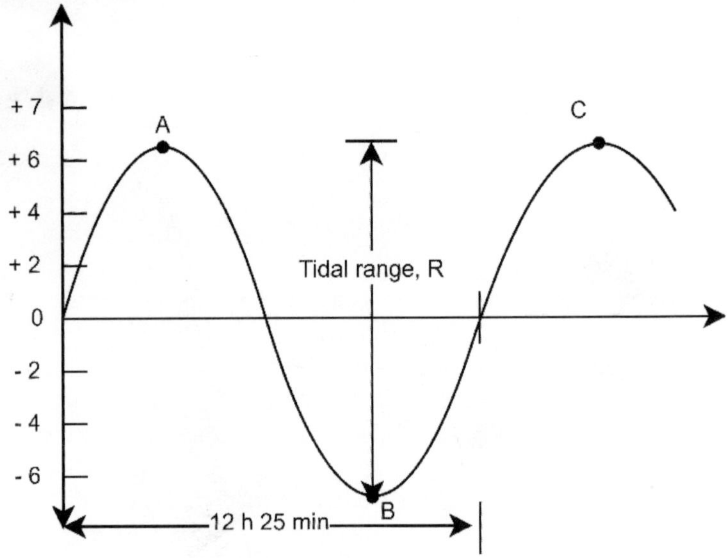

Fig. 22.1 Tidal wave

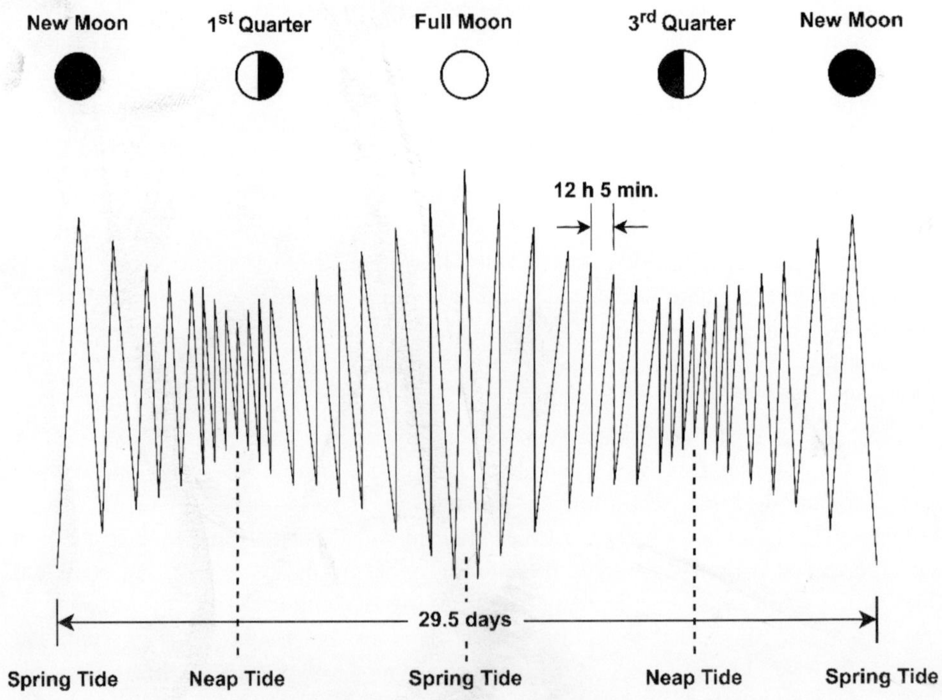

Fig. 22.2 Variations in tidal range during lunar month

22.2 Tidal Power Generation Methods

Tidal power systems use the energy of seawater in motion to generate electricity. The schemes for exploiting the energy of tides may be classified in to two based on the energy transformation. They are basically designed to either convert the (i) potential energy of water to electricity or (ii) the kinetic energy of water directly. The systems which transform potential energy harness the daily movement of tides towards and away from coastal land of backwaters. The systems based on potential energy transformation require tidal barrages as shown in Fig. 22.3.

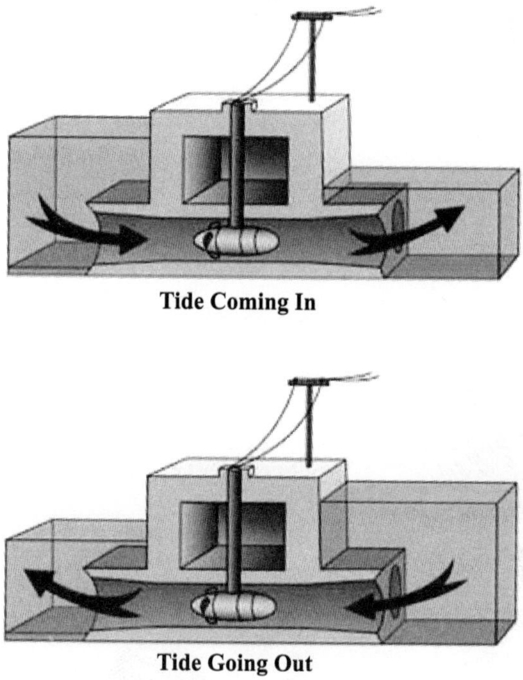

Fig. 22.3 Principle of a tidal barrage power generation system *(Courtesy: NASA)*

Tidal stream generators can make use of the kinetic energy of fast-moving currents in underwater channels. They are similar to wind turbines in operation. Among the several types of tidal power generating systems, tidal stream generators are the cheapest and least affecting the ocean ecology.

22.3 Tidal Barrage Power Generation Systems

There are three fundamental components of a tidal barrage power plant:

(i) The powerhouse
(ii) The barrage or dam
(iii) Sluice-ways and flood-gates

The powerhouse of a tidal power plant comprises of a hydro-turbine, electric generator and other auxiliary equipment. A purpose of dam is to create a barrier between the sea and the basin, or between one basin and another. Barrage and dam are nearly identical terms; however, the term 'barrage' has been proposed as relatively more precise for tidal power generation systems. A barrage should have reinforced concrete channels for the turbines shielded by flood-gates. The flood-gates are used to fill the basin during the high-tide and to discharge it at low-tide.

Different methodological approaches can be utilized to tap tidal energy. The easiest method consists of a single-basin-system which works in only one direction. A dam separates a bay from the open sea, although it is still connected to the sea via a floodgate and a turbine. The turbine and floodgate control systems ensure that water can only enter the bay through the floodgate and can only escape the bay through the turbine. To generate power, this turbine is attached to an electrical generator.

22.3.1 Classification of tidal barrages power plants

For generating useful electric power from a water resource, a hydraulic head (difference in water levels) is the basic requirement and hence various concepts have been proposed to fulfil the same in the context of tidal power generation. Electricity production from tidal range utilizes flow between an artificially developed basin and the sea connected via sluice-ways. In order to operate in a more or less continuous manner, this basic concept may have two or more basins. Therefore, based on the number of basins involved, tidal power plants can fundamentally be classified as, (i) Single-basin type and (ii) Double/Multi-basin type.

A. Single-basin arrangement: The most common method of tidal power generation is using a single basin with a retaining wall/barrage to trap the sea water at high-tide. The sea and barrage are separated and the flow between the two is regulated by flood-gates and sluice-ways equipped with low-head turbines provided along the barrage. Potential hydraulic head is accrued during rise and fall of tidal water levels by blocking the entrance of a long estuary with a barrage across it, creating an artificial pool.

The power generation through a single-basin tidal system could be attained in either of the following systems:

(a) Single ebb-cycle system
(b) Single tide-cycle system
(c) Dual cycle system

(a) Single ebb-cycle system: In this system, sluices are opened as the high-tide occurs, to allow sea water into the basin when the hydro-turbines are turned off. As the water level rises in the basin, potential energy stowed in the form of tidal range offers a hydraulic head during low tide. Electricity generation commences, when the flood-gates are closed and the basin's water runs over the turbine into the ebbing (flowing out) sea during low-tide. Turbine generators are kept operational until the next high-tide makes the water level risen in the sea and a proper head for working of turbines no more exists. A point at which the falling reservoir level and rising sea level lessen the effective hydraulic head across the turbines where it can no longer generate power efficiently, the turbines are shut off. Once the turbines are shut off, the flood-gates are opened again, to reiterate the process in cyclic manner. The power generation from a single ebb-cycle system is intermittent and extremely variable since the turbines are built for one-way operation as illustrated in Fig. 22.4 to Fig. 22.6.

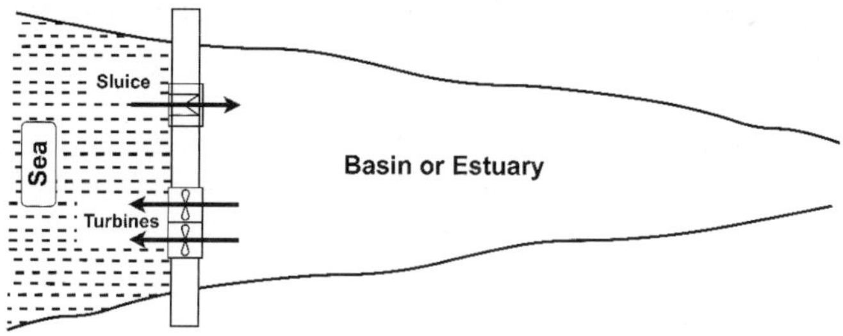

Fig. 22.4 Tidal plant operation as seen from top (Plan)

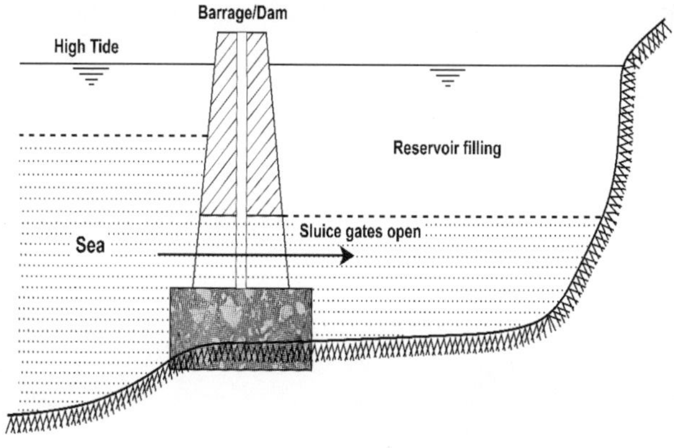

Fig. 22.5 Sluice gates open, Turbines shut-off (sectional view)

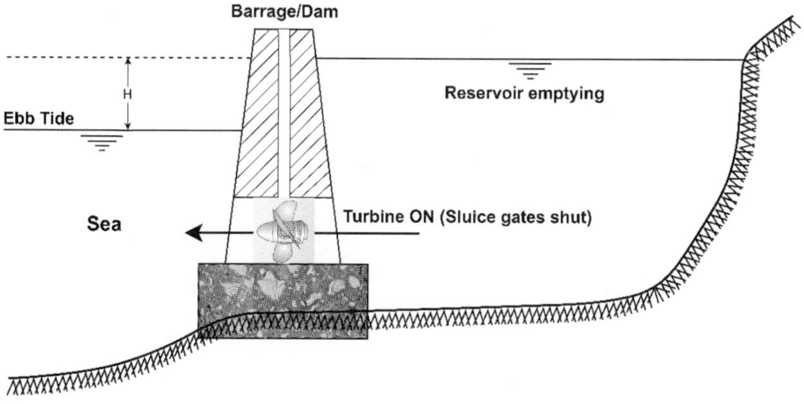

Fig. 22.6 Ebb-generation – Sluice gates shut, Turbines in operation (sectional view)

(b) Single tide-cycle system: In a single tide-cycle system, power generation commences at the onset of high-tide. Seawater is brought into the basin over the turbine generators and hence sometimes called 'Flood generation'. Once the high-tide phase is completed and the sea level commences to fall, the turbines are shut off and the basin water is discharged into the sea via sluice ways. Requirement of larger plant size and shorter duration of power generation makes this system less efficient than the ebb-tide system. One reason for low efficiency is due to the low operational head. In ebb-cycle power generation process, hydraulic head between the basin and the sea-side of the barrier decreases faster than it does in flood generation. Moreover, the power output through a single tide-cycle system is also intermittent in nature.

(c) Dual-cycle system: The dual or double cycle system utilizes reversible turbines, which allows power generation throughout flood-tide as well as ebb-tide. During ebb and flood tides, the direction of flow through the turbine alternates, however, the specially designed reversible turbines generate power in either direction of flow as shown in Fig. 22.7. Though it uses reversible turbines for generating power during two cycles (basin filling and emptying cycles), it is not continuous in operation. Based on lunar day of 24 hours and 50 minutes, this system generates electricity through two short periods over each tidal period of 12 hours, 25 minutes or once in every 6 hours, 12.5 minutes.

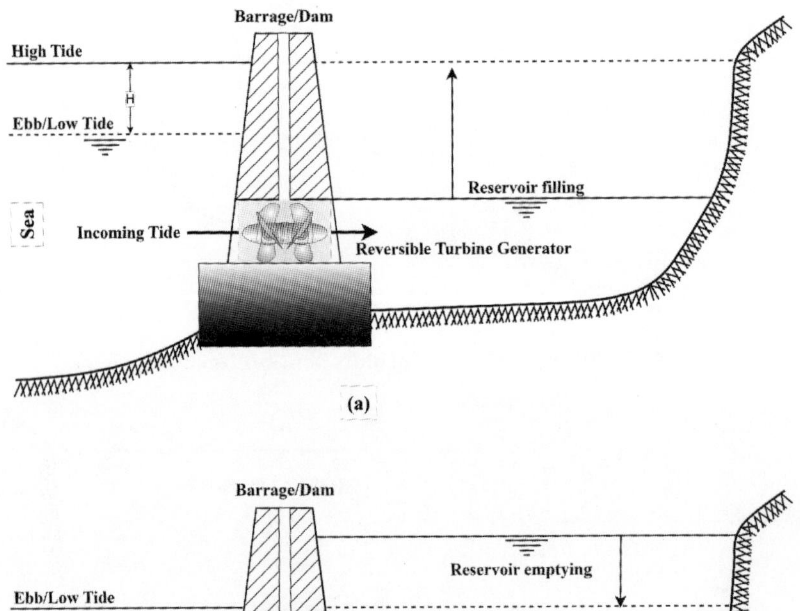

(a)

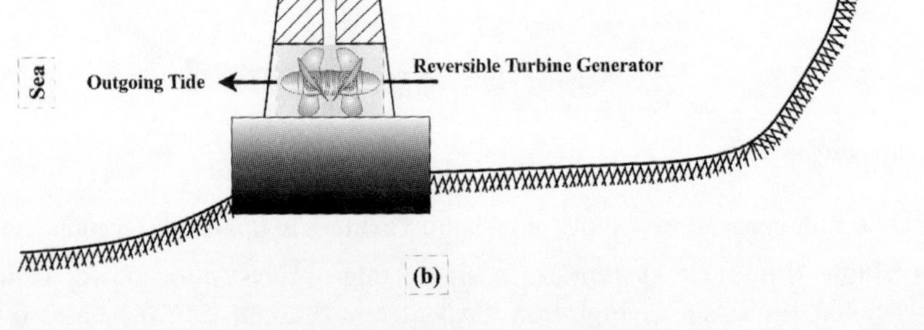

(b)

Fig. 22.7 Dual-cycle system

22.3.2 Double-basin arrangement

Apart from simple design, construction and operation, the fundamental disadvantage of single basin tidal power plants is that power generation can only be achieved during comparatively short time periods. This could be improved using a reversible or cross-flow turbine which is often useful. Additionally, tidal power plants could be designed as a twin-basin system (Fig. 22.8). The first basin is situated at a higher altitude compared to the second one (which is at relatively higher altitude than the low-tide level). Another turbine is built in between the two basins. This turbine is constructed within a dam or a connecting canal between two basins that are also connected to the sea by turbines. Water intake and outflow are regulated by the fact that water enters one basin at high tide and departs the other basin at low tide after passing through the corresponding turbines.

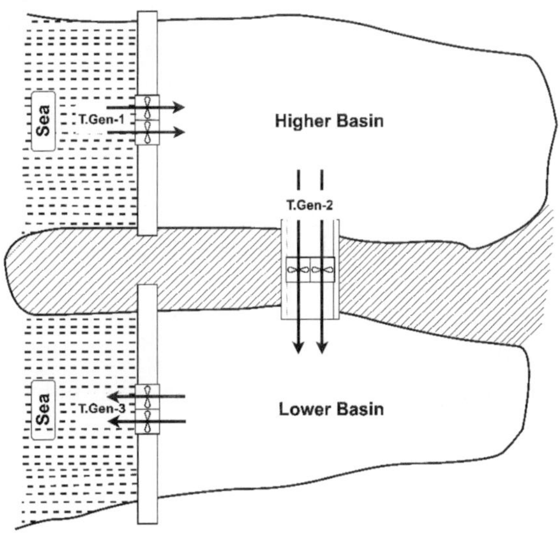

Fig. 22.8 Double basin tidal power plant schematics

22.4 Tidal Turbines

A collective effect of gravitational forces of the sun and moon could create semidiurnal movements in the shallow sea waters, predominantly in places with the appropriate coastal morphology. Most often the flow velocities are not sufficient enough to produce significant quantities of power. But at certain 'pinch points', the currents could accelerate with considerably higher speeds as the tidal flow is constrained by land and seabed topography. Strong currents and tidal waves are produced by these occurrences, which are typically observed in the British Isles and many other parts of the world with comparable circumstances. In such cases, tidal energy can be subjugated directly from these sea currents caused by natural systems formed around the sea foreland and among some islands. In locations with velocities in the range 2 to 3 $m \cdot s^{-1}$, the energy density becomes sufficient enough to install gadgets to tap the kinetic energy.

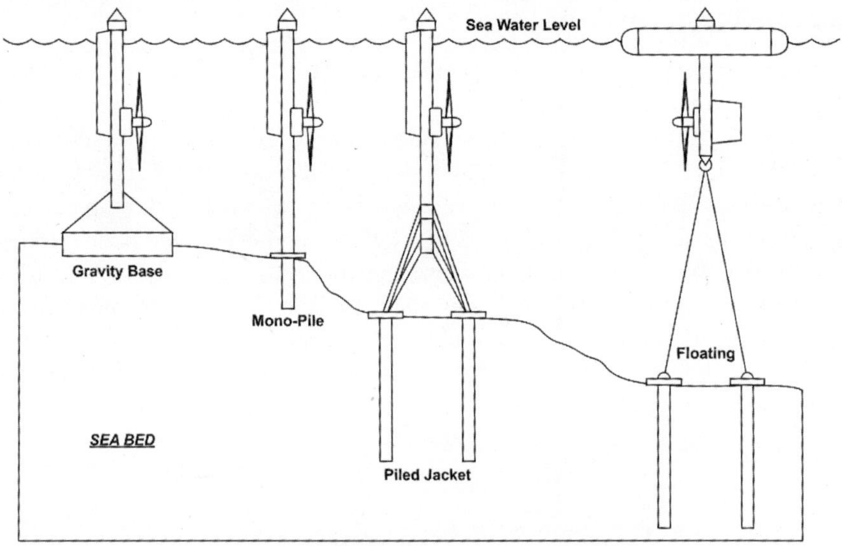

Fig. 22.9 Tidal Stream Energy converter

A tidal stream generator, sometimes denoted to as '*tidal energy converter*' is a type of turbine harnessing the energy from the moving water during tides (Fig. 22.9). Similar to the windmills, kinetic energy of these currents is thus transformed into electricity by employing a turbine power generator across suitable locations. These turbines positioned close to the lowermost part of the water column, can be horizontal, vertical, open or ducted.

The power generation potential of an individual tidal turbine is comparatively greater than that of a similar wind turbine. This is due to the fact that the density of water is much higher relative to air (about 800 times that of air). Hence, a single tidal turbine generator can produce significant power even at low tidal flow velocities in contrast to comparable wind speeds for wind turbines. Because the power obtainable from the kinetic energy of a fluid is proportional to the density of the fluid and the cube of flow velocity, water flowing at speeds of approximately one-tenth the speed of wind provides the same power from the same size of turbine. As the energy availability is extremely sensitive to the flow velocity, doubling the velocity results in eight times the energy density. The energy density in a tidal stream at a flow velocity of 2 ms^{-1} is 4.1 $kW \cdot m^{-2}$, but an increase to 3 ms^{-1} represents an energy density of 13.9 $kW \cdot m^{-2}$.

22.5 Tidal Power Generation – World Scenario

The amount of energy available from tidal energy varies based on the location and time of year. Available energy is basically proportional to the square of mean tidal wave amplitude. Tides are only suited for energy extraction in places where energy is concentrated in massive waves or in geographical positions that make construction and installation of the power plant easier. Such locations have been discovered in the United Kingdom, France, Eastern Canada, Russia's Northern and Eastern Coasts, Korea, China, Mexico, and Chile. Other locations have been discovered in Argentina's Patagonian coast, in Western Australia, and in Western India.

The Rance Tidal Power Station (average tidal range is 8.5 m, 24 turbines, installed capacity of 240 MW and approx. power density of 2.6 W·m^{-2}) situated on the Rance River's estuary in Brittany, France is the world's first of its kind (Fig. 22.10). The power facility was inaugurated in 1966 and is currently being operated by Électricité de France, supplying nearly 0.12 per cent of the power demand of France. For 45 years, it was the world's largest tidal power station with respect to installed capacity until the South Korean Sihwa Lake Tidal Power Station (254 MW installed capacity) surpassed it in 2011.

Fig. 22.10 Rance Tidal Power Station, Brittany, France

(Author: Dani 7C3, Dam of the tidal power plant on the estuary of the Rance River, Bretagne, France, Available at https://commons.wikimedia.org/wiki/File:Rance_tidal_power_plant.JPG, GNU Free Documentation License CC-BY-SA/3.0 via Wikimedia Commons)

M/s. Crown Estate, a UK-based company, manages the seabed surrounding England, Wales, and Northern Ireland, and collaborates with stakeholders to maximize the potential of different renewable energy projects. They grant the seabed rights, encourage innovation, and promote the long-term growth of offshore wind, wave, and tidal energy systems. M/s. Marine Current Turbines Ltd. is a firm established in the United Kingdom that develops tidal stream generators. Off the coast of Lynmouth in UK, it has a 300-kW experimental installation known as 'Seaflow' (Fig. 22.11).

Fig. 22.11 Tidal turbines installed at Lynmouth, UK

(Author:Fundy, The SeaFlow tidal stream generator prototype with rotor raised, Available at https://commons.wikimedia.org/wiki/File:Seaflow_raised_16_jun_03.jpg, GNU Free Documentation License CC-BY-SA/3.0 via Wikimedia Commons)

CHAPTER 23

Geothermal Energy

The word *geothermal* is originated from combination of the two Greek words '*ge*' i.e., Earth (geo) and '*therme*' i.e., heat (thermal) which means heat generated and stored within the planet. In broad terms, geothermal energy is the thermal energy stored in the Earth's crust. The earth's thermal energy is dispersed between its constituent host rocks and the natural fluid trapped in its fissures and pores at temperatures above ambient. However, the term is now commonly used to refer to the amount of Earth's heat that can (or might) be collected and used as a sustainable energy source.

The occurrence of volcanoes, hot springs and other geothermal phenomena in ancient times must have caused our ancestors to believe that the Earth's interior was hot. However, it wasn't until the 16th and 17th centuries, when the first mines were dug to a depth of a few hundred metres below ground level, that people observed the Earth's warmth rising with depth.

The earliest thermometer observations were most likely made in a mine near Belfort, France, around 1740 (Bullard, 1965) and by 1870 contemporary scientific methods were being employed to examine the Earth's thermal regime. It was in the twentieth century we finally grasped the phenomenon of the Earth's heat balance, thanks to the discovery of the role performed by radiogenic heat. In reality, all contemporary thermal models of the Earth must account for the heat created from decay of the radioactive isotopes of Uranium (U^{238}, U^{235}), Thorium (Th^{232}), and Potassium (K^{40}) found in the Earth's crust (Lubimova, 1969) and the primordial energy of planetary accretion. It was revealed in the 1980s that the equilibrium does not exist between the heat generated in the Earth's interior and the heat radiated into space by the Earth, and that our planet is gradually cooling.

It is at the vapor dominated field at Larderello, Italy, electricity was first produced using geothermal steam in 1904. Other hydrothermal developments since then, such as the steam field at "The Geysers in California", and the hot water systems at Wairakei in New Zealand, Cerro Prieto in Mexico, and Reykjavik, Iceland as well as those in Indonesia and the Philippines, have resulted in an installed global electrical production capacity of nearly 10,000 MWe and a direct use, nonelectric potential of more than 100,000 MWt (megawatts of thermal power) at the turn of 21st century.

23.1 Mechanics of Geothermal Phenomenon

The source and transport mechanisms of geothermal heat are unique for various sites. Heat flows through the crust of the Earth at an average rate of almost 59 $mW{\cdot}m^{-2}$. The intrusion of large masses of molten rock can increase this normal heat flow locally; but for most of the continental crust, the heat flow is due to two primary processes:

(i) Upward convection and conduction of heat from the Earth's mantle and core, and

(ii) Heat generated by the decay of radioactive elements in the crust, particularly isotopes of uranium, thorium and potassium.

23.1.1 Geothermal activity and tectonic plate theory

The temperature rises gradually with depth, as well as volcanoes, geysers, hot springs etc., along with other phenomena that are less observable by natural sense. The study related to these phenomena is collectively called as the plate tectonics theory, which describes the dynamics of Earth's outer shell with respect to its inner core.

The Earth consists of a*crust*, which varies in thickness of about 20 to 65 km in continental areas and about 5 to 6 km in oceanic zones, a *mantle*, which is coarsely 2900 km thick, and a *core*, about 3470 km in radius (Fig. 23.1). The physical and chemical physiognomies of the crust, mantle and core differ from the exterior of the Earth to its centre. The outermost shell of the Earth, known as the *lithosphere (*Greek words *'lithos'* i.e., rocky + *'sphaira'* i.e., sphere*)*, is made up of the crust and the upper layer of the mantle. It ranges in thickness from less than 80 km in oceanic zones to over 200 km in continental plates; behaving as a rigid body. The next zone beneath the lithosphere is referred as the *asthenosphere* (Greek word '*asthenes*' meaning 'weak'), less rigid or more plastic in behavior, ranging 200 to 300 km in thickness. In fact, on a geological scale, where time is measured in millions of years, this shell of the Earth behaves similar as a fluid in certain processes.

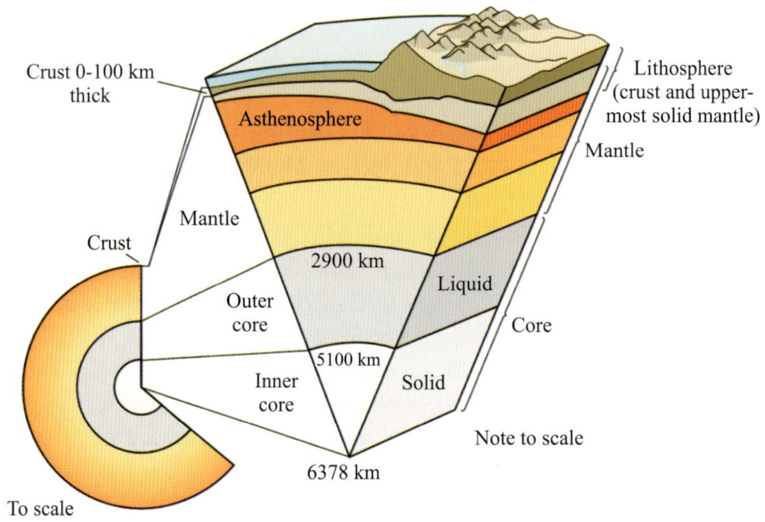

Fig. 23.1 Cutaway views showing the internal structure of the Earth
(Courtesy: United States Geological Survey)

The temperature gradient between the various layers of the asthenosphere, leads to the slow convective movements. Their exceptionally slow activity (a few millimetres to centimetres per year) is sustained by the heat generated from continuous decay of the radioactive elements and the upward thermal convection from the Earth's core. The gigantic volumes of deep, scorching rocks, lighter than the relatively colder adjacent material, rise towards the surface. On other hand, the cooler, denser and heavier rocks near the surface tend to sink, re-heat and rise again to the surface in the so-called *convective currents* or*cycles* . This phenomenon is mutually analogous to the water being boiled in a pot, however, on the extremely slow rate of movement with respect to the geological time.

The lithosphere is pushed upwards and broken (in the regions of lesser thickness, especially in the oceanic zones) by the sweltering, partially molten mass ascending from the asthenosphere. This is the mechanism that has created and still creates the rims spreading more than 60,000 km underneath the oceans and also creeping between the continents. A comparatively small fraction of the molten magma surging from the asthenosphere emerges through the crests of these ridges and solidifies into a new oceanic crust after it come in contact with seawater. However, underneath the lithosphere, much of the material ascending from the asthenosphere splits into two branches that flow in opposite directions. The continual pull of these two opposite flows causes the ridges to shear perpendicularly by massive fractures, called *the transform faults*.

The lithosphere folds downward in subduction zones, plunging beneath the neighboring lithosphere and sink to the deeper zones, where it is 'digested' by the extremely hot mantle. Part of the digested lithospheric material re-surfaces in a molten state through fractures in the crust, and the cycle continues. Consequently, the magmatic arcs with several volcanoes are formed parallel to the oceanic trenches, in the opposite direction of the ridges (Fig. 23.2).

Process of slab pull

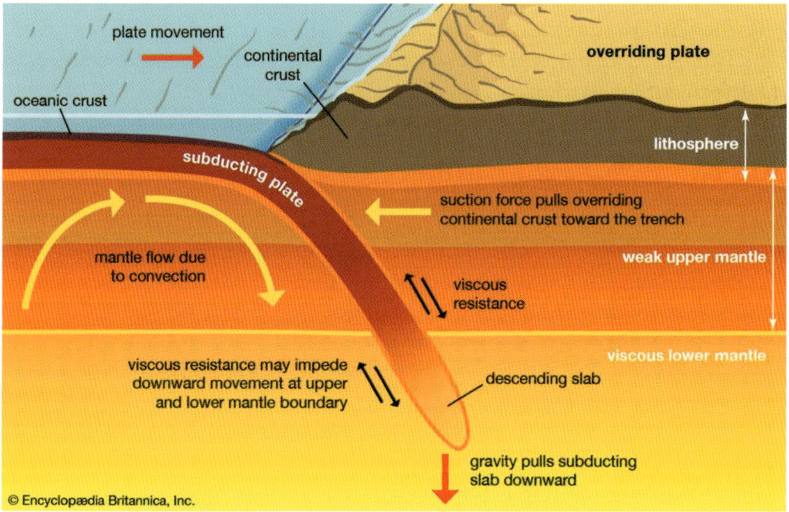

Fig. 23.2 Tectonic activity between different layers of Earth

(Courtesy: Encyclopedia Britannica, Inc.)

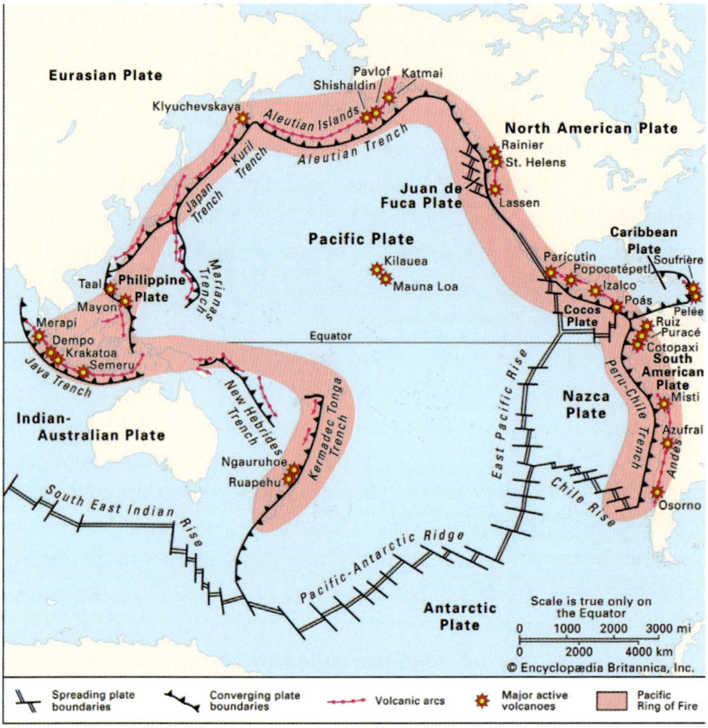

Fig. 23.3 World pattern of tectonic plates, oceanic ridges, oceanic trenches, subduction zones and the major geothermal and tectonic activity field (series of dots) – The Ring of Fire.
(Courtesy: Encyclopædia Britannica, Inc.)

A huge network of spreading ridges, transform faults, and subduction zones splits our world into seven massive and multiple smaller tectonic plates. These plates move gently up against one another, altering position constantly, due to the enormous stresses caused by the Earth's thermal engine and the imbalance of the zones creating and consuming lithospheric material. The plate boundaries correspond to weak, extensively fractured crustal zones marked by severe seismicity, a huge number of volcanoes, and a high terrestrial heat flux due to the ascent of very hot materials towards the surface. The most important geothermal zones, as illustrated in Fig. 23.3, are found along plate boundaries. Major tectonic plates of the Earth:

(i) *African Plate* – A key tectonic plate of the East African Crack – 61,300,000 km^2

(ii) *Antarctic Plate* – A tectonic plate encompassing the Antarctica and ranging outward beneath the adjacent oceans – 60,900,000 km^2

(iii) *Eurasian Plate* – Encompasses a major portion of the Eurasian continent – 67,800,000 km^2

(a) *Indo-Australian Plate* – is formed by the union of the Indian and Australian plates – 58,900,000 km^2 frequently referred as two plates:

(b) *Australian Plate* – A formerly part of the ancient continent of Gondwana – 47,000,000 km^2

(v) *Indian Plate* – A minor tectonic plate detached from Gondwana – 11,900,000 km^2

(vi) *North American Plate* – Large tectonic plate covering most of the North America, Greenland and part of Siberia. – 75,900,000 km^2

(vii) *Pacific Plate* – An oceanic tectonic plate under the Pacific Ocean – 103,300,000 km^2

(viii) *South American Plate* – Includes most of South America and a large part of the south Atlantic – 43,600,000 km^2

23.2 Geothermal Energy Resource: Location and Viability

The location (depth and position) and quality (fluid chemistry and temperature) of a given geothermal resource are influenced by local and regional geology and tectonic processes. For instance, tectonic plate borders and areas of geologically recent igneous activity and/or volcanic eruptions are linked with greater heat fluxes than usual (predominantly younger than about 1 million years). This has prompted many people to associate geothermal energy with places where such conditions exist, such as Iceland, New Zealand, or Japan (plate borders), or Yellowstone National Park in the United States. Moreover, certain other aspects must be satisfied, before one defines a particular location as a viable geothermal resource.

(a) *Accessibility:* This is normally accomplished by drilling to the desired depths, often employing technologies similar to those used to extract oil and gas from subsurface reserves. The ease and expense to explore the resource collectively referred as accessibility.

(b) *Sufficient reservoir productivity:* For enabling a long-term production in hydrothermal systems, substantial volumes of hot, natural fluids should be trapped in an aquifer with high natural rock permeability and porosity. with economic feasibility. In case of inadequate natural recharge to the hydrothermal system (which is normally the case), a reinjection system is requisite to maintain the productivity. The heat extraction method must be designed to work well within limitations imposed by in situ hydrologic, lithologic, and geologic circumstances (Fig. 23.4).

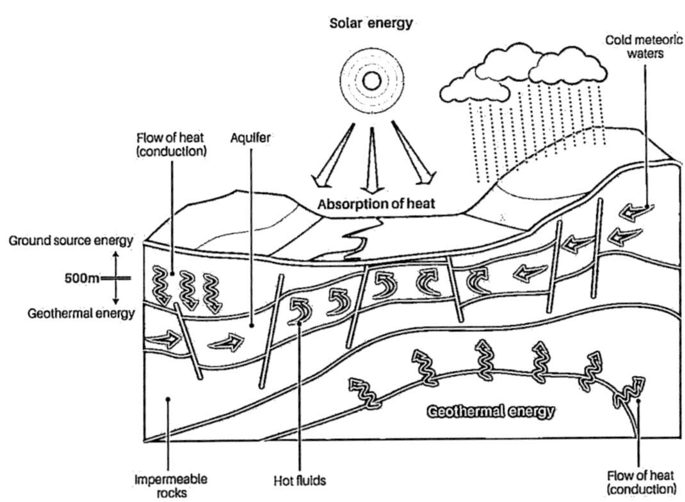

Fig. 23.4 Geothermal energy resource

(Courtesy: British Geological Survey, NERC)

23.3 Geothermal Systems for Power Generation

Coupled transport methods (i.e., convective heat transfer in permeable and/or fractured segments of rock and conduction through the rock itself) are used to extract thermal energy from the reservoir. In the hydrothermal power plant, hot water or steam (125°C < T < 225°C) is produced and its energy is converted mainly into electricity via intermediate power conversion mechanism, such as steam turbine generators. Any waste products (flue gases and particulate matter) must be properly treated and safely disposed-off to complete the process.

According to the US Department of Energy, Enhanced Geothermal Systems (EGS) are constructed reservoirs that have been built to extract affordable volumes of heat from poor permeability and/or porosity geothermal resources. High-grade hydrothermal resources would be excluded, but conduction dominated, low permeability resources in sedimentary and basement formations, as well as geopressured, magma, and low grade, unproductive hydrothermal resources, would be included. EGS designs would collect thermal energy from underlying rocks by constructing or gaining access to a network of open, linked fractures through which water might be injected, heated by contact with the rocks, and then returned to the surface via production wells, forming a complete loop (Fig. 23.5). The concept is a straightforward generalization of naturally existing hydrothermal circulation systems –those are currently producing commercial power and heat for direct use in 71 nations throughout the world.

The mechanism underlying EGS is governed by the concept of thermal fluid convection. Fluid convection takes place due to heating and consequent thermal expansion of fluids subjected to the gravity field. The heat, which is absorbed at the base of the circulation system, itself, drives the system. The lighter and hot fluid tends to rise which is replaced by reinjecting the colder fluid of high density. Another distinction between geothermal systems is that based on reservoir equilibrium state (Nicholson, 1993), which takes into account reservoir fluid movement and heat transport mechanisms. The reservoir in dynamic systems is constantly replenished by hot water, which is subsequently released from the reservoir either to the surface or into subsurface permeable formations. Convection and fluid motion transport heat throughout the system. High-temperature (>150°C) and low-temperature (150°C) systems are included in this category. There is only a small or no replenishment to the reservoir in static systems (also known as static or storage systems), and heat is solely transported through conduction. Low-temperature and geopressured systems fall into this group.

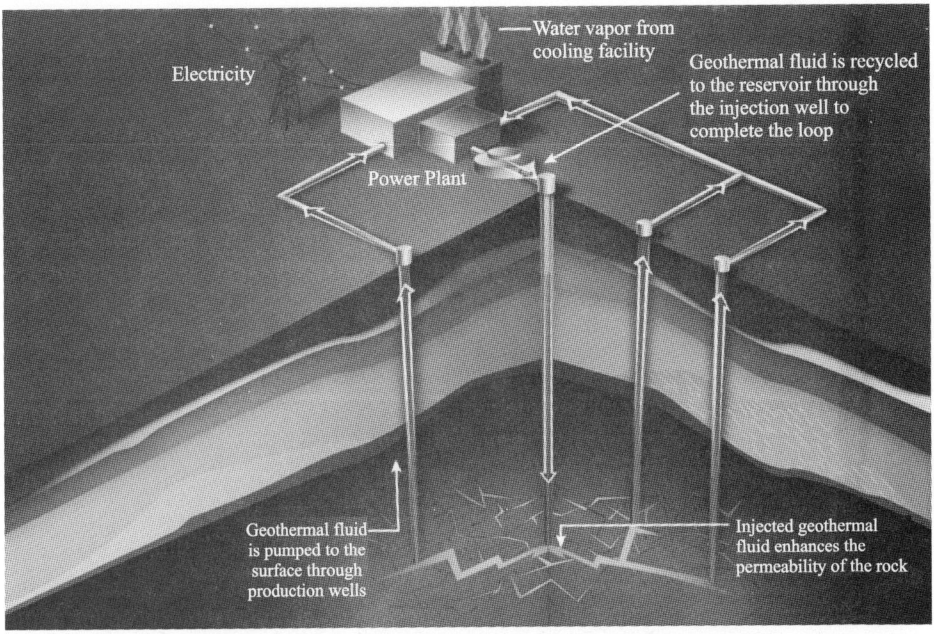

Fig. 23.5 Schematic of a conceptual EGS in hot rock in a low permeability crystalline basement formation *(Courtesy: U.S. Department of Energy)*

Geopressured systems are commonly found at depths of 3 to 7 km in major sedimentary basins (e.g., the Gulf of Mexico, USA). The geopressured reservoirs are made up of permeable sedimentary rocks embedded in impermeable low-conductivity layers that hold pressurized hot water that was trapped when the sediments were deposited. Methane can be found in large quantities in geopressured reservoirs. Thermal and hydraulic energy (highly pressurized water) as well as methane gas could be produced via geopressured systems.

23.3.1 Types of heat-engine based power plants

The nature of geothermal resource (temperature and fluid chemistry) influences the design and production process of the proposed power plant. Though, basic principle of the power conversion mechanism remains the same i.e., thermo-mechanical electricity production or heat engine; interactions of the thermal fluid, mode of heat exchange and utilization defines the topology of particular power plant.

(a) *Vapor-dominated systems:* The simplest type of system to harness geothermal energy is called a *vapor-dominated system*. The power plants operational at vapor-dominated sites are usually more efficient than the other types of geothermal power plants. In nature, vapor-dominated geothermal sites are quite rare. Such geothermal resource produces *dry steam*, i.e., hot steam, but no liquid water. Vapor-dominated systems can produce great amounts of dry steam. In order to utilize such resource, conduits are connected to the producer wellhead to maintain pressure and to provide a path from the wellhead to the turbine-generator (Fig. 23.6).

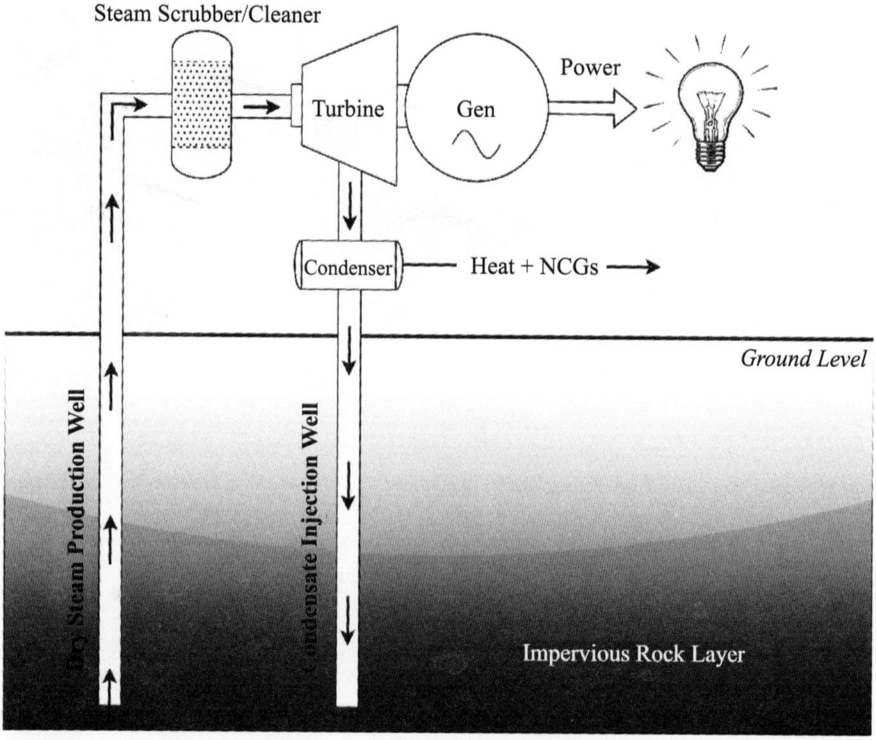

Fig. 23.6 Vapor-dominated system

The dry steam emerging from the well contains impurities like dirt particles and other gases, which must be separated out. Then, as discussed earlier, the hot steam under pressure flows through a directional valve that leads it counter to the blades of the turbine. The expanding hot steam drives the turbine coupled to the generator, producing electricity. Once past the turbine reaction chamber, the steam is routed to a condenser, where it condenses into the liquid phase. The condenser, which causes a pressure drop downstream of the turbine, enhances the efficiency of the energy conversion process and facilitates the separation of the water from any other gases that may be present. Finally, the liquid, or at least some of it, is reinjected into the source to maintain the supply of steam. Vapor-dominated geothermal systems are mechanically simpler and more efficient than other types of geothermal systems (10 to 21 per cent). The Geysers (the world's largest geothermal power plant complex) operated by Calpine Corporation; California is a vapor-dominated system. It has an installed capacity of 1,517 MWe comprising 22 power plants, out of which 15 are presently active to output 725 MWe, enough to electrify 7,25,000 homes.

(b) *Flash systems:* Geothermal hot-water resources, which are found more often in the nature than the vapor-dominated or dry steam resource, are the thermal reservoirs that are saturated with liquid water rather than water vapor. The power generation systems installed on such resources are called as *flash systems*. In fact, the emergent water is still very hot, sometimes more than twice the boiling point of water at atmospheric pressure (100°C). Since, the pressure deep beneath Earth's surface is much higher than at the surface, such that, the water remains in a liquid state, even at blazing temperatures. As the water rises up the well toward the surface, the pressure begins to drop, and the water begins to boil within the well, and part of it turns into steam. What arrives at the wellhead is a mixture of hot liquid water and fairly high-pressure steam. The pressure on this mixture is reduced further in a component called '*flash tank*', which allows more of the water to change into steam. The steam component in the mixture of steam and liquid water (still under substantial pressure) is separated and the liquid water is pumped into the injection well. The steam now drives the turbine-generator pair in just the same way as in a vapor dominated system, and once past the turbine, the steam is cooled in a condenser until it becomes a liquid. The part of water from the condenser, is routed back into the thermal reservoir to complete the cycle having an average conversion efficiency in the range of 12 per cent. Based on this technology, for example, the Coso Hot Springs geothermal power plant complex in California is operated by Coso Operating Company LLC., to output net 145 MWe (Fig. 23.7).

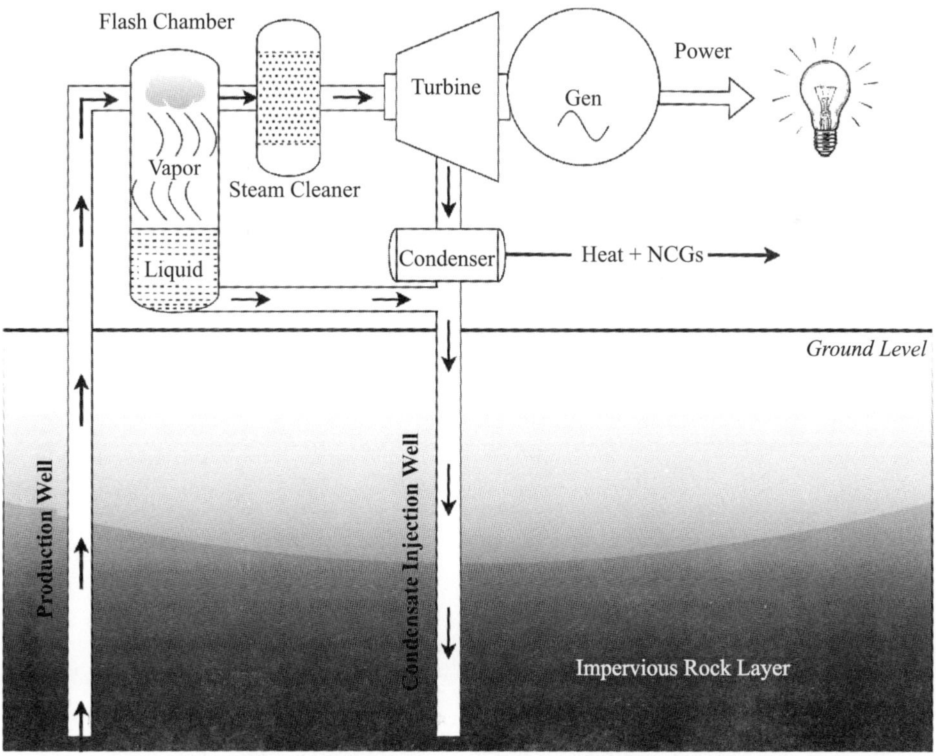

Fig. 23.7 Single-flash system

(c) *Binary systems:* The third method of converting geothermal energy into electricity is used when the water emerging from production well is not hot enough (<150°C) to form large amounts of steam. Even though temperature of the water is well above the boiling point of water at atmospheric pressure, the conventional power plants struggle to deliver the steam at a fairly high pressure to drive the turbine. Such systems are referred as *moderate-temperature systems* and unsuitable for driving a steam turbine directly. However, even at a moderate temperature, the water still carries significant amount of thermal energy. To make use of this thermal energy for creating a relatively high-pressure vapor, the solution is to use a so-called *binary system*, which uses a thermal exchange fluid other than water. Heat engines have been utilizing ammonia, carbon dioxide, helium and other exotic liquids and gases.

A typical binary system uses two closed loops, each having a different thermal fluid. The first loop in a binary geothermal system circulates the hot water produced from the well. This fluid flows up the well to a heat exchanger, where a portion of the thermal energy carried by the water is transferred to the secondary working fluid. After passing through the heat exchanger, the heat exhausted water is injected back into the rock formation, where it is reheated in cycles. On the other side of the heat exchanger the working fluid absorbs the heat from the water and changes from a liquid to a high-pressure gaseous phase, which drives the turbine-generator.

Once past the turbine, the vapor passes through a condenser, where it again becomes a liquid, and recirculated to the heat exchanger, where the cycle reiterates. As a consequence, more water, when measured by mass, must be circulated in order to produce the same amount of electricity. Therefore, the maximum conversion efficiency attainable by the binary system is the lowest of the three systems, typically below 10 per cent (Fig. 23.8).

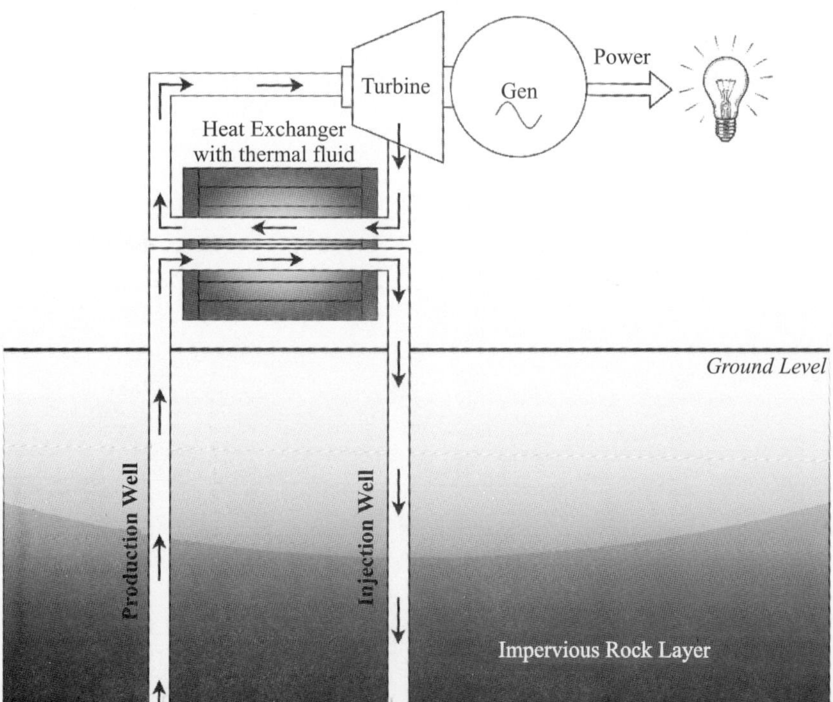

Fig. 23.8 Binary geothermal power plant

23.4 Direct Heat Utilization

The simplest way to use water that is warm, but not warm enough for generating electricity, is space heating. This involves pumping the hot water from the geothermal site, circulate it through a radiator or other type of heat exchanger, and dispose or reinject the heat exhausted water back into the source. The value of the resource depends on the initial temperature of the water, the distance between the well where it is produced and the location where it is used, and the amount of hot water that can be sustainably extracted (Fig. 23.9).

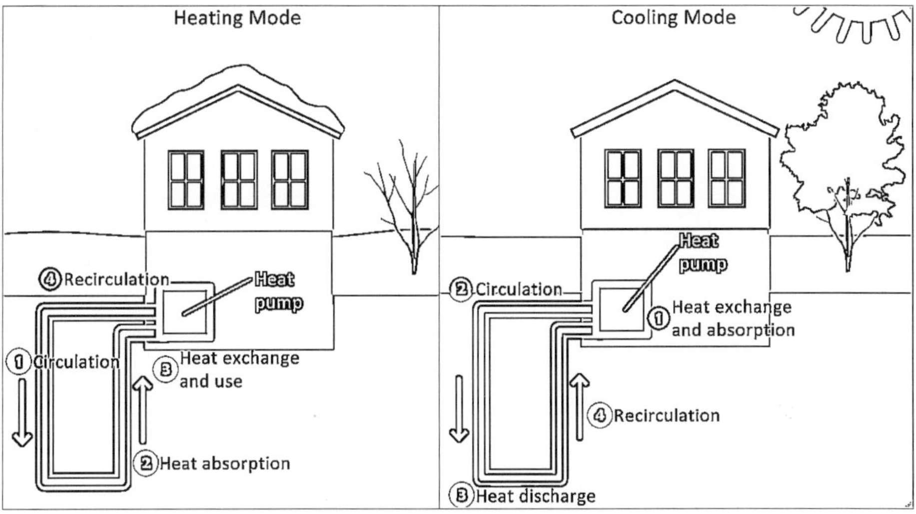

Fig. 23.9 Geothermal heat pump system for space heating and cooling
(Courtesy: U.S. Environmental Protection Agency)

23.5 Prospects of Geothermal Energy

Geothermal energy is typically categorized as renewable and sustainable. The term renewable refers to a characteristic of an energy source, whereas sustainable refers to how the resource is used. The word '*geothermal field*' refers to a region having one or more geothermal systems, regardless of whether or not they are being utilized.

The rate of energy recharge is the most important factor in classifying geothermal energy as a renewable energy source. The recharging of energy occurs during the exploitation of natural geothermal systems by advection of thermal water on the same time scale as the resource's output. This demonstrates the geothermal energy resource's renewable nature. Energy is replenished exclusively through thermal conduction in hot, dry rocks and some hot water aquifers in sedimentary basins; however, due to the slowness of the latter process, hot dry rocks and some sedimentary reservoirs should be regarded as limited energy resources (Stefansson, 2000). A resource's consumption sustainability is determined by its initial amount, consumption rate and regeneration rate. Consumption can clearly be sustained for as long as a resource is being generated quicker than it is being destroyed. The term sustainable development is used by the

World Commission on Environment and Development to mean development which "*meets the needs of the present generation without compromising the needs of future generations.*"

Geological Survey of India (GSI) with CSIR - National Geo-physical Research Institute (NGRI) carried out preliminary resource assessment for the exploration and utilization of geothermal resources in 1970s and 1980s in the country. According to early research conducted by the GSI, India has roughly 300 geothermal hot springs. The majority of these geothermal hot springs are in the middle (100°C to 200°C) and low (100°C) potential zones. The most prospective geothermal power generating locations in the country are listed in Table 23.1.

Table 23.1 Geothermal sites identified in India

Electric power generation	Direct heat utilization
Puga Valley and Chummathang – Jammu & Kashmir	Rajgir – Bihar
Cambay – Gujarat	Manikaran– Himachal Pradesh
Tattapani – Chhattisgarh	Surajkund – Jharkhand
Khammam – Telangana	Tapoban – Uttarakhand
Ratnagiri – Maharasthra	Sohana region – Haryana

CHAPTER

24

Clean Development Mechanism and Renewable Energy

Even though there were lone voices airing their concern on the degradation of environment due to over use of energy for development, it was in 1972 that a global conference was held by the United Nations at Stockholm to seriously discuss these issues. Even though 100 countries participated in the event, there were only two heads of nations who made their presence in the United Nations Conference on the Human Environment, among which one was Smt. Indira Gandhi, the prime minister of India. It was noteworthy that the Indian leader took an affirmative stand on the disparity of countries in combating the environmental issues by stating that *"poverty is the worst polluter"*. In her speech she advocated for the poor nations in the world by asking the developed world, *"...how can we speak to those who live in villages and in slums about keeping the oceans, the rivers and the air clean, when their own lives are contaminated at the source?"*

24.1 United Nations Framework Convention on Climate Change

Twenty years later, the United Nations Conference on Environmental and Development (UNCED), popularly referred as the 'Earth Summit', was held at Rio de Janeiro in Brazil during 3-14 June 1992. The conference was attended by more than 100 heads of nations, showing an increased global awareness of the serious environmental issues. It should be noted that there existed very less evidences of human induced climate change during the early nineties. But there was a general consensus and sense of cooperation of world leaders on climate change which led to the historic event in which the United Nations Framework Convention on Climate Change (UNFCCC) was formally established. An international environmental treaty was signed by 154 countries with the objective to resist *'dangerous human interference with the climate system'*. It was on 21st of March 1994 that the UNFCCC came into force and as on today 197 countries are members of the convention who ratified it. A comprehensive action plan called Agenda 21 was formulated in the earth summit. It was decided to take up this agenda at global, national and local levels in every sphere of human activities which had an impact on environment. In order to ensure the implementation of Agenda 21, a 'Commission on Sustainable Development' was also created in December 1992.

24.2 Kyoto Protocol

Subsequent to the successful conduct of the Earth Summit there was a general belief that global warming is real and is anthropogenic. The cause was excessive greenhouse gas (GHG) emission by human activities, especially carbon dioxide. Further extension of the UNFCCC held the COP 3 at Kyoto, Japan which led to the signing of Kyoto protocol on December 11, 1997. According to the provisions contained in Article 24 of the protocol, it was open from 16 March 1998 to 15 March 1999 at United Nations Headquarters, New York for the countries to sign. Eighty-four nations signed the Kyoto Protocol, but it took a long time for the ratification process. For example, even though the president of the United States of America, Mr. Bill Clinton signed the agreement in 1998, he could not get it ratified by the US congress. His successor Mr. George W. Bush withdrew from the protocol. The 2002 United Nations Climate Change Conference was the 6th in the series (COP 6) which took place from 23rd October to 1st November 2002, in New Delhi, India. Finally, the protocol could be enforced on 16th February 2005, which is regarded as a major milestone in the history of climate change mitigation (Fig. 24.1).

As per this agreement the developed countries as well as countries in transition were grouped as 'Annex B' countries. Kyoto protocol could set the mandatory limits of emission by Annex B countries based on an emission bench mark as on 1990. The Kyoto period was regarded as 2008-12 and the regulations insisted the Annex B countries to limit their GHG emissions 5.2 per cent below their total emission in 1990. The protocol listed seven greenhouse gases (Annex A), which were carbon dioxide (CO_2), methane (CH_4), nitrous oxide (N_2O), hydrofluorocarbons (HFCs), perfluorocarbons (PFCs), and sulphur hexafluoride (SF_6).

To achieve the above target there were different flexible mechanisms outlined for Annex B as well as non-Annex B countries. '*Emission Trading*', the possibility to trade emissions of a country to another was an important aspect. There were also facilities provided for '*Joint Implementation*' by Annex B countries.

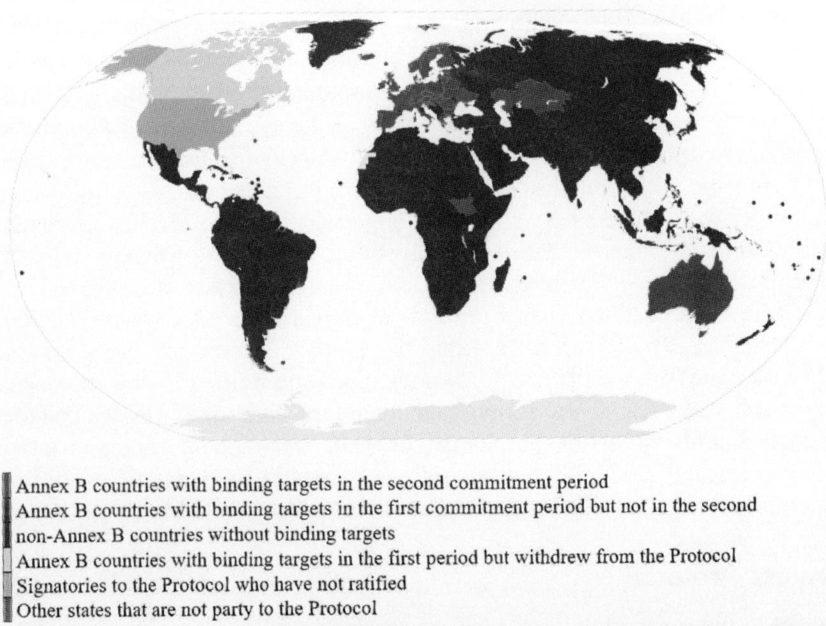

Fig. 24.1 Map of the world showing parties in the Kyoto Protocol *(Source: Wikipedia)*

Another important mechanism was '*Clean Development Mechanism*' (CDM), which was considered as a major feature of Kyoto protocol.

The Doha Amendment to the Kyoto Protocol formulated in 2012 decided to extend the commitment period for the protocol to 2020, known as the second commitment period. The 37 countries with bound targets have a goal of reducing greenhouse gas (GHG) emissions by 18 per cent compared to 1990 levels.

24.3 The Paris Climate Accords

The UNFCCC could evolve a land mark Agreement on climate change mitigation on 12th December 2015 in the COP 12 held at Paris. This agreement often referred to as the Paris Climate Accords, was a notable achievement in the international efforts to mitigate climate change, which covered a wide range of issues like mitigation, adaptation and finance.

The major aim of the Paris agreement was to boost the global awareness on climate change which could augment positive reaction among countries and communities. The target was to minimise the drastic effects of climate change by keeping a global temperature rise below 2 degrees Celsius above pre-industrial levels in the current century. It also encouraged the parties to strengthen their efforts to limit the temperature increase to 1.5 degrees Celsius, which would render the earth much safer than the targeted 2° C increase. Further, the UNFCCC states that, "the agreement aims to increase the ability of countries to deal with the impacts of climate change, and at making finance flows consistent with a low GHG emissions and climate-resilient pathway".

24.4 Glasgow Climate Change Conference

COP 26 of the UN Climate Change Conference, hosted by the UK in partnership with Italy, was held from 31 October to 13 November 2021 in Glasgow, UK. The achievements could be broadly comprehended as the strengthening of the three major aspects viz. adaptation, mitigation and finance.

With respect to mitigation, the long prevailing gaps in emissions reduction targets could be identified properly and Parties of the COP 26 could reach an agreement so as to reduce that gap. This is expected to boost the global efforts in ensuring that the rise in average temperature is limited to 1.5 degrees. The commitment of parties to strengthen their emissions reductions so as to fulfil their national climate action pledges in accordance with the Paris Agreement was also reiterated.

One key outcome of Glasgow conference was the finalisation of the so-called 'Paris rulebook' after six years of negotiations, which is related to the fundamental norms on carbon markets required to make the Paris agreement operative.

Financial implications of the mitigation efforts were discussed in detail leading to a consensus in the need to offer support to developing countries. A reaffirmation was there on the commitment of developed countries to provide 100 billion dollars annually to developing countries.

24.5 Clean Development Mechanism (CDM)

Article 12 of the Kyoto protocol had delineated the specific objectives of the CDM listed as:

(i) Assistance in the achievement of sustainable development

(ii) Contribute to the attainment of goals outlined in the UNFCCC

(iii) To provide assistance to Annex B countries to fulfil their emission reduction commitments.

A notable advantage for developing (non-Annex B) countries was inherent in CDM. A developed country facing difficulty in meeting the emission reduction target can invest in a project in a developing country so as to reduce the emission from an enterprise such as electric power production. The emission credits can be owned by the firm in the developing country and can be sold. Emission reduction from such projects, now called 'CDM projects' are quantified as 'Certified Emissions Reduction' (CER). Each CER is equivalent to 1 tonne of carbon dioxide. The developed countries can use these CERs to get them counted in their emission reduction commitments. By this mechanism, both parties are benefitted. The developing countries get advanced technology from the developed countries, whereas it is less costly for the developed countries to invest in less developed areas of the world to get equivalent CERs. Highly developed countries can thus meet their commitments more easily by undertaking simple and low-cost emission reduction activities such as providing efficient cook stoves in the rural sector (Fig. 24.2).

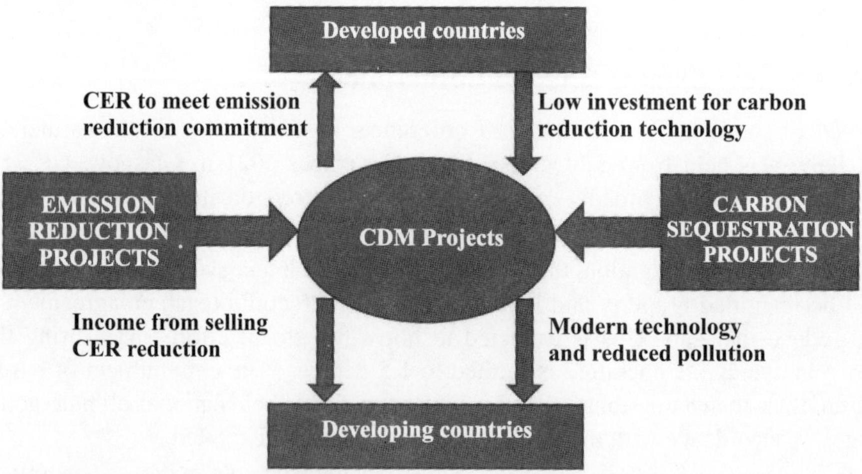

Fig. 24.2 Salient features of CDM projects

There can also be large scale CDM projects in developing countries, like modernization of coal fired thermal plants or replacement of fossil fuel with renewable energy in power production facilities or other industries so as to reduce the carbon emission from these ventures significantly. The capital invested can be recovered profitably using the provisions of emission trading. The developing countries are benefitted as the industries get better and modern technology without much investment. Additionally, the local pollution hazard is also minimized, which has got an indirect bearing on the expenditure on health of the local population. The developed country gets the advantage that it gets the required emission reduction at a much lower cost than required in their own country. There is a CDM Executive Board (CDM EB) constituted for supervising the CDM projects. The basis for qualifying a project under CDM should be well documented in the project document submitted to the CDM EB for its registration. There are approved methodologies for CDM projects and they are required to follow these methodologies. If such an approved methodology is not existing for a specific project, the new methodology should be got approved by the CDM EB. About 100 approved methodologies have been included in the category of small-scale projects, 92 in the category of large-scale projects and 25 approved consolidated methodologies.

24.6 CDM and Renewable Energy Projects

Energy sector is the largest contributor in the increased GHG emission. As CDM allows trading of emission reductions, renewable energy (RE) projects are highly sought after due to their ability to readily reduce GHG emissions to earn CER. They can be easily implemented in developed as well as developing countries without much techno-economic and legal complications. But there is also an apprehension that the CER from renewable energy projects is not economical as the sale price of electricity is less in many countries. Another aspect is that the CER obtainable from reduction of CO_2 is less compared to other CDM projects in which CH_4, HFC or N_2O are involved due to their high global warming potential compared to CO_2.

However, energy conservation and renewable energy development in developing countries could contribute much to arrest CO_2 emission thus helping to prevent global warming coupled with sustainable development through energy security in developing countries. There is a popular notion that RE projects should be so designed to provide incentives for developing countries failing which they cannot promote energy conservation and use of renewable energy. A major share of the approved methodologies for CDM projects are renewable energy and energy conservation projects. The CDM methodology booklet released in 2020 by UNFCCC is available for the public (https://cdm.unfccc.int/methodologies/documentation/meth_booklet.pdf#ACM0002). The 'simplified modalities and procedures' for the small-scale RE projects are also outlined by UNFCCC. In general, the following category of RE projects as listed in Table 24.1 comes under the approved list of CDM.

Table 24.1 Categories of renewable Energy projects suitable under CDM

Major category	Project type
Solar	Photo-voltaic power production (Grid connected) Photo-voltaic power production (Off-grid) Solar water heating Solar thermal power production
Wind	Large on-shore wind power generation Large off shore wind power generation Small scale wind power (electric/mechanical)
Hybrid	Hybrid mini grid RE systems (Hybrid with fossil fuels like Diesel)
Biogas	Farm scale plants Institutional and community biogas plants
Biomass	Biomass gasification Improved cooking stoves Vegetable oil/ Biodiesel as fuels
Waste to energy	Power and heat production from waste Collection and usage of land fill gas Anaerobic waste water treatment Solid waste management for energy Briquetting of crop residues
Energy from water & Ocean	Hydro electricity generation Tidal power Wave power
Geothermal energy	For power and heat

CHAPTER 25

Energy Management and Conservation

The management of any resource involves systematic implementation of a conscious and continuous methodology to use the available resources to achieve the set goals. Energy is one among the most crucial resource for human development. Apart from that, the environmental concern on energy use is currently widely discussed globally. The management of energy is associated with a number of concepts such as balancing energy supply and demand, conservation through better utilization, minimization of wastage, energy recovery, waste recycling, energy conversion of wastes and production of energy from renewable sources.

All sectors of economy are sensitive to energy costs and it is imperative that wasteful practices in energy use is curtailed. Energy conserved at the end point can be regarded as renewable energy produced at that point. The actual benefit is much higher and almost double when we consider the losses occurring in transmission and distribution. Hence conservation of energy is regarded as the most benign concept when we consider the environmental and economic advantages. On a broader sense, the efforts to minimize energy use without sacrificing the GDP of the country can be called 'Energy Conservation' at national level. In other words, energy conservation is the judicious consumption of energy resources with a view to conserve it. This is achieved through energy management which includes various measures like improving energy efficiency in processes and operations. Utilization of waste heat in processes or selection of less energy intensive processes are all steps to minimize energy consumption. Different sectors like industry, transport and agriculture require different strategies and techniques.

25.1 Energy Conservation in Industries

Industry is the largest energy consuming sector and the product cost has a direct bearing on the energy cost, which varies between 10 to 50 per cent of the product cost depending on the energy intensity in its production. As early as 1981, the National Productivity Council conducted a study in 200 industries in India which revealed that there is possibility of saving up to 25% of energy use in industrial sector. Another agency concerned with conserving petroleum products is the 'Petroleum Conservation Research Association' and they had proposed various measures to conserve petroleum. In industries, the purchased energy and its management are of concern as energy is one among the crucial inputs. Depending on the sector the techniques to be followed also varies. Even though the overall objective is to use energy judiciously and minimize its use wherever possible, the strategies are highly specific with each industry. In general, the various steps to conserve energy in industries may be listed as below:

(i) Improving the utilization percentage of installed capacity: This involves strategies to use the full capacity of the plant as far as possible so that the overall energy use per unit of product will be less.

(ii) Promotion of efficient operational and maintenance practices with energy efficient regulation of various processes.

(iii) Minimizing heat losses through better insulation.

(iv) Heat recovery measures

(v) Replacement of inefficient equipment and processes with efficient ones.

(vi) Recycling of wastes and refuse

(vii) Co-generation and use of combined heat and power plants, if applicable.

(viii) To conserve electricity, the recommended measures should be followed for the specific industry. The load in different phases should be balanced and the load factor of motors used for various equipment should match. Proper instrumentation is necessary for good maintenance practices. Any leakage of energy should be avoided by constant monitoring.

(ix) Use of automatic controls for boilers help to save fuel in industries where steam is used. Improving the efficiency of combustion is very important along with thorough and routine inspection of boiler house. Conservation can also be ensured by proper maintenance of steam distribution and recovery systems.

25.2 Energy Accounting

Energy accounting is the primary step by which energy management can be undertaken. This is an essential requirement to plan and implement the energy conservation approaches for the specific system. It comprises of the following steps (Fig. 25.1):

(i) *Fixing the objectives:* The objective may be assessment of overall energy flow or for a specific unit or subsection.

(ii) *Selecting the boundary:* The boundary of the system studied need to be fixed so as to identify the inputs and outputs within the system.

(iii) *Preparation of flow diagrams of processes:* The typical process flow charts are necessary for the accounting procedure.

(iv) *Identification of material inputs:* Mass flow of materials in the process should be noted.

(v) *Identification of energy inputs:* Energy inputs by various sources and different forms need to be understood.

(vi) *Quantification of inputs:* Properly measuring the quantities of mass flow and energy flow in various forms are required to account them.

(vii) *Identifying the outputs:* Both material output and energy output need to be identified.

(viii) *Quantification of mass and energy inputs:* Measurement of both material output and energy output if any is also a requisite for accounting.

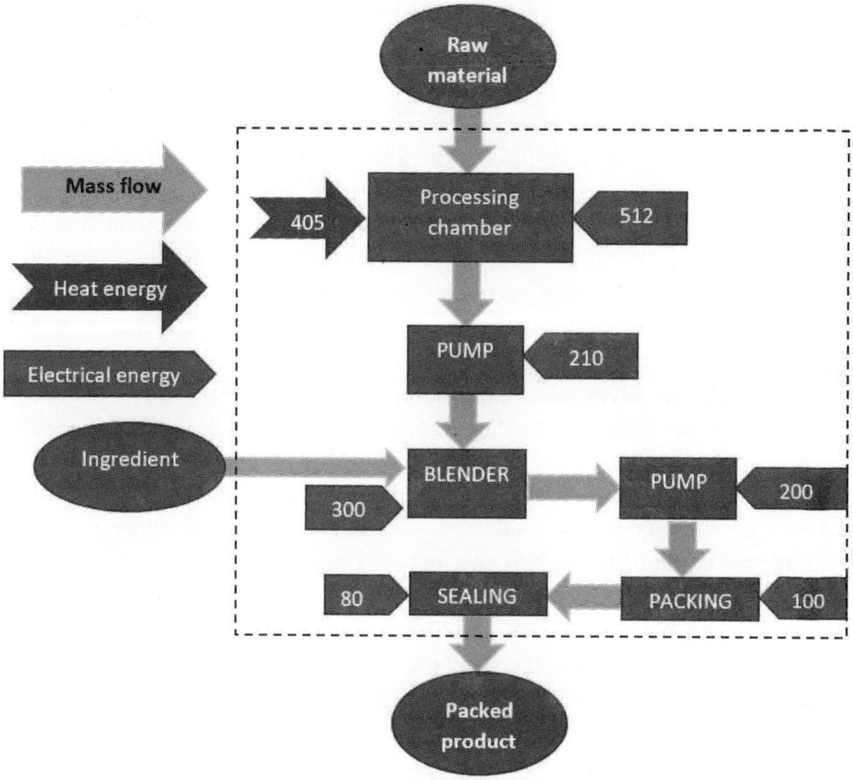

Fig. 25.1 Energy accounting diagram of a food processing plant

25.3 Energy Audit

Energy audit is an important tool to keep track of the performances in energy conversation. The audit procedure involves measuring the actual energy used in a processing facility or a factory or any unit in comparison to the minimum energy required. It aids to identity all energy flows in and out of a unit. It enables to identity the points of energy wastage and locate the points where energy savings potential exists. Energy audit gives answer to questions such as what all purposes energy is used, how much is used, how it is used, why it is used and what is the cost for each form of energy. It gives comprehensive information on all aspects of energy and suggests possible conservation measures and means to improve energy efficiency. The main objectives are:

(i) To review energy consumption in relation to production and compare with the recommended target or to recommend target.

(ii) Carry out energy analysis to identify key areas which need attention.

(iii) To recommend additional facilities.

(iv) To evolve an energy monitoring system and implement it.

(v) To analysis energy efficiency of the overall process and prepare a detailed report.

(vi) To make available necessary guidelines for energy management.

Energy audits are usually categorised into three viz. Walk-through audit, mini-audit and maxi-audit. In a walk-through audit, a quick inspection of the factory/unit is made to assess the energy use pattern and to find possibilities for energy saving opportunities. Additionally, basic information is gathered to determine if a detailed analysis is required. In mini-audit, certain tests and measurements are done to quantity energy uses and losses and determine the possible economic advantage in incorporating modifications and the cost involved. Maxi-audit consists of a detailed analysis of energy use in operation / process. The analytical results show clear pictures of energy use patterns along with required forecasts for various aspects of energy use with a view to arrest wastage and maximise process efficiency. There are essentially two parts for energy audit, (i) Data acquisition and (ii) Data analysis. Data acquisition will need collection of necessary bills for all forms of energy used in the unit to have a document on the quantity as well as cost of energy. The data analysis involves a detailed interpretation of the data with a list of 'Energy Conservation Opportunities'. Next step is the preparation of energy use profiles in the case of an industrial plant or food processing plant. Different methods are used to represent energy flows and energy consumption. To represent energy use from different sources, a pie chart can be used. A Sankey diagram representing flow of different forms of primary energy may also be prepared (Fig. 25.2).

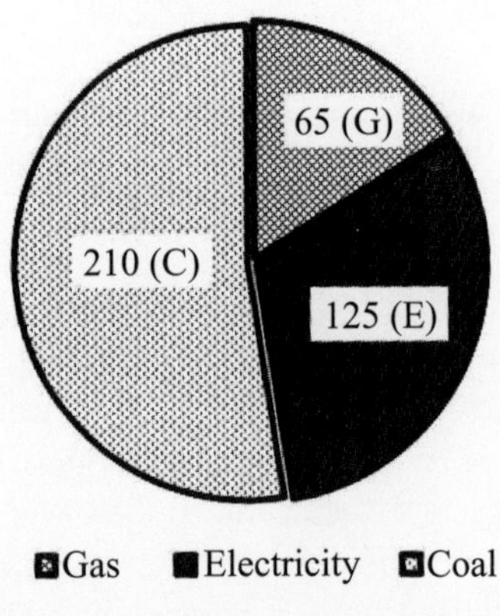

Fig. 25.2 Pie diagram representing consumption of different sources

25.4 Energy Efficiency in Indian Agriculture

The early man used most of his energy in the form of food whereas modern man needs energy in various forms for food production. In agricultural production, we use external energy inputs to augment the natural photosynthetic process so that the food production per unit area is increased. This is required to feed the growing population. Many professional bodies like International Energy Agency have predicted that population expansion and economic growth will increase the global demand for energy and water by 40 percent by 2030. Practices required to get increased food production involve increased consumption of energy which come from various sources like human labour, electricity, chemical energy of fossil fuels, renewable energy or mechanical energy derived from liquid fuel fired engines.

The pattern of commercial energy use in India indicate that agriculture is a prominent consumer of energy after industry, transport and residential sectors. In the last decade, almost 18 per cent of the total electricity consumption in India was by the agricultural sector. About 15 to 17 per cent of the total diesel fuel consumption in India also is by the agricultural sector. Modern agriculture is increasingly becoming more energy intensive in India and energy conservation in agriculture is highly warranted. Apart from direct use of energy in the form of petroleum products and electricity, indirect use of energy in the form of chemical fertilizers and other inputs are also important in agricultural systems. The increased farm power availability has indicated a direct relationship with agricultural production. More and more energy will be used in agriculture and hence efforts to increase energy efficiency in agriculture is important.

25.4.1 Energy efficiency in irrigation

The major energy use in agriculture is for irrigation and electricity is the most sought-after energy form to power water pumps. Even though the attainable efficiency in water pumping by electric pump sets are about 60 per cent, the actual efficiency achieved in the fields is much low. On an average, diesel engine powered pump sets in India work at a very low efficiency of about 10 to 12 per cent. The major problem for energy losses in agricultural water pumping is the wrong selection of pumps, low efficiency of pumps, under sized pipelines with high friction losses, inefficient foot valves of centrifugal pumps, inefficient power transmission belts, unnecessarily high delivery head and too long suction pipes. The agricultural pumps can be classified based on their energy efficiency as shown in Table 25.1.

Table 25.1 Categorisation of pump sets based on energy efficiency

Sl. No.	Category based on energy consumption	Electricity per unit irrigation work* (kWh)	Diesel per unit irrigation work* (Litre)
1.	Ideal	0.50 – 0.70	0.150 – 0.200
2.	Marginally high	0.71 – 0.90	0.201 – 0.250
3.	Very high	0.91 – 1.20	0.251 – 0.350
4.	Excessive	1.21 – 1.50	0.351 – 0.450
5.	Exorbitant	Above 1.50	Above 0.450

*Lifting 100m^3 water through 1 metre

Since the power for irrigation is free or subsidised in many Indian states, most farmers are not bothered about the energy use. There is also lack of knowledge among farmers on efficient irrigation equipment and there has been suggestions to implement

CDM projects for agricultural pump sets. The improvement of water use efficiency of crops by the use of modern irrigation techniques could decrease the water and energy use to considerable extends. Better soil and water conservation measures can address the issue of lowering of water table to some extent, which also has a favourable effect on energy use as the suction lift is reduced. Rain water harvesting is another promising technique to reduce pumping energy consumption. Agronomic measures like mulching to prevent evaporation losses and improving water holding capacity of soil by improving the organic matter content of soil are also crucial in energy conservation as they help to reduce need of pumped water.

25.4.2 Energy efficient implements for crop production

Crop production activities like tillage consumes lot of energy, especially for crops like rice where repeated ploughing operation may be required to make the field ready for sowing/ planting. There is a significant increase in the rate of increase of tractor population in India, and hence introduction of energy efficient matching implements is highly relevant. The implement may not match with tractor power in many occasions resulting in wastage of energy. The conventional tillage implements like the tractor drawn cultivator are less efficient and need to be replaced with active implements like rotovators or power harrows (Fig. 25.3). Both rotovators and power harrow are rotary tillage implements which pulverise the soil better than a soil stirring type tillage implement.

Passive implements like an ordinary cultivator derive the power from the pulling of tractor achieved through the drive wheels. Slippage of drive wheels can occur in many situations which results in energy wastage. On the contrary, active implements take their power directly from the tractor engine and hence do not depend on the drive wheels. Rotary tillage implements mentioned above also have an increased field efficiency as they manipulate the soil better than the drag type conventional implements. The quality of tilth obtained from a single pass of rotary tillage implement is better than two passes by a drag type cultivator. This has a direct impact on the time required for the operation as well as the fuel consumed for the specific operation.

Fig. 25.3 Power harrow

In rice cultivation, transplanted systems and direct (wet) seeded systems require puddling of the fields. Puddling is the mechanical manipulation of soil in presence of standing water to create a least pervious soil strata so that water is maintained in the field after transplanting the seedlings. Most often no special implements other than 'cage wheels' attached to tractor are used for puddling. '*Puddlers*' are energy efficient implements which can be attached to tractors for puddling of paddy lands. The tractor drawn Helical Blade Puddler developed by Kerala Agricultural University (Fig. 25.4) is a simple implement which was certified to be energy efficient by the Energy Management Centre of Kerala state.

Fig. 25.4 Demonstration of KAU Helical Blade puddler

25.5 Energy Ratio of Agricultural Production Systems

The introduction of high yielding varieties of major crops in India paved the way for green revolution resulting in increased use of chemical inputs in the form of fertilizers as well as plant protection chemicals. Plant protection chemicals include insecticides, fungicides, herbicides etc. These indirect sources of energy do not release energy directly but the inherent energy content should be considered. The energy used for the production of that input is referred as equivalent energy coefficient as given in Table 25.2. This indirect energy input should be accounted in assessing the energy efficiency of agricultural production systems.

The energy efficiency of crop production systems can be estimated from the specific energy used for production of unit mass of agricultural produce. The total input energy to unit area of land for cultivation of a specific crop can be estimated from data collected through field observations. Each input quantity may be multiplied by the corresponding equivalent energy coefficient to get the energy input by way of specific inputs. Sum of all these energy quantities will give the total energy input. The total output also can be estimated in a similar way. The energy ratio indicates the overall efficiency of the production system and is given by,

$$\text{Energy Ratio} = \frac{\text{Energy output, MJ}}{\text{Energy input, MJ}} \quad \text{...(25.1)}$$

The total energy input per unit quantity of the agricultural product (MJ·kg^{-1} produce) is often referred as Specific energy of the production system.

$$\text{Specific Energy (MJ kg}^{-1}\text{)} = \frac{\text{Total energy input, MJha}^{-1}}{\text{Production from unit area, kg ha}^{-1}} \quad \text{...(25.2)}$$

Farmers in most parts of India are using energy inputs like fertilizers and chemicals sometimes without following the scientific recommendations. The dependence on chemical energy should be minimized for sustainable crop production. The over use of chemicals in fields has resulted is numerous environmental hazards which directly and indirectly affects us. High level of pesticides and chemicals have been identified in fruits, vegetables and even milk. While pollution directly affects human health, the killing of beneficial insects upsets the balance of the ecosystem causing pest outbreaks. Some of these pests gain resistance over the chemicals and farmers are compelled to use more and more deadly pesticides at heavy doses. Over use of fertilizers also affects the soil and environment in several ways. Pollution of drinking water sources is the direct impact and soil degradation is the indirect impact. The use of chemical fertilizers can be partially substituted by farmyard manure and biogas slurry.

Several studies have shown that there is wasteful use of energy in crop production activities. It was observed that to increase the yield of rice by one tonne, an additional input of 8000 MJ of commercial energy is required. It is noteworthy that by increasing the organic carbon in the soil by 0.1 per cent there could be a net saving of 1620 MJ of commercial energy.

Table 25.2 Equivalent energy coefficients of various agricultural inputs

Sl. No.	Particulars		Equivalent energy coefficient, MJ
A.	**INPUT**		
1.	Human labour, hour	Adult man	1.96
		Woman	1.57
2.	Animal, pair-hour	Large bullocks, > 450 kg	14.1
		Medium weight, 350 to 450 kg	10.1
		Small < 350 kg	8.1
3.	Agricultural machinery, kg*		62.7
4.	Chemical fertilizers, kg	(a) Nitrogen	60.6
		(b) P_2O_5	11.1
		(c) K_2O	6.7
5.	Farm yard manure, kg (dry)		0.3
6.	Seed, kg	(a) Cereals and pulses	14.7
		(b) Oil seeds	25.0
7.	Agro-chemicals, kg	(a) Superior	120.0
		(b) Inferior	10.0
8.	Electricity, kWh		11.93
B.	**OUTPUT**		
1.	Cereals and pulses, kg		14.7
2.	Oil seeds, kg		25.0
3.	Sugarcane, kg		5.3
4.	Tuber crops, kg	(a) High food value (e.g., Tapioca)	5.6
		(b) Medium food value	3.6
		(c) Low food value	1.6
5.	Vegetables, kg		Variable from 0.8 to 1.9
6.	Straw, kg		12.5
7.	Fodder crops like maize, kg		18.0

* Distribute the weight of the machinery equally over the total life span hours (kg per hour). Then find the hours of machine required for particular operation.

CHAPTER 26

Energy Conversion of Municipal Solid Waste

It is well known that the environment of the world is alarmingly deteriorating from many economic activities of modern human civilization. The anthropometric discharge of solid, liquid and gaseous wastes resulted in pollution of the atmosphere, pollution of water bodies and is posing imminent threat to world biodiversity. Many species of aquatic and terrestrial life have been extinct and many more are in the verge of extinction. Management of solid wastes is a serious issue in the urban areas in almost all countries. Any material once discarded as waste is a menace which is likely to be harmful to human life and cause damage to the environment. The amounts of waste produced keep on increasing year after year with corresponding escalation of the related problems, if not tackled.

The present era may be called the 'Age of Wastes' and this is true in the case of many developing societies with high population. Many cities in developing countries are facing serious issues with urban waste, with compounding problems of what to do with the large quantities generated every day. Unfortunately, in many developing countries as not enough is being done, the problems are multiplying. Waste discharged to land and water without proper treatment can become just as big a killer as anything, through contamination of drinking water as well as disease outbreak. Municipal governments in many countries collect the wastes, which we call 'Municipal Solid Waste (MSW)' but then there is the problem of what to do with it. Another area of concern involves sewage treatment. When sewage is treated by conventional processes, large quantities of sludge is produced and dealing with the sludge is again yet another problem. It is very important to understand how to handle MSW for its utilization with minimum environmental issues.

Mismanaged wastes lead to creation of more waste and the failure to salvage and reuse these waste materials cause an avoidable depletion of valuable natural resources as well. The present unhealthy methods of waste disposal have led to further deterioration of the environment due to GHG emission from landfill sites, and black smoke and acid rain due to incineration. The conventional methods of waste disposal are mainly concentrated at disposal of waste only and often do not include methods of utilising it for useful applications like energy generation or conversion to fertiliser. It is possible to extract energy from MSW and there is also a possibility to convert the organic fraction into fertiliser. Different aspects of waste management and feasible technologies for energy conversion are discussed here.

26.1 MSW Generation and its Composition

Municipal Solid Waste (MSW) comprise of the solid or liquid materials rejected due to domestic as well as community or commercial activities in the urban or municipal area. All types of wastes from households, hotel and restaurants, markets, public offices, commercial establishments as well as community institutions like schools, places of worship, auditoriums and cinema halls contribute to MSW.

Waste generation and their composition in different countries are contingent upon many factors like levels and nature of urbanization and industrialization, socio-economic status of households, geographical location etc. Nature, quality and quantity of wastes generated also varies widely. The per capita waste generation is comparatively less in developing countries in comparison to the developed countries and the national per capita waste generation in India is likely to be about 0.4 kg per day. The composition of MSW includes paper, wood, metal, glass, plastic, rubber, leather, construction demolition wastes, food wastes, rejected clothes, biowaste like leaves, street sweepings, rags, etc. The composition can vary widely with respect to the economic status of the people as well as the population. For example, components such as paper, rubber, glass and plastics increase as the population of the city increases. In most of the developed countries, MSW has 70 to 80 per cent of organic matter while Indian MSW contains only 40 to 60 per cent of organic matter. A general classification of components of MSW are shown in Table 26.1.

26.2 Waste to Energy (WTE) Projects

The organic matter in the MSW can be converted into energy. Hence the MSW is actually an energy rich material. The sewage also contains only 2 to 4 per cent of solid content and hence it can also give energy even though in smaller quantities. It was estimated by Ministry of New and Renewable Energy of the Government of India that there is a potential of about 5690 MW of power from municipal and industrial wastes in our country and is increasing with overall increase in population as well as economic and industrial development. There is also a huge potential for energy generation from agricultural residues and the potential is likely to increase further with increasing agricultural development.

Table 26.1 Classification of urban refuse materials

Terminology	Contents	Source
Garbage	Wastes from food preparation and serving	Restaurants, Places of worship, auditoriums for social and religious functions and households.
Rubbish	Market refuse, packings, wastes from handling, storage and sale of various products. Combustible components (mostly organic and plastics), Non-combustible components (Metals, glass and other mineral materials)	Commercial establishments, markets etc
Ashes	Residues from burning solid fuels like coal and biomass	Industrial or domestic
Bulky wastes	Large auto parts like tyres, discarded furniture, refrigerators, crates, trees, branches etc	Individual, Commercial, public utilities
Street wastes	Street sweepings containing both organic and in organic fractions	Streets
Abandoned vehicles	All types of two wheelers and other automobiles	Private or public
Construction wastes	Broken concrete, abandoned roofing materials, plaster, conduits, pipes and other scraps	Private and public
Industrial refuse	Solid wastes both organic and inorganic from industrial processes, manufacturing or food processing.	Industries, power plants and processing units
Special wastes	Hospital wastes, hazardous wastes, explosives, radioactive wastes, pathogenic wastes	Households, hospitals, industries, institutions etc.
Animal and agricultural wastes	Dead animals. Manures, crop residues, chicken wastes, waste meat and animal parts	Domestic, farms, poultry units, slaughter houses, fish markets
Sewage treatment wastes	Sludges, screenings, grit	Sewage treatment plants and septic tanks

26.2.1 Waste collection and pre-processing

The waste generated is generally classified into combustible and non-combustible wastes for the purpose of energy conversion. Combustible wastes include materials like paper, cardboard, paper derived wastes, plastics, wood or wood-based wastes, yard wastes like leaves and tree branches, food and vegetative wastes, rubber, leather, etc. Non-combustible wastes include soil, rocks, ashes, dust and construction / demolition wastes, commercial and industrial wastes, hazardous and toxic wastes, ferrous and non-ferrous metals, glass, etc. The common practice is to pick up all these wastes together and transport to the dumping yard.

Only combustible wastes can be used for the WTE projects and hence need to be sorted out before conversion depending on the technology used. In many cities, high calorific wastes like paper and plastics are removed by rag pickers and they may also collect glass and metallic items as they could be recycled. Segregation of waste at source will be more economical as well as convenient. Separate collection bags with different colouring are provided to households for source sorting of wastes in many developed societies.

26.2.2 Technologies for energy conversion

The conventional method of disposal of MSW is landfilling and composting with the main objective of waste disposal and hence no energy is produced. Composting is the biodegradation of the constituents in organic wastes, by mostly aerobic microbial activity. During composting organic matter is decomposed into a stable, humus like substance and the outcome of the composting process will be manure which can be used in agriculture. For small quantities of wastes, vermi-composting can also be attempted. As the organic manures can replace the energy intensive chemical fertilizers, this may also be considered as a renewable energy conversion method. However, increased quantity of wastes, the inherent slowness of composting process, scarcity of land, objection of neighbourhood due to foul smell and insect problems, etc. render this method un-popular and unsuitable in a developed society.

Landfill is a dump yard where waste materials are dumped outside the city and left out without any treatment for a long time. Due to biochemical reactions in presence of moisture and oxygen, the organic matter degrades. As most landfills are unprotected, rain water dissolves chemicals and other pollutants which pass over to the ground water in the meantime. Besides the harmful leachates, gases like methane and CO_2 are evolved. It should be remembered that methane is explosive and also responsible for global warming and its Global Warming Potential is 24 times that of CO_2. Landfills are banned at many developed countries in European union and in US. However, sanitary landfills are scientifically managed and hence permitted in many developed countries also. Ordinary landfills without a gas recovery system do not produce energy and hence cannot be included under WTE projects.

Technical options for waste to energy

The following are the common technologies for conversion of MSW to energy:

Sanitary landfills with gas recovery

Anaerobic digestion to produce biogas

Bioreactor landfills for anaerobic composting

Incineration with waste heat recovery

Gasification and pyrolysis

Production of RDF

Alcohol production

Sanitary Landfills with gas recovery

Sanitary landfill is an 'engineered method' of disposing of MSW on land with proper precautionary measures to prevent pollution. In this method, the waste is spread in thin layers and compacted to a small volume, subsequently covering it with compacted soil at frequent intervals as required to keep the landfill hazard free. As the organic wastes are buried underground without proper contact with atmospheric air, anaerobic condition prevails enabling the growth of methane producing anaerobic bacterial consortium. Methane is combustible and can be recovered. Normally, the concept is called Landfill Gas (LFG) utilisation program and involves introduction of vertical holes or vertical perforated pipes that reach down 90 per cent of the depth of the landfill. These are distributed throughout the landfill area. The vertical pipes are attached to a gas collection system to which a partial vacuum is applied by the use of blowers. Due to the negative pressure, landfill gas generated within the landfill is sucked through the collection pipes and reach a central collection chamber from where it can be used in an IC engine to produce electricity (Fig. 26.1).

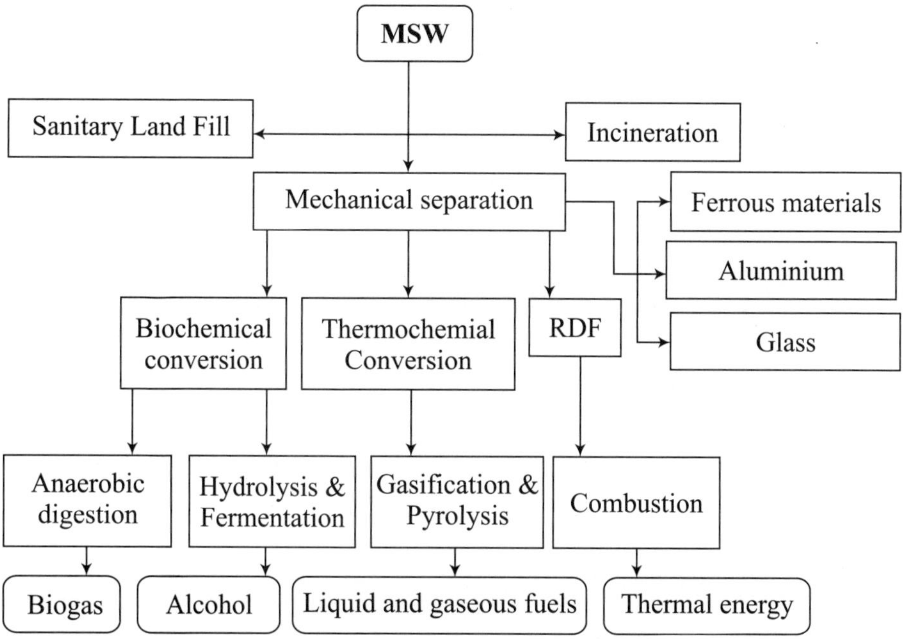

Fig. 26.1 Energy conversion options for MSW

LFG may contain incombustible CO_2 as well as nitrogen and other minor components like water vapour. It is a low-to-medium calorific value fuel gas and there can be daily or seasonal variations in the calorific value and flow rate. LFG generally contain 40 to 50 per cent methane, 30 to 40 per cent CO_2, 10 to 20 per cent nitrogen, less than 1 per cent oxygen and traces of other gases including H_2S. During the landfill process, a moisture content above 40 per cent is maintained so as to have better microbial growth and faster degradation. The LFG also may contain significant amount of water vapour, as the waste material contain substantial amount of moisture and is almost saturated with water vapour. The fact that gas recovery occurs usually at ambient temperatures above 25°C also is another reason for high moisture in LFG. The liquid fraction that percolates through the waste in landfills, called 'leachates' are periodically collected at the bottom and is pumped out to drying beds prepared for this purpose. The digested and stabilised solids can be retrieved and can be used as compost to manure crops, if such a facility is there in the site (Fig. 26.2).

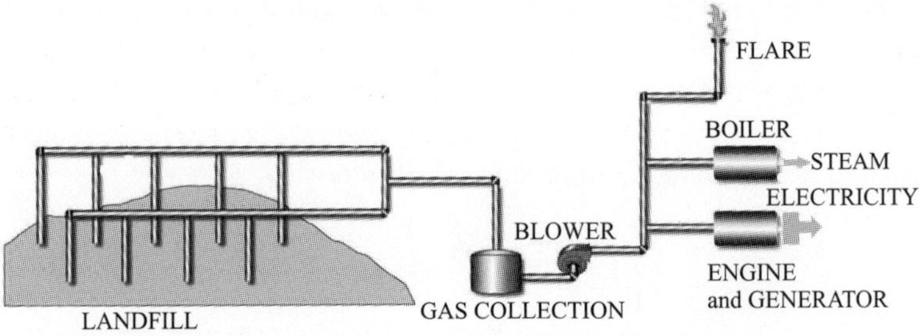

Fig. 26.2 Schematic view of a landfill gas recovery system

26.3 Anaerobic Digestion of Organic Fraction of MSW

Biogas is the combustible gas produced during anaerobic digestion of organic matter. Even though anaerobic digestion of MSW is more capital intensive than aerobic composting, the major advantage is that a renewable fuel can be produced. As the solid content in MSW is likely to be higher than that in the ordinary animal manure-based plants, care should be taken in the design and operation of MSW digesting plants. The biogas generation is highly variable, depending on the characteristics of the waste, weather factors, nutrient availability etc. It can be in a wide range of 10-200 m^3/ton of the organic fraction of MSW. A portion of the energy produced is required for running the plant and the consumption will be more if heating is required. The net electrical energy that can be produced from biogas-based engine generator system is in the range of 100 to 150 kWh per tonnne of MSW (organic fraction). The commercial use of high solids (30 per cent or more) anaerobic digesters for MSW treatment has been in use in Europe with a daily feeding rate of 40 to 300 tonnnes of mixed (source separated) MSW per day capacity at 18 to 25 days HRT. The high-rate anaerobic bioreactors are generally used for organic liquid wastes and may be used also for energy recovery from sewage. Sometimes, the sludge produced is dried and converted into fuel pellets for direct combustion.

Anaerobic digestion of MSW has the major advantages that energy is produced in the process and the solid residue is a value-added compost for use as an organic manure. Most pathogens are destroyed in the process and is cost effective as it requires less land compared to landfills. No GHG emission occurs and is more hygienic as there is no air pollution in the surrounding areas. But it has the limitation that only the organic fraction of MSW can be treated and the non-biodegradable fraction need to be handled separately. Other disadvantages are the inherent slowness of anaerobic process due to the long retention time, the inability of digesters to withstand shock loading and the fact that non-biodegradable organic fraction cannot be treated.

26.4 Bioreactor Landfills for Anaerobic Composting

Anaerobic composting in modified landfills designated as 'bioreactor landfill' or 'controlled landfill' is one among the newer developments in landfill design and management. Bioreactor for anaerobic composting is more suitable to developing countries where the organic solid content of MSW is high. Basically, this is a sanitary landfill designed under optimum conditions for bioconversion of organic fraction of MSW to a fuel gas with methane as the major combustible constituent. The technology involves recycling of the leachate in a continuous manner.

A bioreactor landfill has a modular design with cells or modules designed to hold 0.25 to 1.0 million tonnes of MSW. They are provided with a system to collect all leachate at the base of the cell which in turn is pumped back to the surface and distributed across the top of the piles of MSW. This technology has two advantages; preventing the leachate which has a high pollution potential from contaminating nearby soil and ground water and secondly, the organic wastes are kept moist all the time which promotes bacterial growth by supplying moisture and nutrients. This circulation of leachates retains the landfill in a wet condition which is congenial to bacterial growth, rendering it as a high-solids anaerobic bioreactor. In the long run, the pollution load of leachates drops in the order of about 10,000 to one tenth (BOD). When the pollution potential of leachate is within the safe ranges (as per the prevailing norms), they can be disposed of into regular sewage system. Bioreactor landfills are capable of reducing the volume of organic portion of MSW significantly at a fast rate. If the organic fraction is approximately 50%, it can get reduced to less than 2 per cent within a few years due to the high level of microbial activity enhancing waste degradation. Due to fast waste conversion, the modules can be emptied to take out the stabilised material to be processed and used for agricultural use. This modular operation enables refilling of the emptied cells with fresh charge of MSW. The overall advantage is that we require only smaller landfills. This method is capable of disposing of all types of biodegradable solid wastes, irrespective of size, moisture content and other characteristics.

26.5 Incineration with Waste Heat Recovery

Incineration refers to controlled burning of MSW in large centralised facilities generally referred as 'Municipal Waste Combustion' (MWC) systems. The systems with waste heat recovery are grouped under 'Waste to Energy' systems and most of these mass burn facilities use boilers to produce steam. The high temperature gases produced during combustion are circulated through a chamber lined with water filled tubes which are designed to generate high pressure steam. The steam may be utilized for electricity production using a steam turbine.

The advantage of such a facility with energy recovery over an LFG recovery facility is that much higher quantity of energy can be produced. But it has the disadvantage of significantly high air pollution. Another advantage of incineration is the size reduction to 90 per cent by volume, even though the problem of ash disposal is there which is only about 10 per cent of the original MSW. The MSW can be used in an incinerator system even without much fuel preparation. Even biologically hazardous waste can be safely disposed of as complete sterilisation can be achieved by complete combustion of the material. But the problem of air pollution cannot be completely avoided even in advanced systems. The requirement of additional cost for pollution control equipment will increase the cost of installation and these plants need more maintenance due to chances of fouling and slagging. Incineration with mixed waste input is a popular thermochemical conversion technology for MSW and is conveniently done in Combined Heat and Power (CHP) plants.

Even though the combustion of waste to produce energy seems to be a simple and practical technological option for convenient and safe disposal of waste, these plants are expensive to construct and operate. They are not economical to produce electricity, if they cannot charge a fee for each tonne of waste they accept. These kinds of plants will be economically feasible, if this fee is lower than that for landfilling. The fee levied for accepting wastes combined with the income from energy production may render the plant economical.

26.6 Thermochemical Gasification and Pyrolysis of MSW

The thermochemical processes of Gasification and Pyrolysis can be used to alter the physical and chemical character of the municipal solid waste (MSW). These two processes can convert MSW to gaseous and liquid fuels that can be used in different combustion systems. Pyrolysis is the process of heating MSW in the absence of oxygen to get liquid fuel. Usually, no other reacting material is introduced into the reactor and complex organic molecules in the wastes break into smaller and simpler molecules due to high temperature. Gasification process introduces air or other gases into the reactor for the production of mostly gaseous components. Common gases introduced include oxygen, steam or hydrogen that causes reaction. Gasification reaction usually takes place at higher temperature than required for pyrolysis. It is more advantageous to use pyrolysis as a process for treatment of MSW as a replacement for incineration. But these systems are more complicated in their design and operation and they require sufficient experience for operating commercial systems.

26.7 Production of RDF from MSW

Many developed countries promote more environment friendly waste management options, by reducing the quantity of biodegradable wastes disposed of in landfills. Production of 'Refuse derived fuel' (RDF) by processing MSW convert MSW to a fuel that can be sold or used for various combustion applications such as boiler fuel. The first step to produce RDF is shredding of wastes, which should be done carefully and then sorted to remove all non-combustible materials such as glass, metal, and stone. The shredded combustible portion of the MSW is further processed into pellets that can be sold as fuel. Shredding and separation are done in a series of energy intensive mechanical processes. RDF plants use a number of equipment like flail mills, rotating screens, air blowers, etc. to shred and separate out components of incoming MSW. This process requires substantial amount of electrical energy (80-100 kWh/ton) with another 110-130 kWh for drying the fuel (Fig. 26.3).

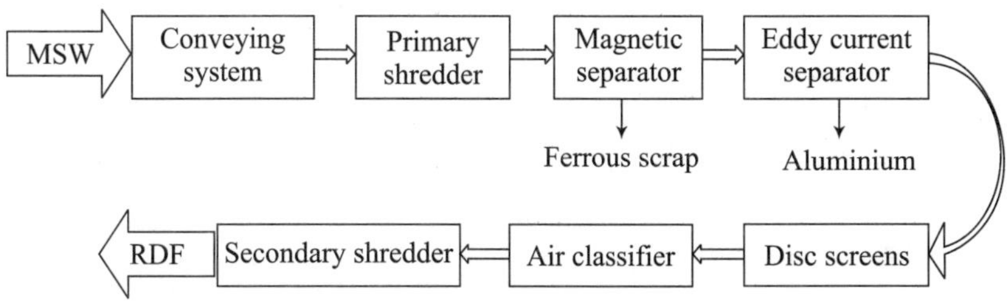

Fig. 26.3 Process flow in production of RDF from MSW

Originally, RDF was intended to be used as a compatible fuel suitable for mixing with coal in coal-fired power plants. As this eventually led to combustion problems, now the strategy is to burn the fuel in specially designed combustion systems. Other alternatives are to mix the RDF with biomass like crop residues or to use RDF in gasifiers.

26.8 Alcohol Production from MSW

As ethanol is being widely used as a renewable fuel, the conversion of organic fraction of MSW into alcohol is also being tried. The principles of continuous acid hydrolysis of cellulosic materials may be utilised to produce a sugar rich product, which in turn is fermented. In laboratory experiments 50 to 60 per cent glucose was recovered from cellulosic materials in MSW at temperatures about 230°C.

A report of the U.S. Department of Energy in 2009 described a combined process for gasification and alcohol synthesis from MSW. A flow diagram for the MSW to ethanol process is shown in Fig. 26.4. MSW is sorted first to remove recyclable materials followed by shredding and milling. It is then dried prior to gasification to produce syngas, which is a mixture of H_2 and CO with some CH_4. The syngas is cleaned by passing through a tar reformer and scrubbing system. The cleaned syngas free of tars and particulates is further sent to a sulphur removal unit. The sulphur free syngas is sent to a steam reformer to convert the methane to hydrogen and carbon monoxide. The syngas is now required to be adjusted for proper ratio of H_2 and CO and compressed which is required for synthesis of mixed alcohols. The products are passed to a cooling system to condense the vapours and the unconverted gaseous products are separated. The moisture in liquid product is then removed using a molecular sieve and sent to a fractionating column to separate it into ethanol and higher alcohols. Steam generated in the processes is utilised by sending to the steam cycle for power generation. Process steam is derived from the turbines for use in the gasifier, steam reformer and for thermal applications in various processes. The entire process is assumed to be installed in a landfill site so as to make use of the MSW available there. The capital involved in producing alcohols from MSW is comparatively higher, due to the need to process the MSW first into RDF.

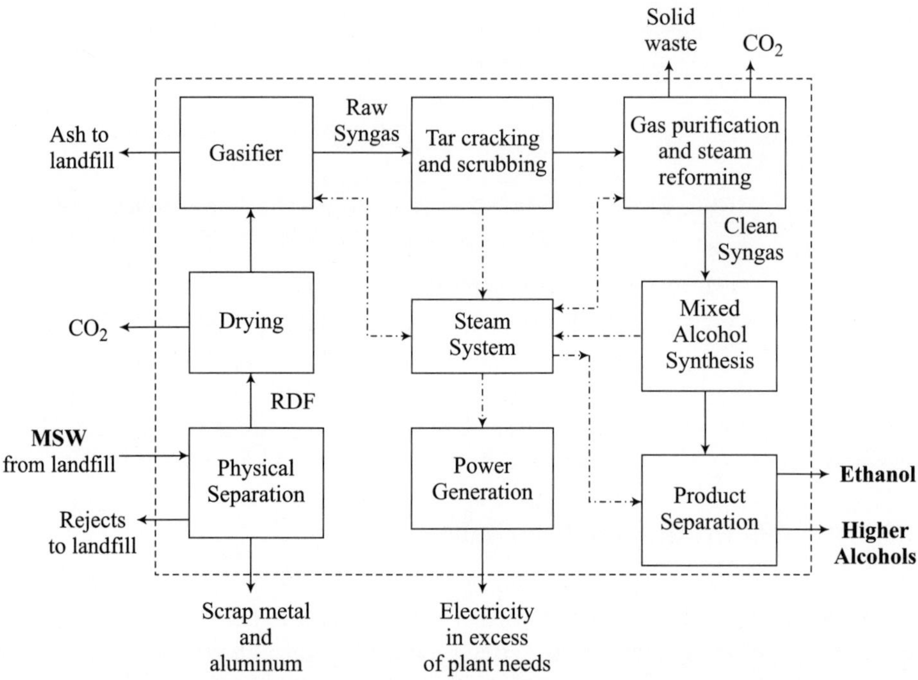

Fig. 26.4 Process flow diagram for synthesis of mixed alcohols from MSW

26.9 Environmental Implications of Waste to Energy Plants

The conversion of MSW to energy appears to be an environment friendly concept. But mass processing of MSW to produce energy can have a negative impact on the environment as well. There is often a public unrest if the waste-to-energy plant site is located in populated areas. There is a possibility of pollution due to dust and bad odours where waste is stored and sorted. The transport of high volumes of waste to processing plant is associated with considerable traffic movement in already congested city roads, especially in developing countries. When the waste material is burnt in CHP system, there is a health hazard due to emission of many toxic substances. These may include dangerous heavy metals such as mercury, cadmium and thallium and other hazardous organic compounds like dioxins and furans. These toxic gases can cause disorders in humans as well as animals. Hence these plants can work in a restricted mode only so as to meet stringent limits on the emission of harmful compounds to the surrounding atmosphere.

BIBLIOGRAPHY

Abrahamse, W., and Schuitema, G. (2020). Psychology and energy conservation: contributions from theory and practice. In M. Lopes, C. H. Antunes, and K. B. Janda (Eds.), *Energy and Behaviour* (pp. 19–44). Academic Press. https://doi.org/10.1016/B978-0-12-818567-4.00001-6

Acharya, B. (2018). Cleaning of product gas of gasification. In P. Basu (Ed.), *Biomass Gasification, Pyrolysis and Torrefaction (3rd Edition)* (pp. 373–391). Academic Press. https://doi.org/10.1016/B978-0-12-812992-0.00010-8

Acharya, J. (2011). *Clean Development Mechanism: Overview*. Asian Development Bank. https://www.adb.org/sites/default/files/publication/29055/cdm-brief-01-overview.pdf

Adams, P., Bridgwater, T., Lea-Langton, A., Ross, A., and Watson, I. (2018). Biomass conversion technologies. In P. Thornley and P. Adams (Eds.), *Greenhouse Gas Balances of Bioenergy Systems* (pp. 107–139). Academic Press. https://doi.org/10.1016/B978-0-08-101036-5.00008-2

Adhikari, S., Abdoulmoumine, N., Nam, H., and Oyedeji, O. (2017). Biomass gasification producer gas cleanup. In F. Dalena, A. Basile, and C. Rossi (Eds.), *Bioenergy Systems for the Future* (pp. 541–557). Woodhead Publishing. https://doi.org/10.1016/B978-0-08-101031-0.00016-8

Adlak, K., Chandra, R., Vijay, V. K., and Pant, K. K. (2021). Physicochemical activation and palletisation of Azadirachta indica wood carbons for increased biomethane adsorbed energy storage. *Journal of Analytical and Applied Pyrolysis*, *155*, 105102. https://doi.org/10.1016/j.jaap.2021.105102

Afsal, A., David, R., Baiju, V., Suhail, N. M., Parvathy, U., and Rakhi, R. B. (2020). Experimental investigations on combustion characteristics of fuel briquettes made from vegetable market waste and saw dust. *Materials Today: Proceedings*, *33*, 3826–3831. https://doi.org/10.1016/j.matpr.2020.06.222

Afzal, I., Shahid, A., Ibrahim, M., Liu, T., Nawaz, M., and Mehmood, M. A. (2017). Microalgae: A promising feedstock for energy and high-value products. In K. M. Zia, M. Zuber, and M. Ali (Eds.), *Algae Based Polymers, Blends, and Composites* (pp. 55–75). Elsevier. https://doi.org/10.1016/B978-0-12-812360-7.00003-3

Agenbroad, J., DeFoort, M., Kirkpatrick, A., and Kreutzer, C. (2011). A simplified model for understanding natural convection driven biomass cooking stoves - Part 1: Setup and baseline validation. *Energy for Sustainable Development*, *15*(2), 160–168. https://doi.org/10.1016/j.esd.2011.04.004

Aggarwal, S. (2000). Estimation of convective mass transfer in active single slope solar still. In A. A. M. Sayigh (Ed.), *World Renewable Energy Congress VI* (pp. 2186–2189). Pergamon. https://doi.org/10.1016/B978-008043865-8/50471-2

Aghaei, M., Ebadi, H., de Oliveira, A. K. V., Vaezi, S., Eskandari, A., and Castañón, J. M. (2020). New concepts and applications of solar PV systems. In S. Gorjian and A. Shukla (Eds.), *Photovoltaic Solar Energy Conversion* (pp. 349–390). Academic Press. https://doi.org/10.1016/B978-0-12-819610-6.00011-9

Aghaei, M., Kumar, N. M., Eskandari, A., Ahmed, H., de Oliveira, A. K. V., and Chopra, S. S. (2020). Solar PV systems design and monitoring. In S. Gorjian and A. Shukla (Eds.), *Photovoltaic Solar Energy Conversion* (pp. 117–145). Academic Press. https://doi.org/10.1016/B978-0-12-819610-6.00005-3

Aguilar-Rivera, N., Arora, R., Barbera, E., Behera, S. S., Bertucco, A., Bhattarai, T., Blandino, A., Buckeridge, M. S., Caro, I., Díaz, A. B., de Farias Silva, C. E., de Souza Abud, A. K., Dutta, A., Eibes, G., Escamilla-Alvarado, C., Ferrari, M. D., Gómez-Vanegas, N. A., García-Torreiro, M., Grandis, A., Webb, C. (2019). In R. C. Ray and S. Ramachandran (Eds.), *Bioethanol Production from Food Crops* (pp. xi–xii). Academic Press. https://doi.org/10.1016/B978-0-12-813766-6.00027-8

Aguilar-Rivera, N., Serna-Lagunes, R., Michel-Cuello, C., and Trujillo-Mata, A. (2019). Upgrading comparative and competitive advantages for ethanol fuel production from agroindustrial crops in developing countries: Mexico as a case study. In R. C. Ray and S. Ramachandran (Eds.), *Bioethanol Production from Food Crops* (pp. 401–415). Academic Press. https://doi.org/10.1016/B978-0-12-813766-6.00020-5

Akhlaghi, N., and Najafpour-Darzi, G. (2020). A comprehensive review on biological hydrogen production. *International Journal of Hydrogen Energy*, *45*(43), 22492–22512. https://doi.org/10.1016/j.ijhydene.2020.06.182

Akradecki, A. R. (2005). *Solar Energy Generating Systems solar power plants III-VII at Mojave Desert, California*. https://commons.wikimedia.org/wiki/File:Solarplant-050406-04.jpg

Al Garni, H. Z., and Awasthi, A. (2018). Solar PV power plants site selection: A review. In I. Yahyaoui (Ed.), *Advances in Renewable Energies and Power Technologies* (pp. 57–75). Elsevier. https://doi.org/10.1016/B978-0-12-812959-3.00002-2

Alam, M. S., and Tanveer, M. S. (2020). Conversion of biomass into biofuel: a cutting-edge technology. In L. Singh, A. Yousuf, and D. M. Mahapatra (Eds.), *Bioreactors* (pp. 55–74). Elsevier. https://doi.org/10.1016/B978-0-12-821264-6.00005-X

Alcorn, R. (2014). Wave energy. In T. M. Letcher (Ed.), *Future Energy (2nd Edition)* (pp. 357–382). Elsevier. https://doi.org/10.1016/B978-0-08-099424-6.00017-X

Alibrahim, H. A., SeedAhmed, S., Ahmed, U., and Zahid, U. (2019). Comparative analysis of gasification and reforming technologies for the syngas production.

In A. A. Kiss, E. Zondervan, R. Lakerveld, and L. Özkan (Eds.), *29th European Symposium on Computer Aided Process Engineering* (Vol. 46, pp. 1759–1764). Elsevier. https://doi.org/10.1016/B978-0-12-818634-3.50294-0

AlNouss, A., Mckay, G., and Al-Ansari, T. (2018). Optimum utilization of biomass for the production of power and fuels using gasification. In A. Friedl, J. J. Klemeš, S. Radl, P. S. Varbanov, and T. Wallek (Eds.), *28th European Symposium on Computer Aided Process Engineering* (Vol. 43, pp. 1481–1486). Elsevier. https://doi.org/10.1016/B978-0-444-64235-6.50258-8

AlNouss, A., McKay, G., and Al-Ansari, T. (2019). Superstructure optimization for the production of fuels, fertilizers and power using biomass gasification. In A. A. Kiss, E. Zondervan, R. Lakerveld, and L. Özkan (Eds.), *29th European Symposium on Computer Aided Process Engineering* (Vol. 46, pp. 301–306). Elsevier. https://doi.org/10.1016/B978-0-12-818634-3.50051-5

Amaral, A. F., Bozzano, G., Pirola, C., and Manenti, F. (2017). Biomass to X: Gasification and pyrolysis integrated. In A. Espuña, M. Graells, and L. Puigjaner (Eds.), *27th European Symposium on Computer Aided Process Engineering* (Vol. 40, pp. 1837–1842). Elsevier. https://doi.org/10.1016/B978-0-444-63965-3.50308-1

Amble. (2014). *Crescent Dunes Solar Energy Project as Seen from an Airliner*. https://en.wikipedia.org/wiki/Crescent_Dunes_Solar_Energy_Project#/media/File:Crescent_Dunes_ Solar_December_2014.JPG

Amoo, L. M., and Layi Fagbenle, R. (2020). Climate change in developing nations of the world. In R. O. Fagbenle, O. M. Amoo, S. Aliu, and A. Falana (Eds.), *Applications of Heat, Mass and Fluid Boundary Layers* (pp. 437–471). Woodhead Publishing. https://doi.org/10.1016/B978-0-12-817949-9.00023-2

Anaya-Lara, O. (2016). Offshore wind farm arrays. In Chong Ng and L. Ran (Eds.), *Offshore Wind Farms* (pp. 389–417). Woodhead Publishing. https://doi.org/10.1016/B978-0-08-100779-2.00012-X

Annamalai, N., Elayaraja, S., Oleskowicz-Popiel, P., Sivakumar, N., and Bahry, S. Al. (2020). Volatile fatty acids production during anaerobic digestion of lignocellulosic biomass. In V. K. Gupta, H. Treichel, R. C. Kuhad, and S. Rodriguez-Cout (Eds.), *Recent Developments in Bioenergy Research* (pp. 237–251). Elsevier. https://doi.org/10.1016/B978-0-12-819597-0.00012-X

Arafa, S., Farag, M., Abul-Naga, B., and Foote, E. (1988). Preparation of fuel briquettes for agricultural residues for village communities. In W. H. Bloss and F. Pfisterer (Eds.), *Advances in Solar Energy Technology* (pp. 2330–2334). Pergamon. https://doi.org/10.1016/B978-0-08-034315-0.50434-1

Archer, R. (2020). Geothermal energy. In T. M. Letcher (Ed.), *Future Energy (3rd Edition)* (pp. 431–445). Elsevier. https://doi.org/10.1016/B978-0-08-102886-5.00020-7

Argun, H., Gokfiliz, P., and Karapinar, I. (2017). Biohydrogen production potential of different biomass sources. In A. Singh and D. Rathore (Eds.), *Biohydrogen*

Production: Sustainability of Current Technology and Future Perspective (pp. 11–48). Springer India. https://doi.org/10.1007/978-81-322-3577-4_2

Arnórsson, S., Thórhallsson, S., and Stefánsson, A. (2015). Utilization of geothermal resources. In H. Sigurdsson (Ed.), *The Encyclopedia of Volcanoes (2nd Edition)* (pp. 1235–1252). Academic Press. https://doi.org/10.1016/B978-0-12-385938-9.00071-7

Ashley Dace. (2011). *Cromford Baryte Crusher.* https://www.geograph.org.uk/photo/2297009

Ashok, B., and Nanthagopal, K. (2019). Eco friendly biofuels for CI engine applications. In K. Azad (Ed.), *Advances in Eco-Fuels for a Sustainable Environment* (pp. 407–440). Woodhead Publishing. https://doi.org/10.1016/B978-0-08-102728-8.00015-2

Ashter, S. A. (2018). *Technology and Applications of Polymers Derived from Biomass* (pp. 75–110, 213–221). William Andrew Publishing.

Ashworth, J. (1996). *Mining the urban waste stream for energy: Options, technological limitations and lessons from the field.* Diane Pub Co.

Aslam, M., Ahmad, R., Yasin, M., Khan, A. L., Shahid, M. K., Hossain, S., Khan, Z., Jamil, F., Rafiq, S., Bilad, M. R., Kim, J., and Kumar, G. (2018). Anaerobic membrane bioreactors for biohydrogen production: Recent developments, challenges and perspectives. *Bioresource Technology*, *269*, 452–464. https://doi.org/10.1016/j.biortech.2018.08.050

Awasthi, A., and Bhaskar, T. (2019). Combustion of lignocellulosic biomass. In A. Pandey, C. Larroche, C.-G. Dussap, E. Gnansounou, S. K. Khanal, and S. Ricke (Eds.), *Biofuels: Alternative Feedstocks and Conversion Processes for the Production of Liquid and Gaseous Biofuels (2nd Edition)* (pp. 267–284). Academic Press. https://doi.org/10.1016/B978-0-12-816856-1.00011-7

AWS Scientific, Inc. (1997). *Wind Resource Assessment Handbook, Fundamentals for Conducting a Successful Monitoring Program.* https://www.nrel.gov/docs/legosti/fy97/22223.pdf

Aydin, E. S., Yucel, O., and Sadikoglu, H. (2018). Numerical investigation of fixed-bed downdraft woody biomass gasification. In I. Dincer, C. O. Colpan, and O. Kizilkan (Eds.), *Exergetic, Energetic and Environmental Dimensions* (pp. 323–339). Academic Press. https://doi.org/10.1016/B978-0-12-813734-5.00018-4

Babarit, A. (2017). *Wave Energy Conversion* (pp. 37–200). Elsevier.

Babu, D., and Anand, R. (2019). Biodiesel-diesel-alcohol blend as an alternative fuel for DICI diesel engine. In A. K. Azad and M. Rasul (Eds.), *Advanced Biofuels* (pp. 337–367). Woodhead Publishing. https://doi.org/10.1016/B978-0-08-102791-2.00014-3

Bach, Q. V., Gye, H. R., and Lee, C. J. (2018). Process modeling for steam biomass gasification in a dual fluidized bed gasifier. In M. R. Eden, M. G. Ierapetritou, and G. P. Towler (Eds.), *13th International Symposium on Process Systems*

Engineering (PSE 2018) (Vol. 44, pp. 343–348). Elsevier. https://doi.org/10.1016/B978-0-444-64241-7.50052-5

Bailey, B. H. (2016). Wind resources for offshore wind farms: Characteristics and assessment. In Chong Ng and L. Ran (Eds.), *Offshore Wind Farms* (pp. 29–58). Woodhead Publishing. https://doi.org/10.1016/B978-0-08-100779-2.00003-9

Bajpai, P. (2020). *Biomass to Energy Conversion Technologies* (pp. 1–204). Elsevier, 264p.

Barone, G., Buonomano, A., Forzano, C., and Palombo, A. (2019). Solar thermal collectors. In F. Calise, M. D. D'Accadia, M. Santarelli, A. Lanzini, and D. Ferrero (Eds.), *Solar Hydrogen Production* (pp. 151–178). Academic Press. https://doi.org/10.1016/B978-0-12-814853-2.00006-0

Basu, P. (2018). *Biomass Gasification, Pyrolysis and Torrefaction (3rd Edition)* (pp. 29–413). Academic Press.

Bayat, M., and Ozalp, M. (2018). Energy, exergy and exergoeconomic analysis of a solar photovoltaic module. In I. Dincer, C. O. Colpan, and O. Kizilkan (Eds.), *Exergetic, Energetic and Environmental Dimensions* (pp. 383–402). Academic Press. https://doi.org/10.1016/B978-0-12-813734-5.00022-6

Bayod-Rújula, A. A. (2019). Solar photovoltaics (PV). In F. Calise, M. D. D'Accadia, M. Santarelli, A. Lanzini, and D. Ferrero (Eds.), *Solar Hydrogen Production* (pp. 237–295). Academic Press. https://doi.org/10.1016/B978-0-12-814853-2.00008-4

Beck, Earl J. (1976). Ocean Thermal Gradient Power Plant, *US Patent 3967449.* Available at: https://www.freepatentsonline.com/3967449.html

Behling, N. H. (2013). *Fuel Cells*. Elsevier, 619p.

Beig, A. R., and Muyeen, S. M. (2016). Wind energy. In M. H. Rashid (Ed.), *Electric Renewable Energy Systems* (pp. 60–77). Academic Press. https://doi.org/10.1016/B978-0-12-804448-3.00004-9

Belessiotis, V., Kalogirou, S., and Delyannis, E. (2016a). *Thermal Solar Desalination* (pp. 1–326). Academic Press.

Bello, R. S., and Onilude, M. A. (2021). Effects of critical extrusion factors on quality of high-density briquettes produced from sawdust admixture. *Materials Today: Proceedings*, *38*, 949–957. https://doi.org/10.1016/j.matpr.2020.05.468

Belyakov, N. (2019). *Sustainable Power Generation* (pp. 355–500). Academic Press.

Benda, V. (2018). *A Comprehensive Guide to Solar Energy Systems* (pp. 151–213). Academic Press.

Berkowitz, N. (1979). Beneficiation: cleaning, drying, and briquetting. In N. Berkowitz (Ed.), *An Introduction to Coal Technology* (pp. 169–183). Academic Press. https://doi.org/10.1016/B978-0-12-091950-5.50014-6

Bermudez, J. M., and Fidalgo, B. (2016). Production of bio-syngas and bio-hydrogen via gasification. In R. Luque, C. S. K. Lin, K. Wilson, and J. Clark (Eds.), *Handbook of Biofuels Production (2nd Edition)* (pp. 431–494). Woodhead Publishing. https://doi.org/10.1016/B978-0-08-100455-5.00015-1

Bertani, R. (2016). Deep geothermal energy for heating and cooling. In G. Stryi-Hipp (Ed.), *Renewable Heating and Cooling* (pp. 67–88). Woodhead Publishing. https://doi.org/10.1016/B978-1-78242-213-6.00004-7

Bhatia, S. C. (2014). *Advanced Renewable Energy Systems* (pp. 280–522). Woodhead Publishing India.

Bhattacharjee, S. (2021). Wind power technology. In S. Dutta and C. Mustansar Hussain (Eds.), Sustainable Fuel Technologies Handbook (pp. 123–170). Academic Press. https://doi.org/10.1016/B978-0- 12-822989-7.00006-8

Bhuiyan, A. A., Karim, M. R., and Naser, J. (2016). Modeling of solid and bio-fuel combustion technologies. In M. M. K. Khan and N. M. S. Hassan (Eds.), *Thermofluid Modeling for Energy Efficiency Applications* (pp. 259–309). Academic Press. https://doi.org/10.1016/B978-0-12-802397-6.00016-6

Bi, Z., and He, B. (2020). Biodiesel from microalgae. In E. Jacob-Lopes, M. M. Maroneze, M. I. Queiroz, and L. Q. Zepka (Eds.), *Handbook of Microalgae-Based Processes and Products* (pp. 329–371). Academic Press. https://doi.org/10.1016/B978-0-12-818536-0.00013-0

Bianchi, L., Mannelli, F., Viti, C., Adessi, A., and De Philippis, R. (2010). Hydrogen-producing purple non-sulfur bacteria isolated from the trophic lake Averno (Naples, Italy). *International Journal of Hydrogen Energy*, *35*(22), 12216–12223. https://doi.org/10.1016/j.ijhydene.2010.08.038

Birgen, C., Preisig, H. A., Wentzel, A., Markussen, S., Wittgens, B., Sarkar, U., Ganguly, A., Saha, S., and Baksi, S. (2016). Anaerobic Bio-reactor Modeling. In Z. Kravanja and M. Bogataj (Eds.), *26th European Symposium on Computer Aided Process Engineering* (Vol. 38, pp. 577–582). Elsevier. https://doi.org/10.1016/B978-0-444-63428-3.50101-6

Blackmon, J. B. (2012). Heliostat size optimization for central receiver solar power plants. In K. Lovegrove and W. Stein (Eds.), *Concentrating Solar Power Technology* (pp. 536–576). Woodhead Publishing. https://doi.org/10.1533/9780857096173.3.536

Boateng, A. A. (2020). Energy crops—biomass resources and traits. In A. A. Boateng (Ed.), *Pyrolysis of Biomass for Fuels and Chemicals* (pp. 221–238). Academic Press. https://doi.org/10.1016/B978-0-12-818213-0.00009-6

Boccotti, P. (2015). Design of a wave energy converter. In P. Boccotti (Ed.), *Wave Mechanics and Wave Loads on Marine Structures* (pp. 285–309). Butterworth-Heinemann. https://doi.org/10.1016/B978-0-12-800343-5.00016-0

Boopathi, K., Kushwaha, R., Balaraman, K., Bastin, J., Kanagavel, P., and Reddy Prasad, D. M. (2021). Assessment of wind power potential in the coastal region of Tamil Nadu, India. *Ocean Engineering*, *219*, 108356. https://doi.org/10.1016/j.oceaneng.2020.108356

Borhanipour, M., Karin, P., Tongroon, M., Chollacoop, N., and Hanamura, K. (2014). Comparison study on fuel properties of biodiesel from jatropha, palm and petroleum based diesel fuel. *SAE Technical Papers*, *2*(August). https://doi.org/10.4271/2014-01-2017

Boulouiha, H. M., Allali, A., and Denai, M. (2017). Grid integration of wind energy systems: control design, stability, and power quality issues. In M. G. Rasul, A. kalam Azad, and S. C. Sharma (Eds.), *Clean Energy for Sustainable Development* (pp. 239–335). Academic Press. https://doi.org/10.1016/B978-0-12-805423-9.00009-0

Braun, A. (2018). Hyundai NEXO Hydrogen Fuel Cell Car, Retrieved from: https://commons.wikimedia.org/wiki/File:Hyundai_NEXO_Hydrogen_Fuel_Cell_Car_beim_Autosalon_Genf_2018_y.jpg , CC BY-SA 4.0 International, via Wikimedia Commons, under GNU Free Documentation License.

Breeze, P. (2017). Waste to energy technologies. Academic Press.

Breeze, Paul. (2014). *Power Generation Technologies (2nd Edition)* (pp. 153–349). Newnes.

Breeze, Paul. (2016). In Paul Breeze (Ed.), *Wind Power Generation*. Academic Press, 97p.

Breeze, Paul. (2016). *Solar Power Generation* (pp. 1–34). Academic Press.

Breeze, Paul. (2017). *Fuel Cells*. Academic Press, 93p.

Breeze, Paul. (2018). *Hydropower*. Academic Press, 93p.

Brigham, C., and Bader, J. (2018). From ethanol to biodiesel: A survey of green fuels. In B. Török and T. Dransfield (Eds.), *Green Chemistry* (pp. 861–879). Elsevier. https://doi.org/10.1016/B978-0-12-809270-5.00029-7

British Geological Survey, National Enevironmental Research Council, Geothermal Energy: Fig 1: geothermal energy resources. Available at: https://www.bgs.ac.uk/wp-content/uploads/2020/04/Geothermal_Figure_B.jpg [Access date: 06.12.2022]

Brito Cruz, C. H., Mendes Souza, G., and Barbosa Cortez, L. A. (2013). Biofuels for transport. *Future Energy: Improved, Sustainable and Clean Options for Our Planet*, 215–244. https://doi.org/10.1016/B978-0-08-099424-6.00011-9

Bronicki, L. Y. (2016). Introduction to geothermal power generation. In Ronald DiPippo (Ed.), *Geothermal Power Generation* (pp. 1–3). Woodhead Publishing. https://doi.org/10.1016/B978-0-08-100337-4.00001-2

Brooks, A. E. (2014). Solar energy: photovoltaics. In T. M. Letcher (Ed.), *Future Energy (2nd Edition)* (pp. 383–404). Elsevier. https://doi.org/10.1016/B978-0-08-099424-6.00018-1

Brooks, K. P., Snowden-Swan, L. J., Jones, S. B., Butcher, M. G., Lee, G.-S. J., Anderson, D. M., Frye, J. G., Holladay, J. E., Owen, J., Harmon, L., Burton, F., Palou-Rivera, I., Plaza, J., Handler, R., and Shonnard, D. (2016). Low-Carbon aviation fuel through the alcohol to jet pathway. In C. J. Chuck (Ed.), *Biofuels for Aviation* (pp. 109–150). Academic Press. https://doi.org/10.1016/B978-0-12-804568-8.00006-8

Brownson, J. R. S. (2014). Sun-Earth geometry. In J. R. S. Brownson (Ed.), *Solar Energy Conversion Systems* (pp. 135–178). Academic Press. https://doi.org/10.1016/B978-0-12-397021-3.00006-5

Bryan, K. R., and Power, H. E. (2020). Wave behaviour outside the surf zone. In D. W. T. Jackson and A. D. Short (Eds.), *Sandy Beach Morphodynamics* (pp. 61–86). Elsevier. https://doi.org/10.1016/B978-0-08-102927-5.00004-7

Bullard, E. C. (1965). Historical Introduction to Terrestrial Heat Flow. In Terrestrial Heat Flow, W.H.K. Lee (Ed.). Geophysical Monograph Series, *American Geophysical Union*, 8:1-6. https://doi.org/10.1029/GM008p0001

Bureau of Indian Standards, (1989). IS 12837 : Hydraulic Turbines for Medium and Large Power Houses - Guidelines for Selection. Usage License by CC0 1.0 Universal, Available at: https://archive.org/details/gov.in.is.12837.1989/page/n3/mode/2up

Cabrales, H., Arzola, N., and Araque, O. (2020). The effects of moisture content, fiber length and compaction time on African oil palm empty fruit bunches briquette quality parameters. *Heliyon*, *6*(12), e05607. https://doi.org/10.1016/j.heliyon.2020.e05607

Caillat, S., and Vakkilainen, E. (2013). Large-scale biomass combustion plants: an overview. In L. Rosendahl (Ed.), *Biomass Combustion Science, Technology and Engineering* (pp. 189–224). Woodhead Publishing. https://doi.org/10.1533/9780857097439.3.189

Campanari, S., and Guandalini, G. (2020). Fuel cells: opportunities and challenges. In A. Basile, G. Centi, M. De Falco, and G. Iaquaniello (Eds.), *Catalysis, Green Chemistry and Sustainable Energy* (Vol. 179, pp. 335–358). Elsevier. https://doi.org/10.1016/B978-0-444-64337-7.00018-5

Canakci, M., and Sanli, H. (2008). Biodiesel production from various feedstocks and their effects on the fuel properties. *Journal of Industrial Microbiology and Biotechnology*, *35*(5), 431–441. https://doi.org/10.1007/s10295-008-0337-6

Canuckguy and Danlaycock (2014). Kyoto Protocol parties.svg, Available at: https://commons.wikimedia.org/wiki/File:Kyoto_Protocol_parties.svg

Cardeña, R., Cercado, B., and Buitrón, G. (2019). Microbial electrolysis cell for biohydrogen production. In A. Pandey, S. V. Mohan, J. S. Chang, P. C. Hallenbeck, and C. Larroche (Eds.), *Biohydrogen (2nd Edition)* (pp. 159–185). Elsevier. https://doi.org/10.1016/B978-0-444-64203-5.00007-1

Chakraverty, A. (1989). *Bio-technology and other alternative technologies for utilization of Biomass/Agrl. Wastes*. Wastes. Oxford & IBH Pub. Co. Pvt. Ltd., New Delhi.

Chandolias, K., Richards, T., and Taherzadeh, M. J. (2018). Combined gasification-fermentation process in waste biorefinery. In T. Bhaskar, A. Pandey, S. V. Mohan, D. J. Lee, and S. K. Khanal (Eds.), *Waste Biorefinery* (pp. 157–200). Elsevier. https://doi.org/10.1016/B978-0-444-63992-9.00005-7

Chawla, O. P. (1986). *Advances in biogas Technology*. IARI, New Delhi.

Chelliapan, S., Arumugam, N., Md. Din, M. F., Kamyab, H., and Ebrahimi, S. S. (2020). Anaerobic treatment of municipal solid waste landfill leachate. In L. Singh, A. Yousuf, and D. M. Mahapatra (Eds.), *Bioreactors* (pp. 175–193). Elsevier. https://doi.org/10.1016/B978-0-12-821264-6.00011-5

Chen, C. Y., Liu, C.-H., Lo, Y. C., and Chang, J. S. (2011). Perspectives on cultivation strategies and photobioreactor designs for photo-fermentative hydrogen production. *Bioresource Technology*, *102*(18), 8484–8492. https://doi.org/10.1016/j.biortech.2011.05.082

Chen, C. Y., Yeh, K. L., Aisyah, R., Lee, D. J., and Chang, J. S. (2011). Cultivation, photobioreactor design and harvesting of microalgae for biodiesel production: A critical review. *Bioresource Technology*, *102*(1), 71–81. https://doi.org/10.1016/j.biortech.2010.06.159

Chen, D., and Ding, K. (2016). System stability and control technologies after large-scale wind power integration. In N. Wang, C. Kang, and D. Ren (Eds.), *Large-Scale Wind Power Grid Integration* (pp. 107–184). Academic Press. https://doi.org/10.1016/B978-0-12-849895-8.00004-X

Chen, F., Pao, F., and Yin, H. (2018). Advanced building integrated photovoltaic/thermal technologies. In T. M. Letcher and V. M. Fthenakis (Eds.), *A Comprehensive Guide to Solar Energy Systems* (pp. 299–319). Academic Press. https://doi.org/10.1016/B978-0-12-811479-7.00014-2

Chen, L., Xing, L., and Han, L. (2009). Renewable energy from agro-residues in China: Solid biofuels and biomass briquetting technology. *Renewable and Sustainable Energy Reviews*, *13*(9), 2689–2695. https://doi.org/10.1016/j.rser.2009.06.025

Chen, S., Ding, H., Tang, Z., Hao, S., and Zhao, Y. (2022). Influence of rice straw forming factors on ring die wear and improved wear prediction model during briquetting. *Biosystems Engineering*, *214*, 122–137. https://doi.org/10.1016/j.biosystemseng.2021.12.012

Chen, W.-H., and Lin, B.-J. (2019). Thermochemical conversion of microalgal biomass. In A. Basile and F. Dalena (Eds.), *Second and Third Generation of Feedstocks* (pp. 345–382). Elsevier. https://doi.org/10.1016/B978-0-12-815162-4.00013-6

Cheng, J., Fan, W., and Zheng, L. (2021). Development of a mixotrophic cultivation strategy for simultaneous improvement of biomass and photosynthetic efficiency in freshwater microalga Scenedesmus obliquus by adding appropriate concentration of sodium acetate. *Biochemical Engineering Journal*, *176*, 108177. https://doi.org/10.1016/j.bej.2021.108177

Chidumayo, E. N., and Gumbo, D. J. (2013). The environmental impacts of charcoal production in tropical ecosystems of the world: A synthesis. *Energy for Sustainable Development*, *17*(2), 86–94. https://doi.org/10.1016/j.esd.2012.07.004

Chisti, Y., and Moo-Young, M. (2003). Bioreactors. *Encyclopedia of Physical Science and Technology*, 247–271. https://doi.org/10.1016/b0-12-227410-5/00067-3

Choo, M. Y., Oi, L. E., Ling, T. C., Ng, E. P., Lee, H. V., and Juan, J. C. (2020). Conversion of microalgae biomass to biofuels. In A. Yousuf (Ed.), *Microalgae Cultivation for Biofuels Production* (pp. 149–161). Academic Press. https://doi.org/10.1016/B978-0-12-817536-1.00010-2

Chwieduk, D. (2014). *Solar Energy in Buildings* (pp. 1–62). Academic Press.

Ciolkosz, D. (2020). What's so different about biodiesel fuel? In *Bioenergy* (Second Edi). Elsevier. https://doi.org/10.1016/b978-0-12-815497-7.00037-3

Ciuta, S., Tsiamis, D., and Castaldi, M. J. (2018). Fundamentals of gasification and pyrolysis. In S. Ciuta, D. Tsiamis, and M. J. Castaldi (Eds.), *Gasification of Waste Materials* (pp. 13–36). Academic Press. https://doi.org/10.1016/B978-0-12-812716-2.00002-9

Cleveland, C. J., and Morris, C. (2014). Hydropower. In C. J. Cleveland and C. Morris (Eds.), *Handbook of Energy* (pp. 29–47). Elsevier. https://doi.org/10.1016/B978-0-12-417013-1.00003-0

Collares-Pereira, M., Canavarro, D., and Chaves, J. (2017). Improved design for linear Fresnel reflector systems. In M. J. Blanco and L. R. Santigosa (Eds.), *Advances in Concentrating Solar Thermal Research and Technology* (pp. 45–55). Woodhead Publishing. https://doi.org/10.1016/B978-0-08-100516-3.00003-4

Cooper, P. I. (1969). The absorption of solar radiation in solar stills, *Solar Energy*, 12:333–346.

Córdova, O., and Chamy, R. (2020). Microalgae to biogas: microbiological communities involved. In A. Yousuf (Ed.), *Microalgae Cultivation for Biofuels Production* (pp. 227–249). Academic Press. https://doi.org/10.1016/B978-0-12-817536-1.00015-1

Cousineau, L. (n.d.). *The Kyoto Protocol*. The Climate Change Guide. Retrieved November 7, 2021, from https://www.climate-change-guide.com/kyoto-protocol.html

da Rosa, A. V. (2009). Solar radiation. In A. V. da Rosa (Ed.), *Fundamentals of Renewable Energy Processes (2nd Edition)* (pp. 521–568). Academic Press. https://doi.org/10.1016/B978-0-12-374639-9.00012-9

Daderot. (2014). *Campbell–Stokes sunshine recorder, 2009. Exhibit in the Museum of Science and Industry, Chicago, Illinois, USA*. https://commons.wikimedia.org/wiki/File:Unidentified_object_-_Museum_of_Science_and_Industry_(Chicago)_-_DSC06431.JPG

Dahiya, A. (2020). Bioenergy - biomass to biofuels: an overview. In A. Dahiya (Ed.), *Bioenergy (2nd Edition)* (pp. 1–131). Academic Press.

Damaziak, K., Małachowski, J., and Szafrański, T. (2020). Design of small wind turbine structure with optimized response in frequency domain. In H. R. Karimi

(Ed.), *Vibration Control and Actuation of Large-Scale Systems* (pp. 361–388). Academic Press. https://doi.org/10.1016/B978-0-12-821194-6.00013-5

Damiani, R. R. (2016). Design of offshore wind turbine towers. In Chong Ng and L. Ran (Eds.), *Offshore Wind Farms* (pp. 263–357). Woodhead Publishing. https://doi.org/10.1016/B978-0-08-100779-2.00010-6

Dani, (2004). Dam of the tidal power plant on the estuary of the Rance River, Bretagne, France. Available at: https://commons.wikimedia.org/wiki/File:Rance_tidal_power_plant.JPG

Darmawan, A., and Aziz, M. (2022). Process and products of biomass conversion technology. In A. Darmawan and M. Aziz (Eds.), *Innovative Energy Conversion from Biomass Waste* (pp. 25–60). Elsevier. https://doi.org/10.1016/B978-0-323-85477-1.00004-X

Das, S., Dutta, K., Nessim, G. D., and Kader, M. A. (2020). Introduction to direct methanol fuel cells. In K. Dutta (Ed.), *Direct Methanol Fuel Cell Technology* (pp. 1–12). Elsevier. https://doi.org/10.1016/B978-0-12-819158-3.00001-X

Dave, A., Bhatt, M., and Kumar, M. A. (2020). Approaches for recovering bio-based products from municipal and industrial wastes. In S. Varjani, A. Pandey, E. Gnansounou, S. K. Khanal, and S. Raveendran (Eds.), *Current Developments in Biotechnology and Bioengineering* (pp. 199–216). Elsevier. https://doi.org/10.1016/B978-0-444-64321-6.00010-0

Dawn, S., Tiwari, P. K., Goswami, A. K., Singh, A. K., and Panda, R. (2019). Wind power: Existing status, achievements and government's initiative towards renewable power dominating India. *Energy Strategy Reviews*, *23*, 178–199. https://doi.org/10.1016/j.esr.2019.01.002

Dayton, D. C., and Foust, T. D. (2020). *Analytical Methods for Biomass Characterization and Conversion* (pp. 19–61). Elsevier.

de Jesus, A. C. (2016). Environmental benefits and challenges associated with geothermal power generation. In Ronald DiPippo (Ed.), *Geothermal Power Generation* (pp. 477–498). Woodhead Publishing. https://doi.org/10.1016/B978-0-08-100337-4.00017-6

de la Rubia, M. A., Villamil, J. A., and Mohedano, A. F. (2020). Anaerobic digestion for methane and hydrogen production. In J. A. Olivares, D. Puyol, J. A. Melero, and J. Dufour (Eds.), *Wastewater Treatment Residues as Resources for Biorefinery Products and Biofuels* (pp. 67–83). Elsevier. https://doi.org/10.1016/B978-0-12-816204-0.00004-7

de Melo Ribeiro, G., de Medeiros Costa, H. K., Simões, A. F., and Arlota, C. (2021). The institutional approach of climate change at the multinational level: the new paradigm of the Brazilian legislative experience. In H. K. de Medeiros Costa and C. Arlota (Eds.), *Carbon Capture and Storage in International Energy Policy and Law* (pp. 61–74). Elsevier. https://doi.org/10.1016/B978-0-323-85250-0.00014-1

Debrezion, Y. M., Bovas, J. J. L., Tedla, T. B., James, P. S., James, A., Michael, Y. B., and Jayaraman, S. (2020). *Automation and Robotics for Crisis Management of Agriculture in Developing Countries*. Glasstree Academic Publishing. https://doi.org/10.20850/9781534298972

Deep, S., Sarkar, A., Ghawat, M., and Rajak, M. K. (2020). Estimation of the wind energy potential for coastal locations in India using the Weibull model. *Renewable Energy*, *161*, 319–339. https://doi.org/10.1016/j.renene.2020.07.054

Deng, M., Li, P., Shan, M., and Yang, X. (2020). Optimizing supply airflow and its distribution between primary and secondary air in a forced-draft biomass pellet stove. *Environmental Research*, *184*, 109301. https://doi.org/10.1016/j.envres.2020.109301

Deshmukh, R., Wu, G. C., Callaway, D. S., and Phadke, A. (2019). Geospatial and techno-economic analysis of wind and solar resources in India. *Renewable Energy*, *134*, 947–960. https://doi.org/10.1016/j.renene.2018.11.073

Dhiman, H. S., Deb, D., and Balas, V. E. (2020). In H. S. Dhiman, D. Deb, and V. E. Balas (Eds.), *Supervised Machine Learning in Wind Forecasting and Ramp Event Prediction* (pp. 1–21). Academic Press.

Dhir, R. K., de Brito, J., Lynn, C. J., and Silva, R. V. (2018). Municipal solid waste composition, incineration, processing and management of bottom ashes. In R. K. Dhir, J. de Brito, C. J. Lynn, and R. V Silva (Eds.), *Sustainable Construction Materials* (pp. 31–90). Woodhead Publishing. https://doi.org/10.1016/B978-0-08-100997-0.00003-8

Dincer, I., and Abu-Rayash, A. (2020). *Energy Sustainability* (pp. 1–402). Academic Press.

Dincer, I., and Ozturk, M. (2021). *Geothermal Energy Systems* (pp. 1–503). Elsevier.

Dincer, I., and Siddiqui, O. (2020). *Ammonia Fuel Cells* (pp. 1–76). Elsevier.

Ding, C., Yang, K.-L., and He, J. (2016). Biological and fermentative production of hydrogen. In R. Luque, C. S. K. Lin, K. Wilson, and J. Clark (Eds.), *Handbook of Biofuels Production (2nd Edition)* (pp. 303–333). Woodhead Publishing. https://doi.org/10.1016/B978-0-08-100455-5.00011-4

DiPippo, R. (2016). *Geothermal Power Generation* (pp. 1–492). Woodhead Publishing.

Dong, H., Wang, W., Song, Z., Dong, H., Wang, J., Sun, S., Zhang, Z., Ke, M., Zhang, Z., Wu, W. M., Zhang, G., and Ma, J. (2017). A high-efficiency denitrification bioreactor for the treatment of acrylonitrile wastewater using waterborne polyurethane immobilized activated sludge. *Bioresource Technology*, *239*, 472–481. https://doi.org/10.1016/j.biortech.2017.05.015

Dower, B. (2018). Solar panels in the landscape. In T. M. Letcher and V. M. Fthenakis (Eds.), *A Comprehensive Guide to Solar Energy Systems* (pp. 373–389). Academic Press. https://doi.org/10.1016/B978-0-12-811479-7.00019-1

Duan, Y., and Shi, F. (2014). Bioreactor design for algal growth as a sustainable energy source. In *Reactor and Process Design in Sustainable Energy Technology* (Issue December 2014). https://doi.org/10.1016/B978-0-444-59566-9.00002-8

Dubois, O., Pirelli, T., and Peressotti, A. (2019). Biomass anaerobic digestion and gasification in non-OECD countries - an overview. In M. Materazzi and P. U. Foscolo (Eds.), *Substitute Natural Gas from Waste* (pp. 343–387). Academic Press. https://doi.org/10.1016/B978-0-12-815554-7.00013-1

Earth Science Communications Team, Huge machine harnesses the tides, *Jet Propulsion Laboratory, NASA*. Available at: https://climatekids.nasa.gov/tidal-energy/

El Haj Assad, M., Alhuyi Nazari, M., and Rosen, M. A. (2021). Applications of renewable energy sources. In M. E. H. Assad and M. A. Rosen (Eds.), *Design and Performance Optimization of Renewable Energy Systems* (pp. 1–15). Academic Press. https://doi.org/10.1016/B978-0-12-821602-6.00001-8

Encyclopædia Britannica, Process of slab pull: Earth's layers, Avaiable at: https://www.britannica.com/science/plate-tectonics/Seafloor-spreading#/media/1/463912/229345 [Acces date: 06.12.2022]

Encyclopædia Britannica, Ring of Fire, Available at: https://www.britannica.com/place/Ring-of-Fire#/media/1/118426/156997 [Access date: 06.12.2022]

Ertugrul, N. (2018). Small-Scale PV systems used in domestic applications. In T. M. Letcher and V. M. Fthenakis (Eds.), *A Comprehensive Guide to Solar Energy Systems* (pp. 333–350). Academic Press. https://doi.org/10.1016/B978-0-12-811479-7.00016-6

Esteves, A. F., Almeida, C. J., Gonçalves, A. L., and Pires, J. C. (2020). Microalgae harvesting techniques. In E. Jacob-Lopes, M. M. Maroneze, M. I. Queiroz, and L. Q. Zepka (Eds.), *Handbook of Microalgae-Based Processes and Products* (pp. 225–281). Academic Press. https://doi.org/10.1016/B978-0-12-818536-0.00010-5

Evans, P. G. H. (2020). Conservation actions. In P. G. H. Evans (Ed.), *European Whales, Dolphins, and Porpoises* (pp. 237–275). Academic Press. https://doi.org/10.1016/B978-0-12-819053-1.00007-7

Exoplanet Exploration – NASA. (n.d.). *Milky Way*. Retrieved January 4, 2022, from https://exoplanets.nasa.gov/milky-way-overlay/

Fahmy, T. Y. A., Fahmy, Y., Mobarak, F., El-Sakhawy, M., and Abou-Zeid, R. E. (2020). Biomass pyrolysis: past, present, and future. *Environment, Development and Sustainability*, *22*(1), 17–32. https://doi.org/10.1007/s10668-018-0200-5

Fanun, M. (2017). Alternative fuel. In M. Fanun (Ed.), *Sugar Esters Microemulsions* (pp. 423–434). Elsevier. https://doi.org/10.1016/B978-0-12-811653-1.00012-3

FAO. (1992). *Biogas processes for sustainable development.*

Ferrazza, F. (2018). Crystalline silicon: Manufacture and properties. In S. A. Kalogirou (Ed.), *McEvoy's Handbook of Photovoltaics (3rd Edition)* (pp. 75–93). Academic Press. https://doi.org/10.1016/B978-0-12-809921-6.00004-5

Fletcher, T. H. (2017). Gasification fundamentals. In T. Wang and G. Stiegel (Eds.), *Integrated Gasification Combined Cycle (IGCC) Technologies* (pp. 223–256). Woodhead Publishing. https://doi.org/10.1016/B978-0-08-100167-7.00006-8

Fontanille, P. (2020). Hydrogen generation in bioreactors. In A. Iulianelli and A. Basile (Eds.), *Current Trends and Future Developments on (Bio-) Membranes* (pp. 183–208). Elsevier. https://doi.org/10.1016/B978-0-12-817384-8.00008-X

Fregolente, P. B. L., Fregolente, L. V., and Wolf MacIel, M. R. (2012). Water content in biodiesel, diesel, and biodiesel-diesel blends. *Journal of Chemical and Engineering Data*, *57*(6), 1817–1821. https://doi.org/10.1021/je300279c

Fundy, (2003). The SeaFlow tidal stream generator prototype with rotor raised, Available at https://commons.wikimedia.org/wiki/File:Seaflow_raised_16_jun_03.jpg, GNU Free Documentation License CC-BY-SA/3.0 via Wikimedia Commons.

Furlan, F. F., Giordano, R. C., Costa, C. B. B., Secchi, A. R., and Woodley, J. M. (2015). Process alternatives for second generation ethanol production from sugarcane bagasse. In K. V Gernaey, J. K. Huusom, and R. Gani (Eds.), *12th International Symposium on Process Systems Engineering and 25th European Symposium on Computer Aided Process Engineering* (Vol. 37, pp. 1349–1354). Elsevier. https://doi.org/10.1016/B978-0-444-63577-8.50070-X

Gaba, V. K., and Bhowmick, S. (2019). A CFD-based study of cross-flow turbine for tidal energy extraction. In Kaushik Kumar, D. Zindani, and P. Davim (Eds.), *Sustainable Engineering Products and Manufacturing Technologies* (pp. 177–186). Academic Press. https://doi.org/10.1016/B978-0-12-816564-5.00007-4

Gao, Y., Gao, X., and Zhang, X. (2017). The 2°C global temperature target and the evolution of the long-term goal of addressing climate change—from the united nations framework convention on climate change to the paris agreement. *Engineering*, *3*(2), 272–278. https://doi.org/10.1016/J.ENG.2017.01.022

Garg, H. P., and Kandpal, T. C. (1992). Renewable energy education in developing countries : Indian scenario. In *Renewable Energy, Technology and the Environment* (Vol. 110016). Pergamon Press Ltd. https://doi.org/10.1016/b978-0-08-041268-9.50015-8

Gautam, A., Warudkar, V., and Bhagoria, J. L. (2022). A comparative analysis of wind characteristics for distinct terrains of India. *Sustainable Energy Technologies and Assessments*, *52*, 102089. https://doi.org/10.1016/j.seta.2022.102089

Geng, C., and Cui, Z. (2020). Analysis of spatial heterogeneity and driving factors of capital allocation efficiency in energy conservation and environmental protection industry under environmental regulation. *Energy Policy*, *137*, 111081. https://doi.org/10.1016/j.enpol.2019.111081

Ghasemzadeh, K., Jalilnejad, E., and Basile, A. (2017). Production of bioalcohol and biomethane. In F. Dalena, A. Basile, and C. Rossi (Eds.), *Bioenergy Systems for the Future* (pp. 61–86). Woodhead Publishing. https://doi.org/10.1016/B978-0-08-101031-0.00003-X

Ghassemi, M., Kamvar, M., and Steinberger-Wilckens, R. (2020). Introduction to fuel cells. In M. Ghassemi, M. Kamvar, and R. Steinberger-Wilckens (Eds.),

Fundamentals of Heat and Fluid Flow in High Temperature Fuel Cells (pp. 1–15). Academic Press. https://doi.org/10.1016/B978-0-12-815753-4.00001-4

Gillespie, G. D., Everard, C. D., Fagan, C. C., and McDonnell, K. P. (2013). Prediction of quality parameters of biomass pellets from proximate and ultimate analysis. *Fuel*, *111*, 771–777. https://doi.org/10.1016/j.fuel.2013.05.002

Global Wind Atlas. (n.d.). Retrieved January 2, 2022, from https://globalwindatlas.info/

Gopalakrishnan, B., Khanna, N., and Das, D. (2019). Dark-Fermentative biohydrogen production. In A. Pandey, S. V. Mohan, J.-S. Chang, P. C. Hallenbeck, and C. Larroche (Eds.), *Biohydrogen (2nd Edition)* (pp. 79–122). Elsevier. https://doi.org/10.1016/B978-0-444-64203-5.00004-6

Gorjian, S., and Ebadi, H. (2020). *Photovoltaic Solar Energy Conversion* (pp. 1–190).

Gothika, Ryazanov, M., and Moonraker, O. (2012). *Water turbine*. https://commons.wikimedia.org/w/index.php?curid=18581340

Gould, M. C. (2015). Bioenergy and anaerobic digestion. In A. Dahiya (Ed.), *Bioenergy* (pp. 297–317). Academic Press. https://doi.org/10.1016/B978-0-12-407909-0.00018-3

Grabert, J. (2021). Clean Development Mechanism Methodology (13th Edition). United Nations Framework Convention on Climate Change. 285p. https://cdm.unfccc.int/methodologies/documentation/meth_booklet.pdf#ACM0002

Greaves, P. (2016). Design of offshore wind turbine blades. In Chong Ng and L. Ran (Eds.), *Offshore Wind Farms* (pp. 105–135). Woodhead Publishing. https://doi.org/10.1016/B978-0-08-100779-2.00006-4

Green, M. A. (2018). High-Efficiency silicon solar cell concepts. In S. A. Kalogirou (Ed.), *McEvoy's Handbook of Photovoltaics (3rd Edition)* (pp. 95–128). Academic Press. https://doi.org/10.1016/B978-0-12-809921-6.00005-7

Grillo, L. M. (2013). Municipal solid waste (MSW) combustion plants. In N. B. Klinghoffer and M. J. Castaldi (Eds.), *Waste to Energy Conversion Technology* (pp. 72–97). Woodhead Publishing. https://doi.org/10.1533/9780857096364.2.72

Gu, J. D. (2003). Microbiological deterioration and degradation of synthetic polymeric materials: recent research advances. *International Biodeterioration & Biodegradation*, *52*(2), 69–91. https://doi.org/10.1016/S0964-8305(02)00177-4

Guaitolini, S. V. M., and Fardin, J. F. (2018). Fuel cells: History (short remind), principles of operation, main features, and applications. In I. Yahyaoui (Ed.), *Advances in Renewable Energies and Power Technologies* (pp. 123–150). Elsevier. https://doi.org/10.1016/B978-0-12-813185-5.00013-9

Gualtieri, G. (2019). A novel method for wind farm layout optimization based on wind turbine selection. *Energy Conversion and Management*, *193*, 106–123. https://doi.org/10.1016/j.enconman.2019.04.059

Guan, R., Yuan, H., Yuan, S., Yan, B., Zuo, X., Chen, X., and Li, X. (2022). Current development and perspectives of anaerobic bioconversion of crop stalks to Biogas: A review. *Bioresource Technology*, *349*, 126615. https://doi.org/10.1016/j.biortech.2021.126615

Gude, V. G. (2018). Geothermal source for water desalination—challenges and opportunities. In V. G. Gude (Ed.), *Renewable Energy Powered Desalination Handbook* (pp. 141–176). Butterworth-Heinemann. https://doi.org/10.1016/B978-0-12-815244-7.00004-0

Guisado-Pintado, E. (2020). Shallow water wave modelling in the nearshore (SWAN). In D. W. T. Jackson and A. D. Short (Eds.), *Sandy Beach Morphodynamics* (pp. 391–419). Elsevier. https://doi.org/10.1016/B978-0-08-102927-5.00017-5

Gunasekaran, M., Merrylin, J., Usman, T. M. M., Kumar, G., Kim, S.-H., and Banu, J. R. (2019). Biohydrogen production from industrial wastewater. In A. Pandey, C. Larroche, C.-G. Dussap, E. Gnansounou, S. K. Khanal, and S. Ricke (Eds.), *Biofuels: Alternative Feedstocks and Conversion Processes for the Production of Liquid and Gaseous Biofuels (2nd Edition)* (pp. 733–760). Academic Press. https://doi.org/10.1016/B978-0-12-816856-1.00030-0

Guo, P., Zhu, D., Lam, J., and Li, V. O. K. (2017). The future of wind energy development in china. In T. M. Letcher (Ed.), *Wind Energy Engineering* (pp. 75–94). Academic Press. https://doi.org/10.1016/B978-0-12-809451-8.00005-9

Guo, X., and Lü, X. (2021). The need for biofuels in the context of energy consumption. In X. Lü (Ed.), *Advances in 2nd Generation of Bioethanol Production* (pp. 9–30). Woodhead Publishing. https://doi.org/10.1016/B978-0-12-818862-0.00004-2

Guo, Z., Wu, J., Zhang, Y., Cao, K., Feng, Y., Liu, J., and Li, Ji. (2020). Briquetting optimization method for the lignite powder using response surface analysis. *Fuel*, *267*, 117260. https://doi.org/10.1016/j.fuel.2020.117260

Guo, Z., Wu, J., Zhang, Y., Wang, F., Guo, Y., Chen, K., and Liu, H. (2020). Characteristics of biomass charcoal briquettes and pollutant emission reduction for sulfur and nitrogen during combustion. *Fuel*, *272*, 117632. https://doi.org/10.1016/j.fuel.2020.117632

Gupta, A. (2016). Climate change and kyoto protocol: An overview. In V. Ramiah and G. N. Gregoriou (Eds.), *Handbook of Environmental and Sustainable Finance* (pp. 3–23). Academic Press. https://doi.org/10.1016/B978-0-12-803615-0.00001-7

Gupta, H., and Roy, S. (2007). *Geothermal Energy* (pp. 1–244). Elsevier.

Hagedorn, R., and Neef, E. (1985). The efficiency of photosynthetic energy conversion in continuous and intermittent light. *Journal of Theoretical Biology*, *114*(1), 93–101. https://doi.org/10.1016/S0022-5193(85)80257-5

Hall, S. M. (2018). Energy conservation. In S. M. Hall (Ed.), *Rules of Thumb for Chemical Engineers (6th Edition)* (pp. 345–354). Elsevier. https://doi.org/10.1016/B978-0-12-811037-9.00019-9

Hand, B., and Cashman, A. (2020). A review on the historical development of the lift-type vertical axis wind turbine: From onshore to offshore floating application. *Sustainable Energy Technologies and Assessments*, *38*, 100646. https://doi.org/10.1016/j.seta.2020.100646

Hanelt, D. (2018). Photosynthesis assessed by chlorophyll fluorescence. In D. P. Häder and G. S. Erzinger (Eds.), *Bioassays* (pp. 169–198). Elsevier. https://doi.org/10.1016/B978-0-12-811861-0.00009-7

Hannah, L. (2015). *Climate Change Biology (2nd Edition)* (pp. 365–377, 425–438). Academic Press.

Hansen, A. D. (2017). In T. M. Letcher (Ed.), *Wind Energy Engineering* (pp. 145–184). Academic Press.

Harvey, W., and Wallace, K. (2016). Flash steam geothermal energy conversion systems: Single-, double-, and triple-flash and combined-cycle plants. In Ronald DiPippo (Ed.), *Geothermal Power Generation* (pp. 249–290). Woodhead Publishing. https://doi.org/10.1016/B978-0-08-100337-4.00010-3

Hatfield, J. L., and Dold, C. (2019). Photosynthesis in the solar corridor system. In C. L. Deichman and R. J. Kremer (Eds.), *The Solar Corridor Crop System* (pp. 1–33). Academic Press. https://doi.org/10.1016/B978-0-12-814792-4.00001-2

Hatton, R. A. (2018). Organic photovoltaics. In T. M. Letcher and V. M. Fthenakis (Eds.), *A Comprehensive Guide to Solar Energy Systems* (pp. 255–277). Academic Press. https://doi.org/10.1016/B978-0-12-811479-7.00012-9

Heidenreich, S., Müller, M., and Foscolo, P. U. (2016). *Advanced Biomass Gasification* (pp. 4–10, 98–114). Academic Press.

Heijnen, J. J., Mulder, A., Enger, W., and Hocks, F. (1989). Review on the application of anaerobic fluidized bed reactors in waste-water treatment. *The Chemical Engineering Journal*, *41*, 37-50.

Hong, C. J. (1990). Improvement of anaerobic digestion in porous media packed bioreactor. In: Biomass for energy and industry. *Proc. Vth E.C. Conf. Vol.2*, 2.454-2.457.

Hossain, M. Z., and Charpentier, P. A. (2015). Hydrogen production by gasification of biomass and opportunity fuels. In V. Subramani, A. Basile, and T. N. Veziroğlu (Eds.), *Compendium of Hydrogen Energy* (pp. 137–175). Woodhead Publishing. https://doi.org/10.1016/B978-1-78242-361-4.00006-6

UN Climate Press Release. (2020). Ratification of Multilateral Climate Agreement Gives Boost to Delivering Agreed Climate Pledges and to Tackling Climate Change (02.10.2020, Bonn, New York). https://unfccc.int/news/ratification-of-multilateral-climate-agreement-gives-boost-to-delivering-agreed-climate-pledges-and.

Hu, M., Triulzi, G., and Sharifzadeh, M. (2020). Technological change in fuel cell technologies. In M. Sharifzadeh (Ed.), *Design and Operation of Solid Oxide*

Fuel Cells (pp. 3–41). Academic Press. https://doi.org/10.1016/B978-0-12-815253-9.00001-X

Hu, X., and Gholizadeh, M. (2019). Biomass pyrolysis: A review of the process development and challenges from initial researches up to the commercialisation stage. *Journal of Energy Chemistry*, *39*, 109–143. https://doi.org/10.1016/j.jechem.2019.01.024

Hu, Y., Bassi, A., and Xu, C. (Charles). (2020). Energy from biomass. In T. M. Letcher (Ed.), *Future Energy (3rd Edition)* (pp. 447–471). Elsevier. https://doi.org/10.1016/B978-0-08-102886-5.00021-9

Huang, B. J., Petrenko, V. A., Chang, J. M., Hu, S. H., and Lee, J. B. (2000). A Study of ejector refrigeration system design for solar cooling application. In A. A. M. Sayigh (Ed.), *World Renewable Energy Congress VI* (pp. 1052–1055). Pergamon. https://doi.org/10.1016/B978-008043865-8/50214-2

Hubbard, W. G. (2020). Wood bioenergy. In A. Dahiya (Ed.), *Bioenergy (2nd Edition)* (pp. 69–87). Academic Press. https://doi.org/10.1016/B978-0-12-815497-7.00004-X

Hukseflux Thermal Sensors. (2012a). *Pyranometer*. https://commons.wikimedia.org/wiki/File:SR20_pyranometer_1.jpg

Hukseflux Thermal Sensors. (2012b). *Pyranometer line drawing*. https://commons.wikimedia.org/wiki/File:SR20_pyranometer_linedrawing.pdf

Huq, S., Mahid, Y., and Suliman, N. (2018). Evolution of climate change adaptation policy and negotiation. In Z. Zommers and K. Alverson (Eds.), *Resilience* (pp. 63–75). Elsevier. https://doi.org/10.1016/B978-0-12-811891-7.00005-0

Ifa, L., Yani, S., Nurjannah, N., Darnengsih, D., Rusnaenah, A., Mel, M., Mahfud, M., and Kusuma, H. S. (2020). Techno-economic analysis of bio-briquette from cashew nut shell waste. *Heliyon*, *6*(9), e05009. https://doi.org/10.1016/j.heliyon.2020.e05009

Ileleji, K. E., Martin, C., and Jones, D. (2020). Basics of energy production through anaerobic digestion of livestock manure. In A. Dahiya (Ed.), *Bioenergy (2nd Edition)* (pp. 325–334). Academic Press. https://doi.org/10.1016/B978-0-12-815497-7.00016-6

Iloeje, O. C. (2000). Solar energy refrigeration. In A. A. M. Sayigh (Ed.), *World Renewable Energy Congress VI* (pp. 975–981). Pergamon. https://doi.org/10.1016/B978-008043865-8/50196-3

Immethun, C. M., Henson, W. R., Wang, X., Nielsen, D. R., and Moon, T. S. (2016). Engineering central metabolism for production of higher alcohol-based biofuels. In C. A. Eckert and C. T. Trinh (Eds.), *Biotechnology for Biofuel Production and Optimization* (pp. 1–34). Elsevier. https://doi.org/10.1016/B978-0-444-63475-7.00001-7

Iowa Renewable Fuels Association. (n.d.). *Biodiesel and the Environment*. Retrieved May 11, 2021, from https://iowarfa.org/biodiesel-and-the-environment/

Ismail, S. A. E. A., and Ali, R. F. M. (2015). Physico-chemical properties of biodiesel manufactured from waste frying oil using domestic adsorbents. *Science and Technology of Advanced Materials*, *16*(3). https://doi.org/10.1088/1468-6996/16/3/034602

Jahirul, M. I., Rasul, M. G., Chowdhury, A. A., and Ashwath, N. (2012). Biofuels production through biomass pyrolysis - a technological review. *Energies*, *5*(12), 4952–5001. https://doi.org/10.3390/en5124952

Jahobr. (2016a). *BulbTurbine Sketch.* https://commons.wikimedia.org/wiki/File:BulbTurbineSketch.svg

Jahobr. (2016b). *PeltonSketch.* https://commons.wikimedia.org/wiki/File:PeltonSketch.svg

Jahobr. (2016c). *Tubular Gear Turbine Sketch.* https://commons.wikimedia.org/wiki/File:TubularGearTurbineSketch.svg

Jahobr. (2020a). *Deriaz Turbine Sketch.* https://commons.wikimedia.org/wiki/File:Deriaz_Turbine_Sketch.svg

Jahobr. (2020b). *Turgo Turbine Sketch.* https://commons.wikimedia.org/wiki/File:Turgo_Turbine_Sketch.svg

Jaibiba, P., Naga Vignesh, S., and Hariharan, S. (2020). Working principle of typical bioreactors. In L. Singh, A. Yousuf, and D. M. Mahapatra (Eds.), *Bioreactors* (pp. 145–173). Elsevier. https://doi.org/10.1016/B978-0-12-821264-6.00010-3

Jain, A., Das, P., Yamujala, S., Bhakar, R., and Mathur, J. (2020). Resource potential and variability assessment of solar and wind energy in India. *Energy*, *211*, 118993. https://doi.org/10.1016/j.energy.2020.118993

Jalal, F., Ilavarasi, P. S., and Miranda, L. R. (2011). Fatty methyl esters from vegetable oils for use as a diesel fuel. *1st Conference on Clean Energy and Technology, CET 2011, IEEE*, 125–128. https://doi.org/10.1109/CET.2011.6041472

James, P. A. B., and Bahaj, A. S. (2017). Small-Scale wind turbines. In T. M. Letcher (Ed.), *Wind Energy Engineering* (pp. 389–418). Academic Press. https://doi.org/10.1016/B978-0-12-809451-8.00019-9

James, P. S., Joe Joe, L. B., and Neerackal, G. M. (2005). Production of rubber seed oil biodiesel and its testing in a CI engine. *Bio Energy News*, 15–19.

James, P. S., Joe Joe, L. B., Neerackal, G. M., and Sareena, K. V. (2005). Engine tests with coconut oil biodiesel. *Indian Coconut Journal*, 6–8.

James, P. S., and Kamaraj, S. (2002). Immobilised cell anaerobic bioreactors for energy production from agro industrial waste waters- An introduction. *Bioenergy News*, *6*(3), 10–15.

James, P. S., and Kamaraj, S. (2003). Hybrid anaerobic reactors for energy production from cassava starch factory effluent. *Bioenergy News*, *7*(3), 10–12.

James, P. S., Regina, M. F., and Joe Joe, L. B. (2013). High rate anaerobic bioreactors for treatment and energy conversion of organic effluents from agro processing

Industries. *International Journal of Processing and Post-Harvest Technology*, *4*(1), 13–17.

James, P. S., Regina, M. F., Joe Joe, L. B., and Kamaraj, S. (2013). Development of a hybrid anaerobic bioreactor for treatment and energy conversion of organic effluents. *Journal of Root Crops*, *39*(2), 183–188.

James, P. S., and Regina, F. M. (2011). Improving energy efficiency in agriculture. *Energy Manager*, *4*(4), 20–24.

Janjai, S., and Bala, B.K. (2011). Solar Drying Technology. *Food Engineering Reviews*, 4:16-54.

Jankowska, E., Zieliński, M., Dębowski, M., and Oleśkowicz-Popiel, P. (2019). Anaerobic digestion of microalgae for biomethane production. In A. Basile and F. Dalena (Eds.), *Second and Third Generation of Feedstocks* (pp. 405–436). Elsevier. https://doi.org/10.1016/B978-0-12-815162-4.00015-X

Janssen, M. (2016). Microalgal photosynthesis and growth in mass culture. In J. Legrand (Ed.), *Photobioreaction Engineering* (Vol. 48, pp. 185–256). Academic Press. https://doi.org/10.1016/bs.ache.2015.11.001

Jenkins, N., and Thornycroft, J. (2018). Grid connection of photovoltaic systems: Technical and regulatory issues. In S. A. Kalogirou (Ed.), *McEvoy's Handbook of Photovoltaics (3rd Edition)* (pp. 847–876). Academic Press. https://doi.org/10.1016/B978-0-12-809921-6.00022-7

Jenkins, R. G. (2015). Thermal gasification of biomass – A primer. In A. Dahiya (Ed.), *Bioenergy* (pp. 261–286). Academic Press. https://doi.org/10.1016/B978-0-12-407909-0.00016-X

Jenkins, R. G. (2020). Thermal gasification of biomass - a primer. In A. Dahiya (Ed.), *Bioenergy (2nd Edition)* (pp. 293–324). Academic Press. https://doi.org/10.1016/B978-0-12-815497-7.00015-4

Joe Joe, L. B., and James, P. S. (2016). *High-rate bioreactor for energy production from rice-mill effluent*. LAP LAMBERT Academic Publishing GmbH & Co. KG, Dudweiler Landstr. 99 – D-66123 Saarbrücken.

Johansson, S. A., Stephenson, P. G., Edwards, R. J., Yoshida, K., Moore, C. M., Terauchi, R., Zubkov, M. V, Terry, M. J., and Bibby, T. S. (2020). Isolation and molecular characterisation of Dunaliella tertiolecta with truncated light-harvesting antenna for enhanced photosynthetic efficiency. *Algal Research*, *48*, 101917. https://doi.org/10.1016/j.algal.2020.101917

Jones, S. B., Zhu, Y., and Valkenburg, C. (2009). *Municipal Solid Wastes (MSW) to liquid fuels synthesis, Volume 2: A techno-economic evaluation on the production of mixed alcohols.*

Joseph, A. (2014). The history of measuring ocean currents. In A. Joseph (Ed.), *Measuring Ocean Currents* (pp. 51–92). Elsevier. https://doi.org/10.1016/B978-0-12-415990-7.00002-8

Joseph, A. (2017). Complex wave motions and thermal structure of the oceans. In

A. Joseph (Ed.), *Investigating Seafloors and Oceans* (pp. 139–208). Elsevier. https://doi.org/10.1016/B978-0-12-809357-3.00003-5

Joshi, B., Joshi, J., Bhattarai, T., and Sreerama, L. (2019). Currently used microbes and advantages of using genetically modified microbes for ethanol production. In R. C. Ray and S. Ramachandran (Eds.), *Bioethanol Production from Food Crops* (pp. 293–316). Academic Press. https://doi.org/10.1016/B978-0-12-813766-6.00015-1

Jung, K. W., Kim, D. H., Kim, S. H., and Shin, H. S. (2011). Bioreactor design for continuous dark fermentative hydrogen production. *Bioresource Technology*, *102*(18), 8612–8620. https://doi.org/10.1016/j.biortech.2011.03.056

Junginger, M., Hittinger, E., Williams, E., and Wiser, R. (2020). Onshore wind energy. In M. Junginger and A. Louwen (Eds.), *Technological Learning in the Transition to a Low-Carbon Energy System* (pp. 87–102). Academic Press. https://doi.org/10.1016/B978-0-12-818762-3.00006-6

Kalaiselvam, S., and Parameshwaran, R. (2014). Energy and Energy Management. In S. Kalaiselvam and R. Parameshwaran (Eds.), *Thermal Energy Storage Technologies for Sustainability* (pp. 1–19). Academic Press. https://doi.org/10.1016/B978-0-12-417291-3.00001-3

Kalmikov, A. (2017). Wind Power Fundamentals. In T. M. Letcher (Ed.), *Wind Energy Engineering* (pp. 17–24). Academic Press. https://doi.org/10.1016/B978-0-12-809451-8.00002-3

Kalogirou, S. A. (Ed.). (2009). Appendix 3 - Sun diagrams. In *Solar Energy Engineering* (pp. 717–721). Academic Press. https://doi.org/10.1016/B978-0-12-374501-9.00017-0

Kalogirou, S. A. (2009). *Solar Energy Engineering* (pp. 1-701). Academic Press.

Kamran, M. (2021). Chapter 4 - Solar energy. In M. Kamran and M. R. Fazal (Eds.), *Renewable Energy Conversion Systems* (pp. 109–152). Academic Press. https://doi.org/10.1016/B978-0-12-823538-6.00008-7

Kandiyoti, R., Herod, A., Bartle, K., and Morgan, T. (2017). High-pressure reactor design: Pyrolysis, hydropyrolysis and gasification. In R. Kandiyoti, A. Herod, K. Bartle, and T. Morgan (Eds.), *Solid Fuels and Heavy Hydrocarbon Liquids (2nd Edition)* (Second Edi, pp. 125–203). Elsevier. https://doi.org/10.1016/B978-0-08-100784-6.00004-7

Kapdan, I. K., and Kargi, F. (2006). Bio-hydrogen production from waste materials. *Enzyme and Microbial Technology*, *38*(5), 569–582. https://doi.org/10.1016/j.enzmictec.2005.09.015

Kapfunde, N., Masuku, C. M., and Hildebrandt, D. (2018). Optimization of the thermal efficiency of a fixed-bed gasifier using computational fluid dynamics. In M. R. Eden, M. G. Ierapetritou, and G. P. Towler (Eds.), *13th International Symposium on Process Systems Engineering (PSE 2018)* (Vol. 44, pp. 1747–1752). Elsevier. https://doi.org/10.1016/B978-0-444-64241-7.50286-X

Karthikeyan, N., Kalidasa Murugavel, K., Arun Kumar, S., and Rajakumar, S. (2015). Review of aerodynamic developments on small horizontal axis wind turbine blade. *Renewable and Sustainable Energy Reviews*, *42*, 801–822. https://doi.org/10.1016/j.rser.2014.10.086

Kaur-Sidhu, M., Ravindra, K., Mor, S., and John, S. (2020). Emission factors and global warming potential of various solid biomass fuel-cook stove combinations. *Atmospheric Pollution Research*, *11*(2), 252–260. https://doi.org/10.1016/j.apr.2019.10.009

Kaur, R., Gera, P., Jha, M. K., and Bhaskar, T. (2019). Thermochemical route for biohydrogen production. In A. Pandey, S. V. Mohan, J.S. Chang, P. C. Hallenbeck, and C. Larroche (Eds.), *Biohydrogen (2nd Edition)* (pp. 187–218). Elsevier. https://doi.org/10.1016/B978-0-444-64203-5.00008-3

Keller, M. (2015). Photosynthesis and Respiration. In M. Keller (Ed.), *The Science of Grapevines (2nd Edition)* (p. 125–143). Academic Press. https://doi.org/10.1016/B978-0-12-419987-3.00004-2

Keller, M. (2020). Photosynthesis and respiration. In M. Keller (Ed.), *The Science of Grapevines (3rd Edition)* (pp. 129–148). Academic Press. https://doi.org/10.1016/B978-0-12-816365-8.00004-X

Ken, B. (2011), Sheffield: Abbeydale Industrial Hamlet, Available at https://www.geograph.org.uk/photo/2357009 via Geograph, Under lincense CC-BY-SA2.0

Keong, C. Y. (2021). The United Nations' journey to global environmental sustainability since Stockholm: An assessment. In C. Y. Keong (Ed.), *Global Environmental Sustainability* (pp. 7–61). Elsevier. https://doi.org/10.1016/B978-0-12-822419-9.00002-3

Kerčmar, J., and Pintar, A. (2017). Support material dictates the attached biomass characteristics during the immobilization process in anaerobic continuous-flow packed-bed bioreactor. *Anaerobe*, *48*, 194–202. https://doi.org/10.1016/j.anaerobe.2017.06.007

Keskin, T., Nalakath Abubackar, H., Arslan, K., and Azbar, N. (2019). Biohydrogen production from solid wastes. In A. Pandey, S. V. Mohan, J.-S. Chang, P. C. Hallenbeck, and C. Larroche (Eds.), *Biohydrogen (2nd Edition)* (pp. 321–346). Elsevier. https://doi.org/10.1016/B978-0-444-64203-5.00012-5

Khan, M. U., Ahring, B., Garcia-Perez, T., and Garcia-Perez, M. (2020). Valorization of municipal solid waste in biorefineries for the creation of a circular economy: role of emerging technologies. In R. Kataki, A. Pandey, S. K. Khanal, and D. Pant (Eds.), *Current Developments in Biotechnology and Bioengineering* (pp. 323–347). Elsevier. https://doi.org/10.1016/B978-0-444-64309-4.00014-3

Khanal, S. K., Giri, B., Nitayavardhana, S., and Gadhamshetty, V. (2017). Anaerobic bioreactors/digesters: Design and development. In D. J. Lee, V. Jegatheesan, H. H. Ngo, P. C. Hallenbeck, and A. Pandey (Eds.), *Current Developments in Biotechnology and Bioengineering* (pp. 261–279). Elsevier. https://doi.org/10.1016/B978-0-444-63665-2.00010-2

Khanh Toan, P., Duc Cuong, N., Thi Quy, N., and Kim Sinh, P. (2000). Application of briquetting technology to produce briquettes from agricultural residues and by-products. In A. A. M. Sayigh (Ed.), *World Renewable Energy Congress VI* (pp. 1416–1419). Pergamon. https://doi.org/10.1016/B978-008043865-8/50295-6

Khare, V., Khare, C., Nema, S., and Baredar, P. (2019). *Tidal Energy Systems* (pp. 1-394). Elsevier.

Khushboo, Ankush, Karuna Yadav, Mrinal Kanti Mandal, Supriya Pal, Hirok Chaudhuri, and Kashyap Kumar Dubey. (2020). Bioeconomy of municipal solid waste (MSW) using gas fermentation. In S. Varjani, A. Pandey, E. Gnansounou, S. K. Khanal, and S. Raveendran (Eds.), *Current Developments in Biotechnology and Bioengineering* (pp. 289–304). Elsevier. https://doi.org/10.1016/B978-0-444-64321-6.00015-X

Kiang, Y.-H. (2018). *Fuel Property Estimation and Combustion Process Characterization* (pp. 235–312). Academic Press. https://doi.org/10.1016/B978-0-12-813473-3.00007-6

Killingtveit, Å. (2014). Hydroelectric power. In T. M. Letcher (Ed.), *Future Energy (2nd Edition)* (pp. 453–470). Elsevier. https://doi.org/10.1016/B978-0-08-099424-6.00021-1

Killingtveit, A. (2019). Hydropower. In T. M. Letcher (Ed.), *Managing Global Warming* (pp. 265–315). Academic Press. https://doi.org/10.1016/B978-0-12-814104-5.00008-9

Killingtveit, Å. (2020). Hydroelectric power. In T. M. Letcher (Ed.), *Future Energy (3rd Edition)* (pp. 315–330). Elsevier. https://doi.org/10.1016/B978-0-08-102886-5.00015-3

Kim, J. W., Woo, S. B., Song, J. I., and Kwon, H. K. (2022). An observational study of hydrodynamic impact on water mass transport due to tidal power generation. *Science of The Total Environment*, *807*, 151013. https://doi.org/10.1016/j.scitotenv.2021.151013

Kirk, D. M., and Gould, M. C. (2020). Bioenergy and anaerobic digestion. In A. Dahiya (Ed.), *Bioenergy (2nd Edition)* (pp. 335–360). Academic Press. https://doi.org/10.1016/B978-0-12-815497-7.00017-8

Kirkham, M. B. (2014). Solar time and interception of direct-beam solar radiation. In M. B. Kirkham (Ed.), *Principles of Soil and Plant Water Relations (2nd Edition)* (pp. 533–554). Academic Press. https://doi.org/10.1016/B978-0-12-420022-7.00030-6

Kizito, S., Jjagwe, J., Ssewaya, B., Nekesa, L., Tumutegyereize, P., Zziwa, A., and Komakech, A. J. (2022). Biofuel characteristics of non-charred briquettes from dried fecal sludge blended with food market waste: Suggesting a waste-to-biofuel enterprise as a win–win strategy to solve energy and sanitation problems in slums settlements. *Waste Management*, *140*, 173–182. https://doi.org/10.1016/j.wasman.2021.11.029

Knoema. (n.d.). *Biodiesel production*. Retrieved May 7, 2021, from https://knoema.com/atlas/topics/Energy/Renewables/Biodiesel-production

Kohler, M. (2019). Economic assessment of ethanol production. In A. Basile, A. Iulianelli, F. Dalena, and T. N. Veziroğlu (Eds.), *Ethanol* (pp. 505–521). Elsevier. https://doi.org/10.1016/B978-0-12-811458-2.00019-5

Kostov, G., Iliev, V., Goranov, B., Denkova, R., and Shopska, V. (2020). Immobilized cell bioreactors in fermented beverage production: Design and modeling. In A. M. Grumezescu and A. M. Holban (Eds.), *Biotechnological Progress and Beverage Consumption* (pp. 339–375). Academic Press. https://doi.org/10.1016/B978-0-12-816678-9.00011-4

Kraiem, T., Hassen, A. Ben, Belayouni, H., and Jeguirim, M. (2017). Production and characterization of bio-oil from the pyrolysis of waste frying oil. *Environmental Science and Pollution Research*, *24*(11), 9951–9961. https://doi.org/10.1007/s11356-016-7704-z

Krishna, B. B., Biswas, B., and Bhaskar, T. (2019). Gasification of lignocellulosic biomass. In A. Pandey, C. Larroche, C.-G. Dussap, E. Gnansounou, S. K. Khanal, and S. Ricke (Eds.), *Biofuels: Alternative Feedstocks and Conversion Processes for the Production of Liquid and Gaseous Biofuels (2nd Edition)* (pp. 285–300). Academic Press. https://doi.org/10.1016/B978-0-12-816856-1.00012-9

Kristian Peters-Fabelfroh (2006). Plagiomnium affine laminazellen.jpeg, Retrieved from: https://www.nationalgeographic.org/encyclopedia/chlorophyll/ , CC BYSA3.0 Unported via National Geographic, under GNU Free Documentation License.

Kristoferson, L. A., and Bokalders, V. (1986). *Renewable Energy Technologies* (pp. 43–51,183–231). Pergamon.

Kulikovsky, A. A. (2019). Fuel cell basics. In A. A. Kulikovsky (Ed.), *Analytical Modeling of Fuel Cells (2nd Edition)* (pp. 1–33). Elsevier. https://doi.org/10.1016/B978-0-44-464222-6.00008-3

Kumar, Amit, Tiwari, A. K., and Said, Z. (2021). A comprehensive review analysis on advances of evacuated tube solar collector using nanofluids and PCM. *Sustainable Energy Technologies and Assessments*, *47*, 101417. https://doi.org/10.1016/j.seta.2021.101417

Kumar, Anil, Kumar, N., Baredar, P., and Shukla, A. (2015). A review on biomass energy resources, potential, conversion and policy in India. *Renewable and Sustainable Energy Reviews*, *45*, 530–539. https://doi.org/10.1016/j.rser.2015.02.007

Kumar, Krishna, and Saini, R. P. (2022). A review on operation and maintenance of hydropower plants. *Sustainable Energy Technologies and Assessments*, *49*, 101704. https://doi.org/10.1016/j.seta.2021.101704

Kumar, N. (2019). Study of oxygenated ecofuel applications in CI engine, gas turbine, and jet engine. In A. K. Azad and M. Rasul (Eds.), *Advanced Biofuels* (pp. 405–441). Woodhead Publishing. https://doi.org/10.1016/B978-0-08-102791-2.00016-7

Kumar, N. M., Chopra, S. S., de Oliveira, A. K. V., Ahmed, H., Vaezi, S., Madukanya, U. E., and Castañón, J. M. (2020). Solar PV module technologies. In S. Gorjian and A. Shukla (Eds.), *Photovoltaic Solar Energy Conversion* (pp. 51–78). Academic Press. https://doi.org/10.1016/B978-0-12-819610-6.00003-X

Kumar, P., Subbarao, P. M. V, Kala, L. D., and Vijay, V. K. (2021). Thermogravimetry and associated characteristics of pearl millet cob and eucalyptus biomass using differential thermal gravimetric analysis for thermochemical gasification. *Thermal Science and Engineering Progress*, *26*, 101104. https://doi.org/10.1016/j.tsep.2021.101104

Kumar, S. (2000). Technology options for MSW to energy projects. *TERI Information Monitor on Environment Science*, *5*(1), 2–11.

Kumar, V., Sharma, N., Jaiswal, K. K., Vlaskin, M. S., Nanda, M., Tripathi, M. K., and Kumar, S. (2021). Microalgae with a truncated light-harvesting antenna to maximize photosynthetic efficiency and biomass productivity: Recent advances and current challenges. *Process Biochemistry*, *104*, 83–91. https://doi.org/10.1016/j.procbio.2021.03.006

Küüt, A., Ritslaid, K., Küüt, K., Ilves, R., and Olt, J. (2019). State of the art on the conventional processes for ethanol production. In A. Basile, A. Iulianelli, F. Dalena, and T. N. Veziroğlu (Eds.), *Ethanol* (pp. 61–101). Elsevier. https://doi.org/10.1016/B978-0-12-811458-2.00003-1

L'Orange, C., Volckens, J., and DeFoort, M. (2012). Influence of stove type and cooking pot temperature on particulate matter emissions from biomass cook stoves. *Energy for Sustainable Development*, *16*(4), 448–455. https://doi.org/10.1016/j.esd.2012.08.008

Lam, M. K., Loy, A. C. M., Yusup, S., and Lee, K. T. (2019). Biohydrogen production from algae. In *Biohydrogen*. Elsevier B.V. https://doi.org/10.1016/b978-0-444-64203-5.00009-5

Lamb, J. J., Hillestad, M., Rytter, E., Bock, R., Nordgård, A. S. R., Lien, K. M., Burheim, O. S., and Pollet, B. G. (2020). Traditional routes for hydrogen production and carbon conversion. In J. J. Lamb and B. G. Pollet (Eds.), *Hydrogen, Biomass and Bioenergy* (pp. 21–53). Academic Press. https://doi.org/10.1016/B978-0-08-102629-8.00003-7

Lareo, C., and Ferrari, M. D. (2019). Sweet potato as a bioenergy crop for fuel ethanol production: Perspectives and challenges. In R. C. Ray and S. Ramachandran (Eds.), *Bioethanol Production from Food Crops* (pp. 115–147). Academic Press. https://doi.org/10.1016/B978-0-12-813766-6.00007-2

Lazaro, C. Z., and Hallenbeck, P. C. (2019). Fundamentals of biohydrogen production. In A. Pandey, S. V. Mohan, J.-S. Chang, P. C. Hallenbeck, and C. Larroche (Eds.), *Biohydrogen (2nd Edition)* (pp. 25–48). Elsevier. https://doi.org/10.1016/B978-0-444-64203-5.00002-2

Lee, K., and Lee, I. (2019). Optimization of a heliostat field site in central receiver systems based on analysis of site slope effect. *Solar Energy*, *193*, 175–183. https://doi.org/10.1016/j.solener.2019.09.027

Lemonis, G. (2005). *Wave and Tidal Energy Conversion. In: Background material for Training programme on Renewable energy sources and technologies.*

Letcher, T. M. (2017). Why wind energy? In T. M. Letcher (Ed.), *Wind Energy Engineering* (pp. 3–14). Academic Press. https://doi.org/10.1016/B978-0-12-809451-8.00001-1

Lettinga, G., Velsen, A. F. M., Van Hobma, S. W., Zeeuw, W. J. de, and Klapwijk, A. (1980). Use of the Up-flow Sludge Blacket (USB) reactor concept for biological wastewater treatment. *Biotechnology and Bioengineering*, *22*(1), 699–734.

Liang, S., and Wang, J. (Eds.). (2020). Solar radiation. In *Advanced Remote Sensing (2nd Edition)* (pp. 157–191). Academic Press. https://doi.org/10.1016/B978-0-12-815826-5.00005-2

Lima, D. G., Soares, V. C. D., Ribeiro, E. B., Carvalho, D. A., Cardoso, É. C. V., Rassi, F. C., Mundim, K. C., Rubim, J. C., and Suarez, P. A. Z. (2004). Diesel-like fuel obtained by pyrolysis of vegetable oils. *Journal of Analytical and Applied Pyrolysis*, *71*(2), 987–996. https://doi.org/10.1016/j.jaap.2003.12.008

Lin, C. Y., and Lin, Y. W. (2012). Fuel characteristics of biodiesel produced from a high-acid oil from soybean soapstock by supercritical-methanol transesterification. *Energies*, *5*(7), 2370–2380. https://doi.org/10.3390/en5072370

Lindorfer, J., Rosenfeld, D. C., and Böhm, H. (2020). Fuel cells: Energy conversion technology. In T. M. Letcher (Ed.), *Future Energy (3rd Edition)* (pp. 495–517). Elsevier. https://doi.org/10.1016/B978-0-08-102886-5.00023-2

Liu, J., Lin, H., and Zhang, J. (2019). Review on the technical perspectives and commercial viability of vertical axis wind turbines. *Ocean Engineering*, *182*, 608–626. https://doi.org/10.1016/j.oceaneng.2019.04.086

Louwen, A., and van Sark, W. (2020). Photovoltaic solar energy. In M. Junginger and A. Louwen (Eds.), *Technological Learning in the Transition to a Low-Carbon Energy System* (pp. 65–86). Academic Press. https://doi.org/10.1016/B978-0-12-818762-3.00005-4

Lu, X., and McElroy, M. B. (2017). Global potential for wind-generated electricity. In T. M. Letcher (Ed.), *Wind Energy Engineering* (pp. 51–73). Academic Press. https://doi.org/10.1016/B978-0-12-809451-8.00004-7

Lubimova, E.A., (1968). Thermal history of the Earth, in: The Earth's Crust and Upper Mantle, Amer. Geophys. Un., Geophys. Mon. Ser., 13: 63-77.

Lubimova, E.A. (1969). Thermal History of the Earth. In The Earth's Crust and Upper Mantle, P.J. Hart (Ed.). https://doi.org/10.1029/GM013p0063

Ma, H., Guo, Y., Qin, Y., and Li, Y.-Y. (2018). Nutrient recovery technologies integrated with energy recovery by waste biomass anaerobic digestion. *Bioresource Technology*, *269*, 520–531. https://doi.org/10.1016/j.biortech.2018.08.114

Makareviciene, V., and Skorupskaite, V. (2019). Transesterification of microalgae for biodiesel production. In A. Basile and F. Dalena (Eds.), *Second and Third*

Generation of Feedstocks (pp. 469–510). Elsevier. https://doi.org/10.1016/B978-0-12-815162-4.00017-3

Manirafasha, E., Jiao, K., Zeng, X., Xu, Y., Tang, X., Sun, Y., Lin, L., Murwanashyaka, T., Ndikubwimana, T., Jing, K., and Lu, Y. (2020). Processing of microalgae to biofuels. In A. Yousuf (Ed.), *Microalgae Cultivation for Biofuels Production* (pp. 111–128). Academic Press. https://doi.org/10.1016/B978-0-12-817536-1.00008-4

Markov, S. A. (2012). Hydrogen production in bioreactors: Current trends. *Energy Procedia*, *29*, 394–400. https://doi.org/10.1016/j.egypro.2012.09.046

Markvart, T., and Castañer, L. (2018). Principles of solar cell operation. In S. A. Kalogirou (Ed.), *McEvoy's Handbook of Photovoltaics (3rd Edition)* (pp. 3–28). Academic Press. https://doi.org/10.1016/B978-0-12-809921-6.00001-X

Marsh, G. (2004). Tidal turbines harness the power of the sea. *Reinforced Plastics*, *48*(6), 44–47. https://doi.org/10.1016/S0034-3617(04)00344-3

Martinez-Gracia, A., Arauzo, I., and Uche, J. (2019). Solar energy availability. In F. Calise, M. D. D'Accadia, M. Santarelli, A. Lanzini, and D. Ferrero (Eds.), *Solar Hydrogen Production* (pp. 113–149). Academic Press. https://doi.org/10.1016/B978-0-12-814853-2.00005-9

Marx, S. (2019). Cassava as feedstock for ethanol production: A global perspective. In R. C. Ray and S. Ramachandran (Eds.), *Bioethanol Production from Food Crops* (pp. 101–113). Academic Press. https://doi.org/10.1016/B978-0-12-813766-6.00006-0

Mathur, A. N., and Rathore, N. S. (1992). *Biogas production, management and utilization*. Himanshu Publication. Delhi.

Matsumura, Y. (2015). Hydrothermal gasification of biomass. In A. Pandey, T. Bhaskar, M. Stöcker, and R. K. Sukumaran (Eds.), *Recent Advances in Thermo-Chemical Conversion of Biomass* (pp. 251–267). Elsevier. https://doi.org/10.1016/B978-0-444-63289-0.00009-0

Mattuci. (n.d.). *PEM Fuel Cell*. Retrieved January 2, 2022, from https://upload.wikimedia.org/wikipedia/commons/6/63/Proton_Exchange_Fuel_Cell_Diagram.svg

Mauk, M. G. (2018). Thin crystalline and polycrystalline silicon solar cells. In S. A. Kalogirou (Ed.), *McEvoy's Handbook of Photovoltaics (2nd Edition)* (pp. 159–231). Academic Press. https://doi.org/10.1016/B978-0-12-809921-6.00007-0

Mawire, A. (2015). Solar thermal energy storage for solar cookers. In B. Sorensen (Ed.), *Solar Energy Storage* (pp. 327–358). Academic Press. https://doi.org/10.1016/B978-0-12-409540-3.00014-1

McDonald, A., and Carroll, J. (2016). Design of generators for offshore wind turbines. In Chong Ng and L. Ran (Eds.), *Offshore Wind Farms* (pp. 159–192). Woodhead Publishing. https://doi.org/10.1016/B978-0-08-100779-2.00008-8

McFadden, S., and Basu, B. (2016). Wind turbine gearbox design with drivetrain dynamic analysis. In Chong Ng and L. Ran (Eds.), *Offshore Wind Farms* (pp. 137–158). Woodhead Publishing. https://doi.org/10.1016/B978-0-08-100779-2.00007-6

McKendry, P. (2002). Energy production from biomass (part 3): gasification technologies. *Bioresource Technology*, *83*(1), 55–63. https://doi.org/10.1016/S0960-8524(01)00120-1

McLachlan, A., and Defeo, O. (2018). Management and conservation. In A. McLachlan and O. Defeo (Eds.), *The Ecology of Sandy Shores (3rd Edition)* (pp. 451–495). Academic Press. https://doi.org/10.1016/B978-0-12-809467-9.00017-5

Mehr, G., Durali, M., Khakrand, M. H., and Hoghooghi, H. (2021). A novel design and performance optimization methodology for hydraulic Cross-Flow turbines using successive numerical simulations. *Renewable Energy*, *169*, 1402–1421. https://doi.org/10.1016/j.renene.2021.01.090

MHz`as. (2012). *MODIS ATM solar irradiance*. https://commons.wikimedia.org/wiki/File:MODIS_ATM_solar_irradiance.svg

Miguelsierra (2011). Scheme Chloroplast-en.svg. Available at: https://commons.wikimedia.org/wiki/File:Scheme_Chloroplast-en.svg, CC BY-SA 4.0 International, via Wikimedia Commons, under GNU Free Documentation License.

Ming, T., Liu, W., Liu, C., Zhou, Z., and Yu, X. (2016). In T. Ming (Ed.), *Solar Chimney Power Plant Generating Technology* (pp. 95–220). Academic Press.

Ministry of New and Renewable Energy, Govt. of India (2016). *Potential of Wind Energy in India*. [Online data - 05.12.2022] Retrieved from: https://mnre.gov.in/wind/current-status/

Ministry of Power Govt. of India. (2021). *Power Sector at a Glance All India*. https://powermin.gov.in/en/content/power-sector-glance-all-india

Minseong Kim. (2017). *Hyundai fuel cell bus 1st gen*. Retrieved January 2, 2022, from https://commons.wikimedia.org/wiki/File:Hyundai_fuel_cell_bus_1st_gen.jpg, CC BY-SA 4.0 International, via Wikimedia Commons, under GNU Free Documentation License.

Mital, K. M. (1996). *Biogas systems; Principles and applications*. New Age International (P) ltd. Publishers, New Delhi.

Mitchell, E. J. S., Gudka, B., Whittaker, C., Shield, I., Price-Allison, A., Maxwell, D., Jones, J. M., and Williams, A. (2020). The use of agricultural residues, wood briquettes and logs for small-scale domestic heating. *Fuel Processing Technology*, *210*, 106552. https://doi.org/10.1016/j.fuproc.2020.106552

Miyake, J., and Kawamura, S. (1987). Efficiency of light energy conversion to hydrogen by the photosynthetic bacterium Rhodobacter sphaeroides. *International Journal of Hydrogen Energy*, *12*(3), 147–149. https://doi.org/10.1016/0360-3199(87)90146-7

Miyamoto, K. (1997). Renewable biological systems for alternative sustainable energy production. In *FAO Agricultural Services Bulletin (FAO)*. FAO.

Mofijur, M., Rasul, M. G., Hassan, N. M. S., Masjuki, H. H., Kalam, M. A., and Mahmudul, H. M. (2017). Assessment of physical, chemical, and tribological properties of different biodiesel fuels. *Clean Energy for Sustainable Development: Comparisons and Contrasts of New Approaches*, 441–463. https://doi.org/10.1016/B978-0-12-805423-9.00014-4

Mohammadi, M., Mohammadi, P., Karami, N., Barzegar, A., and Mohamad Annuar, M. S. (2019). Efficient hydrogen gas production from molasses in hybrid anaerobic-activated sludge-rotating biological contactor. *International Journal of Hydrogen Energy*, *44*(5), 2592–2602. https://doi.org/10.1016/j.ijhydene.2018.12.033

Mohan, D., Pittman, C. U., and Steele, P. H. (2006). Pyrolysis of wood/biomass for bio-oil: A critical review. *Energy & Fuels*, *20*(3), 848–889. https://doi.org/10.1021/ef0502397

Mohan, S. V., Rohit, M. V., Subhash, G. V., Chandra, R., Devi, M. P., Butti, S. K., and Rajesh, K. (2019). Algal oils as biodiesel. In *Biofuels from Algae* (Second Edi). Elsevier B.V. https://doi.org/10.1016/b978-0-444-64192-2.00012-3

Mohanty, S. K., and Mishra, S. (2020). Regulatory reform and market efficiency: The case of Indian agricultural commodity futures markets. *Research in International Business and Finance*, *52*, 101145. https://doi.org/10.1016/j.ribaf.2019.101145

Mohd Moiz Khan, Riyaj Uddin Khan, Fahad Zishan Khan, M. A. (2013). Impacts of biodiesel on the environmet. *International Journal of Environmental Enginerring and Management*, *4*(4), 345–350.

Molino, A., Migliori, M., Larocca, V., Marino, T., Figoli, A., Casella, P., Iovane, P., Cerbone, A., Rimauro, J., and Donatelli, A. (2019). Power production by biomass gasification technologies. In A. Basile, A. Cassano, and A. Figoli (Eds.), *Current Trends and Future Developments on (Bio-) Membranes* (pp. 293–318). Elsevier. https://doi.org/10.1016/B978-0-12-813545-7.00012-X

Möllerström, E., Gipe, P., Beurskens, J., and Ottermo, F. (2019). A historical review of vertical axis wind turbines rated 100 kW and above. *Renewable and Sustainable Energy Reviews*, *105*, 1–13. https://doi.org/10.1016/j.rser.2018.12.022

Montañez Hernandez, L. E., García Lozano, M., Peña Garcia, Y., Oropeza Navarro, R., and Balagurusamy, N. (2017). Microbiology and biochemistry of anaerobic treatment. In D.-J. Lee, V. Jegatheesan, H. H. Ngo, P. C. Hallenbeck, and A. Pandey (Eds.), *Current Developments in Biotechnology and Bioengineering* (pp. 231–259). Elsevier. https://doi.org/10.1016/B978-0-444-63665-2.00009-6

Moore, M. C. (2016). Project permitting, finance, and economics for geothermal power generation. In Ronald DiPippo (Ed.), *Geothermal Power Generation* (pp. 499–533). Woodhead Publishing. https://doi.org/10.1016/B978-0-08-100337-4.00018-8

Moral, N. A. (2000). Briquetting of biomass in bangladesh. In A. A. M. Sayigh (Ed.), *World Renewable Energy Congress VI* (pp. 2402–2405). Pergamon. https://doi.org/10.1016/B978-008043865-8/50522-5

More, T. T., Yan, S., Hettiaratchi, J. P., Tyagi, R. D., and Surampalli, R. Y. (2019). Biodiesel impact on environment and health. *Biodiesel Production*, 583–652. https://doi.org/10.1061/9780784415344.ch23

Mörner, N. A. (2016). Sea level changes as observed in nature. In D. J. Easterbrook (Ed.), *Evidence-Based Climate Science (2nd Edition)* (pp. 215–229). Elsevier. https://doi.org/10.1016/B978-0-12-804588-6.00012-4

Morthorst, P. E., and Kitzing, L. (2016). Economics of building and operating offshore wind farms. In Chong Ng and L. Ran (Eds.), *Offshore Wind Farms* (pp. 9–27). Woodhead Publishing. https://doi.org/10.1016/B978-0-08-100779-2.00002-7

Mosier, N. S. (2020). Cellulosic ethanol - biofuel beyond corn. In A. Dahiya (Ed.), *Bioenergy (2nd Edition)* (pp. 217–221). Academic Press. https://doi.org/10.1016/B978-0-12-815497-7.00011-7

Mosier, N. S., and Ileleji, K. E. (2020). How fuel ethanol is made from corn. In A. Dahiya (Ed.), *Bioenergy (2nd Edition)* (pp. 539–544). Academic Press. https://doi.org/10.1016/B978-0-12-815497-7.00026-9

Motta, I. L., Marchesan, A. N., Filho, R. M., and Maciel, M. R. W. (2020). Optimization of biomass circulating fluidized bed gasifier for synthesis applications using simulation and response surface methodology. In S. Pierucci, F. Manenti, G. L. Bozzano, and D. Manca (Eds.), *30th European Symposium on Computer Aided Process Engineering* (Vol. 48, pp. 1585–1590). Elsevier. https://doi.org/10.1016/B978-0-12-823377-1.50265-2

Moustakas, K., Parmaxidou, P., and Vakalis, S. (2020). Anaerobic digestion for energy production from agricultural biomass waste in Greece: Capacity assessment for the region of Thessaly. *Energy*, *191*, 116556. https://doi.org/10.1016/j.energy.2019.116556

Mundon, T. R., and Goldsmith, J. (2014). State of the art and future outlook of integrating wave and tidal energy. In L. E. Jones (Ed.), *Renewable Energy Integration* (pp. 347–358). Academic Press. https://doi.org/10.1016/B978-0-12-407910-6.00027-2

Murray, R., King, G., and Wyse-Mason, R. (2019). Micro-emulsification vs. transesterification: an investigation of the efficacy of methanol use in improving vegetable oil engine performance. *Biofuels*, *0*(0), 1–10. https://doi.org/10.1080/17597269.2019.1598316

Murthy, K. V. S. R., and Raju, M. R. (2011). Electrical energy consumption in Indian agricultural sector- an analysis. *Energy Manager*, *4*(4), 6–11.

Muthu Dinesh Kumar, R., and Anand, R. (2019). Production of biofuel from biomass downdraft gasification and its applications. In A. K. Azad and M. Rasul (Eds.), *Advanced Biofuels* (pp. 129–151). Woodhead Publishing. https://doi.org/10.1016/B978-0-08-102791-2.00005-2

Muthudineshkumar, R., and Anand, R. (2019). Anaerobic digestion of various feedstocks for second-generation biofuel production. In K. Azad (Ed.), *Advances in Eco-Fuels for a Sustainable Environment* (pp. 157–185). Woodhead Publishing. https://doi.org/10.1016/B978-0-08-102728-8.00006-1

Najafpour, G. D. (2015). Immobilization of microbial cells for the production of organic acid and ethanol. In G. D. Najafpour (Ed.), *Biochemical Engineering and Biotechnology (2nd Edition)* (pp. 257–289). Elsevier. https://doi.org/10.1016/B978-0-444-63357-6.00008-0

Nanaki, E. A., and Koroneos, C. J. (2012). Comparative LCA of the use of biodiesel, diesel and gasoline for transportation. *Journal of Cleaner Production*, *20*(1), 14–19. https://doi.org/10.1016/j.jclepro.2011.07.026

Nanda, S, Li, K., Abatzoglou, N., Dalai, A. K., and Kozinski, J. A. (2017). Advancements and confinements in hydrogen production technologies. In F. Dalena, A. Basile, and C. Rossi (Eds.), *Bioenergy Systems for the Future* (pp. 373–418). Woodhead Publishing. https://doi.org/10.1016/B978-0-08-101031-0.00011-9

National Renewable Energy Lab. (2011). *NPC, SERRL, Pyrheliometer Comparisons*. https://www.flickr.com/photos/nrel/6214808068/

Nautiyal, H., and Varun. (2012). Progress in renewable energy under clean development mechanism in India. *Renewable and Sustainable Energy Reviews*, *16*(5), 2913–2919. https://doi.org/10.1016/j.rser.2012.02.008

Navalta, C. J. L. G., Banaag, K. G. C., Raboy, V. A. O., Go, A. W., Cabatingan, L. K., and Ju, Y. H. (2020). Solid fuel from co-briquetting of sugarcane bagasse and rice bran. *Renewable Energy*, *147*, 1941–1958. https://doi.org/10.1016/j.renene.2019.09.129

Navarro Diaz, G. P., Saulo, A. C., and Otero, A. D. (2021). Full wind rose wind farm simulation including wake and terrain effects for energy yield assessment. *Energy*, *237*, 121642. https://doi.org/10.1016/j.energy.2021.121642

Nazir, Y., Halim, H., Prabhakaran, P., Hamid, A. A., and Song, Y. (2020). Microalgae single cell oil. In E. Jacob-Lopes, M. M. Maroneze, M. I. Queiroz, and L. Q. Zepka (Eds.), *Handbook of Microalgae-Based Processes and Products* (pp. 419–444). Academic Press. https://doi.org/10.1016/B978-0-12-818536-0.00016-6

Negi, H., Agrawal, R., Verma, A., and Goel, R. (2019). Municipal solid waste to bioenergy: current status, opportunities, and challenges in indian context. In J. S. Singh and D. P. Singh (Eds.), *New and Future Developments in Microbial Biotechnology and Bioengineering* (pp. 191–203). Elsevier. https://doi.org/10.1016/B978-0-444-64191-5.00014-6

Neill, S. P., and Hashemi, M. R. (2018). *Fundamentals of Ocean Renewable Energy* (pp. 47–106). Academic Press.

Ng, C., and Ran, L. (2016). Introduction to offshore wind energy. In Chong Ng and L. Ran (Eds.), *Offshore Wind Farms* (pp. 3–8). Woodhead Publishing. https://doi.org/10.1016/B978-0-08-100779-2.00001-5

Nguyen, D., Nitayavardhana, S., Sawatdeenarunat, C., Surendra, K. C., and Khanal, S. K. (2019). Biogas production by anaerobic digestion: Status and perspectives. In A. Pandey, C. Larroche, C. G. Dussap, E. Gnansounou, S. K. Khanal, and S. Ricke (Eds.), *Biofuels: Alternative Feedstocks and Conversion Processes for the Production of Liquid and Gaseous Biofuels (2nd Edition)* (pp. 763–778). Academic Press. https://doi.org/10.1016/B978-0-12-816856-1.00031-2

Ni, S.-Q., Ahmad, H. A., and Ahmad, S. (2020). Immobilization of anaerobic ammonium oxidation bacteria for nitrogen-rich wastewater treatment. In M. P. Shah, S. Rodriguez-Couto, and S. S. Şengör (Eds.), *Emerging Technologies in Environmental Bioremediation* (pp. 1–22). Elsevier. https://doi.org/10.1016/B978-0-12-819860-5.00001-8

Nicholson, K. (1993). Geothermal Fluids: Chemistry and Exploration Techniques, 1st Ed., Springer Berlin, Heidelberg. 255p. ISBN: 978-3-642-77846-9. doi: https://doi.org/10.1007/978-3-642-77844-5

Nijssen, R., and de Winkel, G. D. (2016). Developments in materials for offshore wind turbine blades. In Chong Ng and L. Ran (Eds.), *Offshore Wind Farms* (pp. 85–104). Woodhead Publishing. https://doi.org/10.1016/B978-0-08-100779-2.00005-2

Nikitas, G., Bhattacharya, S., and Vimalan, N. (2020). Wind energy. In T. M. Letcher (Ed.), *Future Energy (3rd Edition)* (pp. 331–355). Elsevier. https://doi.org/10.1016/B978-0-08-102886-5.00016-5

Nikitas, G., Bhattacharya, S., Vimalan, N., Demirci, H. E., Nikitas, N., and Kumar, P. (2019). Wind power: A sustainable way to limit climate change. In T. M. Letcher (Ed.), *Managing Global Warming* (pp. 333–364). Academic Press. https://doi.org/10.1016/B978-0-12-814104-5.00010-7

Nikolaisen, L. S., and Jensen, P. D. (2013). Biomass feedstocks: Categorisation and preparation for combustion and gasification. In L. Rosendahl (Ed.), *Biomass Combustion Science, Technology and Engineering* (pp. 36–57). Woodhead Publishing. https://doi.org/10.1533/9780857097439.1.36

Nikoletatos, J., and Halambalakis, G. (2018). Standards, Calibration, and Testing of PV Modules and Solar Cells. In S. A. Kalogirou (Ed.), *McEvoy's Handbook of Photovoltaics (3rd Edition)* (pp. 1155–1182). Academic Press. https://doi.org/10.1016/B978-0-12-809921-6.00033-1

Nobel, P. S. (2020). Photochemistry of photosynthesis. In P. S. Nobel (Ed.), *Physicochemical and Environmental Plant Physiology (5th Edition)* (pp. 259–307). Academic Press. https://doi.org/10.1016/B978-0-12-819146-0.00005-5

Notton, G., and Voyant, C. (2018). Forecasting of intermittent solar energy resource. In I. Yahyaoui (Ed.), *Advances in Renewable Energies and Power Technologies* (pp. 77–114). Elsevier. https://doi.org/10.1016/B978-0-12-812959-3.00003-4

Nurek, T., Gendek, A., Roman, K., and Dąbrowska, M. (2019). The effect of temperature and moisture on the chosen parameters of briquettes made of shredded logging residues. *Biomass and Bioenergy*, *130*, 105368. https://doi.org/10.1016/j.biombioe.2019.105368

O-Thong, S., Mamimin, C., Kongjan, P., and Reungsang, A. (2019). Thermophilic fermentation for enhanced biohydrogen production. In A. Pandey, S. V. Mohan, J. S. Chang, P. C. Hallenbeck, and C. Larroche (Eds.), *Biohydrogen (2nd Edition)* (pp. 123–139). Elsevier. https://doi.org/10.1016/B978-0-444-64203-5.00005-8

Ofosu, E. A., Amo-Boateng, M., Domfeh, M. K., and Andoh, R. (2018). Re-engineering hydropower plant for improved performance. In A. Kabo-Bah and C. J. Diji (Eds.), *Sustainable Hydropower in West Africa* (pp. 189–195). Academic Press. https://doi.org/10.1016/B978-0-12-813016-2.00012-5

Ogunkunle, O., and Ahmed, N. A. (2019). A review of global current scenario of biodiesel adoption and combustion in vehicular diesel engines. *Energy Reports*, *5*, 1560–1579. https://doi.org/10.1016/j.egyr.2019.10.028

Okot, D. K., Bilsborrow, P. E., and Phan, A. N. (2018). Effects of operating parameters on maize COB briquette quality. *Biomass and Bioenergy*, *112*, 61–72. https://doi.org/10.1016/j.biombioe.2018.02.015

Okot, D. K., Bilsborrow, P. E., and Phan, A. N. (2019). Briquetting characteristics of bean straw-maize cob blend. *Biomass and Bioenergy*, *126*, 150–158. https://doi.org/10.1016/j.biombioe.2019.05.009

Ometto, F., Karlsson, A., Ejlertsson, J., Björn, A. V., and Shakeri, S. Y. (2019). Anaerobic digestion: An engineered biological process. In M. Materazzi and P. U. Foscolo (Eds.), *Substitute Natural Gas from Waste* (pp. 63–74). Academic Press. https://doi.org/10.1016/B978-0-12-815554-7.00005-2

Onwosi, C. O., Igbokwe, V. C., Odimba, J. N., and Nwagu, T. N. (2020). Anaerobic bioconversion of poultry industry-derived wastes for the production of biofuels and other value-added products. In N. Krishnaraj Rathinam and R. K. Sani (Eds.), *Biovalorisation of Wastes to Renewable Chemicals and Biofuels* (pp. 113–131). Elsevier. https://doi.org/10.1016/B978-0-12-817951-2.00006-7

Outline, C. (2020). Energy. In *Nuclear Energy* (pp. 3–13). Elsevier Inc. https://doi.org/10.1016/B978-0-12-812881-7.00001-0

Owen, A. (2014). Tidal current energy: Origins and challenges. In T. M. Letcher (Ed.), *Future Energy (2nd Edition)* (pp. 335–356). Elsevier. https://doi.org/10.1016/B978-0-08-099424-6.00016-8

Owen, A. (2020). Tidal current energy: Origins and challenges. In T. M. Letcher (Ed.), *Future Energy (3rd Edition)* (pp. 357–374). Elsevier. https://doi.org/10.1016/B978-0-08-102886-5.00017-7

Özçimen, D., Koçer, A. T., İnan, B., and Özer, T. (2020). Bioethanol production from microalgae. In E. Jacob-Lopes, M. M. Maroneze, M. I. Queiroz, and L. Q. Zepka (Eds.), *Handbook of Microalgae-Based Processes and Products* (pp. 373–389). Academic Press. https://doi.org/10.1016/B978-0-12-818536-0.00014-2

Palm, J., and Thollander, P. (2020). Reframing energy efficiency in industry: A discussion of definitions, rationales, and management practices. In M. Lopes,

C. H. Antunes, and K. B. Janda (Eds.), *Energy and Behaviour* (pp. 153–175). Academic Press. https://doi.org/10.1016/B978-0-12-818567-4.00007-7

Panchasara, H., and Ashwath, N. (2021). Effects of pyrolysis bio-oils on fuel atomisation—a review. *Energies*, *14*(4), 1–22. https://doi.org/10.3390/en14040794

Pang, S. (2019). Advances in thermochemical conversion of woody biomass to energy, fuels and chemicals. *Biotechnology Advances*, *37*(4), 589–597. https://doi.org/10.1016/j.biotechadv.2018.11.004

Panidepu, H., Ram, S. K., Cheernam, V., and Tyagi, R. D. (2019). Biodiesel: Socioeconomic and political aspects. *Biodiesel Production*, 653–663. https://doi.org/10.1061/9780784415344.ch24

Papavinasam, S. (2014). Oil and gas industry network. In *Corrosion Control in the Oil and Gas Industry*. https://doi.org/10.1016/b978-0-12-397022-0.00002-9

Paritosh, K., Yadav, M., Kesharwani, N., Pareek, N., Parthiba Karthyikeyan, O., Balan, V., and Vivekanand, V. (2021). Strategies to improve solid state anaerobic bioconversion of lignocellulosic biomass: an overview. *Bioresource Technology*, *331*, 125036. https://doi.org/10.1016/j.biortech.2021.125036

Parthasarathy, P., Narayanan, K. S., and Arockiam, L. (2013). Study on kinetic parameters of different biomass samples using thermo-gravimetric analysis. *Biomass and Bioenergy*, *58*, 58–66. https://doi.org/10.1016/j.biombioe.2013.08.004

Patel, N. K., and Shah, S. N. (2015). Biodiesel from plant oils. In *Food, Energy, and Water: The Chemistry Connection*. Elsevier Inc. https://doi.org/10.1016/B978-0-12-800211-7.00011-9

Patil, R. A., Kausley, S. B., Joshi, S. M., and Pandit, A. B. (2020). Process intensification applied to microalgae-based processes and products. In E. Jacob-Lopes, M. M. Maroneze, M. I. Queiroz, and L. Q. Zepka (Eds.), *Handbook of Microalgae-Based Processes and Products* (pp. 737–769). Academic Press. https://doi.org/10.1016/B978-0-12-818536-0.00027-0

Pearsall, N. M. (2017). Introduction to photovoltaic system performance. In N. Pearsall (Ed.), *The Performance of Photovoltaic (PV) Systems* (pp. 1–19). Woodhead Publishing. https://doi.org/10.1016/B978-1-78242-336-2.00001-X

Pearson, R. J., and Turner, J. W. G. (2014). Using alternative and renewable liquid fuels to improve the environmental performance of internal combustion engines: key challenges and blending technologies. In R. Folkson (Ed.), *Alternative Fuels and Advanced Vehicle Technologies for Improved Environmental Performance* (pp. 52–89). Woodhead Publishing. https://doi.org/10.1533/9780857097422.1.52

Pereira, L., Concejo, A. V., and Trindade, W. (2020). Tidal modulation. In D. W. T. Jackson and A. D. Short (Eds.), *Sandy Beach Morphodynamics* (pp. 87–101). Elsevier. https://doi.org/10.1016/B978-0-08-102927-5.00005-9

Perkins, G. (2020). Production of electricity and chemicals using gasification of municipal solid wastes. In T. Bhaskar, A. Pandey, E. R. Rene, and D. C. W.

Tsang (Eds.), *Waste Biorefinery* (pp. 3–39). Elsevier. https://doi.org/10.1016/B978-0-12-818228-4.00001-0

Petković, D. (2020). Technology for producing briquettes from wet biomass. In S. Hashmi and I. A. Choudhury (Eds.), *Encyclopedia of Renewable and Sustainable Materials* (pp. 901–907). Elsevier. https://doi.org/10.1016/B978-0-12-803581-8.11122-1

Phuc, N. H. H., Xian, O. Y., and Atsunori, M. (2020). New developments in hydrogen fuel cells. In Sonil Nanda, D.-V. N. Vo, and P. Nguyen-Tri (Eds.), *New Dimensions in Production and Utilization of Hydrogen* (pp. 273–298). Elsevier. https://doi.org/10.1016/B978-0-12-819553-6.00011-8

Plante, R. H. (2014). Simple solar basics. In R. H. Plante (Ed.), *Solar Energy, Photovoltaics, and Domestic Hot Water.*. (pp. 7–25). Academic Press. https://doi.org/10.1016/B978-0-12-420155-2.00002-5

Prado, M., and Polinder, H. (2013). *Electrical Drives for Direct Drive Renewable Energy Systems* (pp. 175–218). Woodhead Publishing.

Pratap Singh, A., Kumar, A., Akshayveer, and Singh, O. P. (2020). Performance enhancement strategies of a hybrid solar chimney power plant integrated with photovoltaic panel. *Energy Conversion and Management*, *218*, 113020. https://doi.org/10.1016/j.enconman.2020.113020

Psomopoulos, C. S. (2014). Residue derived fuels as an alternative fuel for the hellenic power generation sector and their potential for emissions reduction. *AIMS Energy*, *2*(3), 321–341.

Putrus, G., and Bentley, E. (2016). Integration of distributed renewable energy systems into the smart grid. In M. H. Rashid (Ed.), *Electric Renewable Energy Systems* (pp. 487–518). Academic Press. https://doi.org/10.1016/B978-0-12-804448-3.00020-7

Qian, C., Li, Q., Zhang, Z., Wang, X., Hu, J., and Cao, W. (2020). Prediction of higher heating values of biochar from proximate and ultimate analysis. *Fuel*, *265*, 116925. https://doi.org/10.1016/j.fuel.2019.116925

Qiao, Z., and Lü, X. (2021). Industrial bioethanol production: status and bottlenecks. In X. Lü (Ed.), *Advances in 2nd Generation of Bioethanol Production* (pp. 213–227). Woodhead Publishing. https://doi.org/10.1016/B978-0-12-818862-0.00003-0

Queiroz, M. I., Vieira, J. G., and Maroneze, M. M. (2020). Morphophysiological, structural, and metabolic aspects of microalgae. In E. Jacob-Lopes, M. M. Maroneze, M. I. Queiroz, and L. Q. Zepka (Eds.), *Handbook of Microalgae-Based Processes and Products* (pp. 25–48). Academic Press. https://doi.org/10.1016/B978-0-12-818536-0.00002-6

Ragheb, A. M., and Selig, M. S. (2017). Multielement airfoils for wind turbines. In T. M. Letcher (Ed.), *Wind Energy Engineering* (pp. 203–219). Academic Press. https://doi.org/10.1016/B978-0-12-809451-8.00011-4

Ragheb, M. (2017). History of harnessing wind power. In T. M. Letcher (Ed.), *Wind*

Energy Engineering (pp. 127–143). Academic Press. https://doi.org/10.1016/B978-0-12-809451-8.00007-2

Rai, G. D. (2013). *Non-Conventional Energy Sources*. Khanna Publishers, Delhi.

Ram, S. K., Kumar, L. R., Yellapu, S. K., Kaur, R., and Tyagi, R. D. (2019). Economic evaluation of biodiesel production processes. *Biodiesel Production*, *1893*, 527–581. https://doi.org/10.1061/9780784415344.ch22

Ramadhas, A. S., Jayaraj, S., and Muraleedharan, C. (2005). Biodiesel production from high FFA rubber seed oil. *Fuel*, *84*(4), 335–340. https://doi.org/10.1016/j.fuel.2004.09.016

Ramalingam, K., and Indulkar, C. (2017). Solar energy and photovoltaic technology. In G. B. Gharehpetian and S. M. Mousavi Agah (Eds.), *Distributed Generation Systems* (pp. 69–147). Butterworth-Heinemann. https://doi.org/10.1016/B978-0-12-804208-3.00003-0

Ranzi, E., Faravelli, T., and Manenti, F. (2016). Pyrolysis, gasification, and combustion of solid fuels. In K. M. Van Geem (Ed.), *Thermochemical Process Engineering* (Vol. 49, pp. 1–94). Academic Press. https://doi.org/10.1016/bs.ache.2016.09.001

Rao, K. U., and Ramana, P. V. (2001). Bioenergy and climate change- opportunities in india. *Bioenergy News*, *5*(4), 15–19.

Rao, M. N., Sultana, R., and Kota, S. H. (2017). Municipal solid waste. In M. N. Rao, R. Sultana, and S. H. Kota (Eds.), *Solid and Hazardous Waste Management* (pp. 3–120). Butterworth-Heinemann. https://doi.org/10.1016/B978-0-12-809734-2.00002-X

Rao, S., and Parulekar, B. B. (2005). *Energy Technology*. Khanna Publishers, Delhi.

Rashad, A., Kamel, S., and Jurado, F. (2017). The basic principles of wind farms. In G. B. Gharehpetian and S. M. Mousavi Agah (Eds.), *Distributed Generation Systems* (pp. 21–67). Butterworth-Heinemann. https://doi.org/10.1016/B978-0-12-804208-3.00002-9

Ravindranath, N. H., Somashekar, H. I., Nagaraja, M. S., Sudha, P., Sangeetha, G., Bhattacharya, S. C., and Abdul Salam, P. (2005). Assessment of sustainable non-plantation biomass resources potential for energy in India. *Biomass and Bioenergy*, *29*(3), 178–190. https://doi.org/10.1016/j.biombioe.2005.03.005

Reca-Cardeña, J., and López-Luque, R. (2018). Design principles of photovoltaic irrigation systems. In I. Yahyaoui (Ed.), *Advances in Renewable Energies and Power Technologies* (pp. 295–333). Elsevier. https://doi.org/10.1016/B978-0-12-812959-3.00009-5

Redko, A., Redko, O., and DiPippo, R. (2020). *Low-Temperature Energy Systems with Applications of Renewable Energy* (pp. 177–259). Academic Press.

Refaat, A. A. (2012). Biofuels from waste materials. In *Comprehensive Renewable Energy* (Vol. 5). Elsevier Ltd. https://doi.org/10.1016/B978-0-08-087872-0.00518-7

Rehnlund Bjorn. (2008). *Outlook on standardization of Sub task Report October 2008. October*, 59.

Reikard, G. (2017). Wave energy forecasting. In G. Kariniotakis (Ed.), *Renewable Energy Forecasting* (pp. 199–217). Woodhead Publishing. https://doi.org/10.1016/B978-0-08-100504-0.00007-X

Rekioua, D. (2018). Energy management for pv installations. In I. Yahyaoui (Ed.), *Advances in Renewable Energies and Power Technologies* (pp. 349–369). Elsevier. https://doi.org/10.1016/B978-0-12-812959-3.00011-3

Ren, N., Guo, W., Liu, B., Cao, G., and Ding, J. (2011). Biological hydrogen production by dark fermentation: challenges and prospects towards scaled-up production. *Current Opinion in Biotechnology*, *22*(3), 365–370. https://doi.org/10.1016/j.copbio.2011.04.022

Ribeiro, a, Castro, F., and Carvalho, J. (2011). Influence of free fatty acid content in biodiesel production on non-edible oils. *International Conference Waste: Solutions, Treatments and Opportunities*, *c*.

Richardson, Y., Drobek, M., Julbe, A., Blin, J., and Pinta, F. (2015). Biomass gasification to produce syngas. In A. Pandey, T. Bhaskar, M. Stöcker, and R. K. Sukumaran (Eds.), *Recent Advances in Thermo-Chemical Conversion of Biomass* (pp. 213–250). Elsevier. https://doi.org/10.1016/B978-0-444-63289-0.00008-9

Richhariya, G., Kumar, A., and Samsher. (2020). Solar cell technologies. In S. Gorjian and A. Shukla (Eds.), *Photovoltaic Solar Energy Conversion* (pp. 27–50). Academic Press. https://doi.org/10.1016/B978-0-12-819610-6.00002-8

Rimmer, M. (2021). Carbon capture and storage: Intellectual property, innovation policy, and climate change. In H. K. de Medeiros Costa and C. Arlota (Eds.), *Carbon Capture and Storage in International Energy Policy and Law* (pp. 181–203). Elsevier. https://doi.org/10.1016/B978-0-323-85250-0.00012-8

Rincón, B. (2020). Biogas from microalgae. In E. Jacob-Lopes, M. M. Maroneze, M. I. Queiroz, and L. Q. Zepka (Eds.), *Handbook of Microalgae-Based Processes and Products* (pp. 311–328). Academic Press. https://doi.org/10.1016/B978-0-12-818536-0.00012-9

Rizvi, A. A., Danish, S. N., El-Leathy, A., Al-Ansary, H., and Yang, D. (2021). A review and classification of layouts and optimization techniques used in design of heliostat fields in solar central receiver systems. *Solar Energy*, *218*, 296–311. https://doi.org/10.1016/j.solener.2021.02.011

Rodríguez-Sánchez, M. R., Leray, C., Toutant, A., Ferriere, A., and Olalde, G. (2019). Development of a new method to estimate the incident solar flux on central receivers from deteriorated heliostats. *Renewable Energy*, *130*, 182–190. https://doi.org/10.1016/j.renene.2018.06.056

Rosendahl, L. (Ed.). (2013). Woodhead publishing series in energy. In *Biomass Combustion Science, Technology and Engineering* (pp. xiii–xvi). Woodhead Publishing. https://doi.org/10.1016/B978-0-85709-131-4.50015-7

Roshanmanesh, S., Hayati, F., and Papaelias, M. (2020). Tidal turbines. In M. Papaelias, F. P. G. Márquez, and A. Karyotakis (Eds.), *Non-Destructive Testing and Condition Monitoring Techniques for Renewable Energy Industrial Assets* (pp. 143–158). Butterworth-Heinemann. https://doi.org/10.1016/B978-0-08-101094-5.00010-1

Ross, D. (1995). *Power from the waves*. Oxford University Press, Oxford, United Kingdom.

Rostrup-Nielsen, J. R., and Hansen, J. B. (2011). Steam reforming for fuel cells. In D. Shekhawat, J. J. Spivey, and D. A. Berry (Eds.), *Fuel Cells: Technologies for Fuel Processing* (pp. 49–71). Elsevier. https://doi.org/10.1016/B978-0-444-53563-4.10004-5

Ruan, R., Ding, K., Liu, S., Peng, P., Zhou, N., He, A., Chen, P., Cheng, Y., Wang, Y., Liu, Y., Lei, H., Addy, M., and Cobb, K. (2020). Gasification and pyrolysis of waste. In R. Kataki, A. Pandey, S. K. Khanal, and D. Pant (Eds.), *Current Developments in Biotechnology and Bioengineering* (pp. 263–297). Elsevier. https://doi.org/10.1016/B978-0-444-64309-4.00012-X

Sağır, E., and Hallenbeck, P. C. (2019). Photofermentative hydrogen production. In A. Pandey, S. V. Mohan, J.-S. Chang, P. C. Hallenbeck, and C. Larroche (Eds.), *Biohydrogen (2nd Edition)* (pp. 141–157). Elsevier. https://doi.org/10.1016/B978-0-444-64203-5.00006-X

Sakai, Y. (2017). Advanced geothermal steam turbines. In T. Tanuma (Ed.), *Advances in Steam Turbines for Modern Power Plants* (pp. 455–486). Woodhead Publishing. https://doi.org/10.1016/B978-0-08-100314-5.00019-1

Samuel, O. D., Oreko, B. U., Oyejide, J. O., Idi, S., and Fayomi, O. S. I. (2019). Experimental and empirical study of diesel and biodiesel produced from blend of fresh vegetable and waste vegetable oil on density, viscosity, sulphur content and acid value. *Journal of Physics: Conference Series*, *1378*(4). https://doi.org/10.1088/1742-6596/1378/4/042024

Sans, C., Mata-Alvarez, J., Ceechi, F., Pavan, P., and Bassetti, A. (1995). Acidogenic fermentation of organic urban wastes in a pilot plug-flow reactor under thermophilic conditions. *Bioresource Technology*, *54*(1), 105–110.

Santori, G., Di Nicola, G., Moglie, M., and Polonara, F. (2012). A review analyzing the industrial biodiesel production practice starting from vegetable oil refining. *Applied Energy*, *92*, 109–132. https://doi.org/10.1016/j.apenergy.2011.10.031

Santos, F., Eichler, P., de Queiroz, J. H., and Gomes, F. (2020). Production of second-generation ethanol from sugarcane. In F. Santos, S. C. Rabelo, M. De Matos, and P. Eichler (Eds.), *Sugarcane Biorefinery, Technology and Perspectives* (pp. 195–228). Academic Press. https://doi.org/10.1016/B978-0-12-814236-3.00011-1

Santos, J. J. C. S., Rodríguez, C. E. C., Carvalho, M., Barone, M. A., Palacio, J. C. E., and Carrillo, R. A. M. (2018). Geothermal power. In I. Yahyaoui (Ed.), *Advances in Renewable Energies and Power Technologies* (pp. 173–205). Elsevier. https://doi.org/10.1016/B978-0-12-813185-5.00014-0

Santos, T., and Sabbatella, I. (2020). Energy (and climate) challenges in South America: what planning for what transition? In L. Noura Guimarães (Ed.), *The Regulation and Policy of Latin American Energy Transitions* (pp. 59–74). Elsevier. https://doi.org/10.1016/B978-0-12-819521-5.00004-8

Sanz Rodrigo, J., Frías Paredes, L., Girard, R., Kariniotakis, G., Laquaine, K., Stoffels, N., and von Bremen, L. (2017). The role of predictability in the investment phase of wind farms. In G. Kariniotakis (Ed.), *Renewable Energy Forecasting* (pp. 341–357). Woodhead Publishing. https://doi.org/10.1016/B978-0-08-100504-0.00014-7

Saradha Devi, G., Vaishnavi, S., Srinath, S., Dutt, B., and Rajmohan, K. S. (2020). Energy recovery from biomass using gasification. In S. Varjani, A. Pandey, E. Gnansounou, S. K. Khanal, and S. Raveendran (Eds.), *Current Developments in Biotechnology and Bioengineering* (pp. 363–382). Elsevier. https://doi.org/10.1016/B978-0-444-64321-6.00019-7

Saraswat, S. K., Digalwar, A. K., Yadav, S. S., and Kumar, G. (2021). MCDM and GIS based modelling technique for assessment of solar and wind farm locations in India. *Renewable Energy*, *169*, 865–884. https://doi.org/10.1016/j.renene.2021.01.056

Saratale, G. D., Saratale, R. G., Banu, J. R., and Chang, J. S. (2019). Biohydrogen production from renewable biomass resources. In A. Pandey, S. V. Mohan, J. S. Chang, P. C. Hallenbeck, and C. Larroche (Eds.), *Biohydrogen (2nd Edition)* (pp. 247–277). Elsevier. https://doi.org/10.1016/B978-0-444-64203-5.00010-1

Sarbu, I., and Sebarchievici, C. (2017). *Solar Heating and Cooling Systems* (pp. 1–206). Academic Press.

Savla, N., Shinde, A., Sonawane, K., Mekuto, L., Chowdhary, P., and Pandit, S. (2020). Microbial hydrogen production: fundamentals to application. In P. Chowdhary, A. Raj, D. Verma, and Y. Akhter (Eds.), *Microorganisms for Sustainable Environment and Health* (pp. 343–365). Elsevier. https://doi.org/10.1016/B978-0-12-819001-2.00017-6

Schwerz, F., Eloy, E., Elli, E. F., and Caron, B. O. (2019). Reduced planting spacing increase radiation use efficiency and biomass for energy in black wattle plantations: Towards sustainable production systems. *Biomass and Bioenergy*, *120*, 229–239. https://doi.org/10.1016/j.biombioe.2018.11.034

Scully, S. M., and Orlygsson, J. (2019). Biological production of alcohols. In M. Hosseini (Ed.), *Advanced Bioprocessing for Alternative Fuels, Biobased Chemicals, and Bioproducts* (pp. 83–108). Woodhead Publishing. https://doi.org/10.1016/B978-0-12-817941-3.00005-X

Sedighi, A. A., Deldoost, Z., and Karambasti, B. M. (2020). Effect of thermal energy storage layer porosity on performance of solar chimney power plant considering turbine pressure drop. *Energy*, *194*, 116859. https://doi.org/10.1016/j.energy.2019.116859

Sen, P., Nath, A., and Bhattacharjee, C. (2017). Packed-Bed bioreactor and its application in dairy, food, and beverage industry. In C. Larroche, M. Á. Sanromán, G. Du, and A. Pandey (Eds.), *Current Developments in Biotechnology and Bioengineering* (pp. 235–277). Elsevier. https://doi.org/10.1016/B978-0-444-63663-8.00009-4

Sensors, H. T. (2012). *pyrheliometer linedrawing*. https://commons.wikimedia.org/wiki/File:DR01_pyrheliometer_linedrawing.pdf

Seo, D. K., Park, S. S., Hwang, J., and Yu, T.-U. (2010). Study of the pyrolysis of biomass using thermo-gravimetric analysis (TGA) and concentration measurements of the evolved species. *Journal of Analytical and Applied Pyrolysis*, *89*(1), 66–73. https://doi.org/10.1016/j.jaap.2010.05.008

Serrano González, J., Trigo García, Á. L., Burgos Payán, M., Riquelme Santos, J., and González Rodríguez, Á. G. (2017). Optimal wind-turbine micro-siting of offshore wind farms: A grid-like layout approach. *Applied Energy*, *200*, 28–38. https://doi.org/10.1016/j.apenergy.2017.05.071

Seshadri, B., Bolan, N. S., Thangarajan, R., Jena, U., Das, K. C., Wang, H., and Naidu, R. (2016). Biomass energy from revegetation of landfill sites. In M. N. V Prasad (Ed.), *Bioremediation and Bioeconomy* (pp. 99–109). Elsevier. https://doi.org/10.1016/B978-0-12-802830-8.00005-8

Shah, A. (2018). Thin-Film silicon solar cells. In S. A. Kalogirou (Ed.), *McEvoy's Handbook of Photovoltaics (3rd Edition)* (pp. 235–307). Academic Press. https://doi.org/10.1016/B978-0-12-809921-6.00008-2

Shaisundaram, V. S., Chandrasekaran, M., Sujith, S., Praveen Kumar, K. J., and Shanmugam, M. (2021). Design and analysis of novel biomass stove. *Materials Today: Proceedings*, *46*, 4054–4058. https://doi.org/10.1016/j.matpr.2021.02.569

Shakouri, M., Ebadi, H., and Gorjian, S. (2020). Solar photovoltaic thermal (PVT) module technologies. In S. Gorjian and A. Shukla (Eds.), *Photovoltaic Solar Energy Conversion* (pp. 79–116). Academic Press. https://doi.org/10.1016/B978-0-12-819610-6.00004-1

Shan, Y., and Lü, X. (2021). Ending fossil fuels addiction: Prospects for biofuels. In X. Lü (Ed.), *Advances in 2nd Generation of Bioethanol Production* (pp. 229–242). Woodhead Publishing. https://doi.org/10.1016/B978-0-12-818862-0.00008-X

Shandilya, K. K., and Pattarkine, V. M. (2019). Using microalgae for treating wastewater. In M. Hosseini (Ed.), *Advances in Feedstock Conversion Technologies for Alternative Fuels and Bioproducts* (pp. 119–136). Woodhead Publishing. https://doi.org/10.1016/B978-0-12-817937-6.00007-2

Shao, L., Chang, L., Finkelman, R. B., Wang, W., Liu, J., Li, J., Xing, J., and Hou, C. (2020). Distribution of rare earth elements in PM10 emitted from burning coals and soil-mixed coal briquettes. *Journal of Environmental Sciences*, *97*, 96–101. https://doi.org/10.1016/j.jes.2020.05.010

Shen, H., Luo, Z., Xiong, R., Liu, X., Zhang, L., Li, Y., Du, W., Chen, Y., Cheng, H., Shen, G., and Tao, S. (2021). A critical review of pollutant emission factors from fuel combustion in home stoves. *Environment International, 157*, 106841. https://doi.org/10.1016/j.envint.2021.106841

Sheng, S., and O'Connor, R. (2017). Reliability of wind turbines. In T. M. Letcher (Ed.), *Wind Energy Engineering* (pp. 299–327). Academic Press. https://doi.org/10.1016/B978-0-12-809451-8.00015-1

Shi, J., and Erdem, E. (2017). Estimation of wind energy potential and prediction of wind power. In T. M. Letcher (Ed.), *Wind Energy Engineering* (pp. 25–49). Academic Press. https://doi.org/10.1016/B978-0-12-809451-8.00003-5

Shikanai, T. (2020). Regulation of photosynthesis by cyclic electron transport around photosystem I. In T. Hisabori (Ed.), *ATP Synthase in Photosynthetic Organisms* (Vol. 96, pp. 177–204). Academic Press. https://doi.org/10.1016/bs.abr.2020.07.005

Shilpi, S., Lamb, D., Bolan, N., Seshadri, B., Choppala, G., and Naidu, R. (2019). Waste to watt: Anaerobic digestion of wastewater irrigated biomass for energy and fertiliser production. *Journal of Environmental Management, 239*, 73–83. https://doi.org/10.1016/j.jenvman.2019.02.122

Show, K. Y., and Lee, D. J. (2017). Anaerobic treatment versus aerobic treatment. In D. J. Lee, V. Jegatheesan, H. H. Ngo, P. C. Hallenbeck, and A. Pandey (Eds.), *Current Developments in Biotechnology and Bioengineering* (pp. 205–230). Elsevier. https://doi.org/10.1016/B978-0-444-63665-2.00008-4

Show, Kuan-Yeow, Yan, Y. G., and Lee, D. J. (2019a). Biohydrogen production from algae: Perspectives, challenges, and prospects. In A. Pandey, J. S. Chang, C. R. Soccol, D. J. Lee, and Y. Chisti (Eds.), *Biofuels from Algae (2nd Edition)* (pp. 325–343). Elsevier. https://doi.org/10.1016/B978-0-444-64192-2.00013-5

Show, Kuan-Yeow, Yan, Y. G., and Lee, D. J. (2019b). Biohydrogen production: Status and perspectives. In A. Pandey, C. Larroche, C. G. Dussap, E. Gnansounou, S. K. Khanal, and S. Ricke (Eds.), *Biofuels: Alternative Feedstocks and Conversion Processes for the Production of Liquid and Gaseous Biofuels (2nd Edition)* (pp. 693–713). Academic Press. https://doi.org/10.1016/B978-0-12-816856-1.00028-2

Show, Kuan-Yeow, Yan, Y., and Lee, D. J. (2019c). Bioreactor and bioprocess design for biohydrogen production. In A. Pandey, S. V. Mohan, J. S. Chang, P. C. Hallenbeck, and C. Larroche (Eds.), *Biohydrogen (2nd Edition)* (pp. 391–411). Elsevier. https://doi.org/10.1016/B978-0-444-64203-5.00016-2

Siddiqui, O., and Dincer, I. (2018). nergy and exergy analyses of a geothermal-based integrated system for trigeneration. In I. Dincer, C. O. Colpan, and O. Kizilkan (Eds.), *Exergetic, Energetic and Environmental Dimensions* (pp. 213–231). Academic Press. https://doi.org/10.1016/B978-0-12-813734-5.00013-5

Sikarwar, V. S., and Zhao, M. (2017). Biomass gasification. In M. A. Abraham (Ed.), *Encyclopedia of Sustainable Technologies* (pp. 205–216). Elsevier. https://doi.org/10.1016/B978-0-12-409548-9.10533-0

Silva, D. A. L., Filleti, R. A. P., Musule, R., Matheus, T. T., and Freire, F. (2022). A systematic review and life cycle assessment of biomass pellets and briquettes production in Latin America. *Renewable and Sustainable Energy Reviews*, *157*, 112042. https://doi.org/10.1016/j.rser.2021.112042

Silva, V. B. R. E., and Cardoso, J. (2020a). *Computational Fluid Dynamics Applied to Waste-to-Energy Processes* (pp. 87–149). Butterworth-Heinemann.

Silveira, M. H. L., Vanelli, B. A., and Chandel, A. K. (2018). Second generation ethanol production: Potential biomass feedstock, biomass deconstruction, and chemical platforms for process valorization. In A. K. Chandel and M. H. Luciano Silveira (Eds.), *Advances in Sugarcane Biorefinery* (pp. 135–152). Elsevier. https://doi.org/10.1016/B978-0-12-804534-3.00006-9

Silvestre, S. (2018). Review of system design and sizing tools. In S. A. Kalogirou (Ed.), *McEvoy's Handbook of Photovoltaics (3rd Edition)* (pp. 715–734). Academic Press. https://doi.org/10.1016/B978-0-12-809921-6.00018-5

Simeons, C. (1980). *Hydro-Power* (pp. 13–23, 81–163). Pergamon.

Singh, A. K., and Chaudhary, V. P. (2011). Energy efficiency in Indian agriculture. *Energy Manager*, *4*(4), 12–18.

Singh, H., and Das, D. (2020). Biohydrogen from microalgae. In E. Jacob-Lopes, M. M. Maroneze, M. I. Queiroz, and L. Q. Zepka (Eds.), *Handbook of Microalgae-Based Processes and Products* (pp. 391–418). Academic Press. https://doi.org/10.1016/B978-0-12-818536-0.00015-4

Singh, S., Mittal, J. P., Singh, M. P., and Bakhshi, R. (1988). Energy use patterns under various farming systems in Punjab. *Applied Energy*, *30*, 261–268.

Singh, S., Pannu, C. J. S., Singh, S., Singh, I. P., and Kaur, S. (1996). *Energy in Punjab agriculture*.

Singh, V., and Das, D. (2019). Potential of hydrogen production from biomass. In P. E. V de Miranda (Ed.), *Science and Engineering of Hydrogen-Based Energy Technologies* (pp. 123–164). Academic Press. https://doi.org/10.1016/B978-0-12-814251-6.00003-4

Singh, Y. D., Mahanta, P., and Bora, U. (2017). Comprehensive characterization of lignocellulosic biomass through proximate, ultimate and compositional analysis for bioenergy production. *Renewable Energy*, *103*, 490–500. https://doi.org/10.1016/j.renene.2016.11.039

Sinharoy, A., Kumar, M., and Pakshirajan, K. (2020). An overview of bioreactor configurations and operational strategies for dark fermentative biohydrogen production. In L. Singh, A. Yousuf, and D. M. Mahapatra (Eds.), *Bioreactors* (pp. 249–288). Elsevier. https://doi.org/10.1016/B978-0-12-821264-6.00014-0

Sivaramakrishnan, R., Shanmugam, S., Sekar, M., Mathimani, T., Incharoensakdi, A., Kim, S.-H., Parthiban, A., Edwin Geo, V., Brindhadevi, K., and Pugazhendhi,

A. (2021). Insights on biological hydrogen production routes and potential microorganisms for high hydrogen yield. *Fuel*, *291*, 120136. https://doi.org/10.1016/j.fuel.2021.120136

Smith, J. D., Sreedharan, V., Landon, M., and Smith, Z. P. (2020). Advanced design optimization of combustion equipment for biomass combustion. *Renewable Energy*, *145*, 1597–1607. https://doi.org/10.1016/j.renene.2019.07.074

Soares, J. F., Confortin, T. C., Todero, I., Mayer, F. D., and Mazutti, M. A. (2020). Dark fermentative biohydrogen production from lignocellulosic biomass: Technological challenges and future prospects. *Renewable and Sustainable Energy Reviews*, *117*, 109484. https://doi.org/10.1016/j.rser.2019.109484

Soelaiman, T. A. F. (2016). Geothermal energy. In M. H. Rashid (Ed.), *Electric Renewable Energy Systems* (pp. 114–139). Academic Press. https://doi.org/10.1016/B978-0-12-804448-3.00007-4

Solar cookers international. (2015). *SLiCK SM70*. https://solarcooking.fandom.com/wiki/SLiCK_SM70

Solar Cookers international. (2020). *Introduction to solar cooking*. https://solarcooking.fandom.com/wiki/Introduction_to_solar_cooking

Solar system exploration – NASA. (2017). *10 Things, June 12: NASA's First Mission to Touch the Sun*. Retrieved January 3, 2022, from https://solarsystem.nasa.gov/news/268/10-things-june-12-nasas-first-mission-to-touch-the-sun/

Song, Q., Sun, C., Wang, Z., and Cai, K. (2020). Municipal solid waste to electricity development and future trend in China: a special life cycle assessment case study of Macau. In J. Ren (Ed.), *Waste-to-Energy* (pp. 177–212). Academic Press. https://doi.org/10.1016/B978-0-12-816394-8.00007-0

Song, X., Zhang, S., Wu, Y., and Cao, Z. (2020). Investigation on the properties of the bio-briquette fuel prepared from hydrothermal pretreated cotton stalk and wood sawdust. *Renewable Energy*, *151*, 184–191. https://doi.org/10.1016/j.renene.2019.11.003

Sönnichsen, N. (2022). *Demand outlook for selected oil products worldwide from 2020 to 2045*. Statista. https://www.statista.com/statistics/282774/global-product-demand-outlook-worldwide/

Speight, J. G. (2011). Fuels for fuel cells. In D. Shekhawat, J. J. Spivey, and D. A. Berry (Eds.), *Fuel Cells: Technologies for Fuel Processing* (pp. 29–48). Elsevier. https://doi.org/10.1016/B978-0-444-53563-4.10003-3

Speight, J. G. (2014). *Gasification of Unconventional Feedstocks* (pp. 54–614). Gulf Professional Publishing.

Srinivasa Raghavan, D. S., Qiu, G., Song, Y., and Ting, Y. P. (2017). *Current Developments in Biotechnology and Bioengineering* (pp. 205–297). Elsevier.

Stefansson, V. (2000). The renewability of geothermal energy. Proceedings of the World Geothermal Congress 2000, Kyushu-Tohoku, Japan, May-June 2000, 883-888.

Stein, G., and Letcher, T. M. (2018). Integration of PV generated electricity into national grids. In T. M. Letcher and V. M. Fthenakis (Eds.), *A Comprehensive Guide to Solar Energy Systems* (pp. 321–332). Academic Press. https://doi.org/10.1016/B978-0-12-811479-7.00015-4

Stout, B. A. (1983). *Biomass energy profiles*. Rome (Italy) FAO.

Subhakar, D. (1995). Wind energy technology. *In: Background material for Training programme on Renewable energy sources and technologies*.

Suharta, H., Sayigh, A. M., and Abdullah, K. (2000). The solar cooker HS 5521 and HS 5921 tested in Indonesia. In A. A. M. Sayigh (Ed.), *World Renewable Energy Congress VI* (pp. 1143–1148). Pergamon. https://doi.org/10.1016/B978-008043865-8/50236-1

Sukhatme, S. P., and Nayak, J. K. (2012). *Solar Energy: Principles of Thermal Collection and Storage*. Tata Mc-Graw Hill Education Pvt. Ltd., New Delhi.

Sumathy, K. (2000). Solar refrigeration for developing countries using a solid — absorption cycle. In A. A. M. Sayigh (Ed.), *World Renewable Energy Congress VI* (pp. 2212–2215). Pergamon. https://doi.org/10.1016/B978-008043865-8/50477-3

Sundaram, S., Shanks, K., and Upadhyaya, H. (2018). Thin film photovoltaics. In T. M. Letcher and V. M. Fthenakis (Eds.), *A Comprehensive Guide to Solar Energy Systems* (pp. 361–370). Academic Press. https://doi.org/10.1016/B978-0-12-811479-7.00018-X

Sundén, B. (2019). *Hydrogen, Batteries and Fuel Cells* (pp. 123–216). Academic Press.

Sunder Selwyn, T., and Hemalatha, S. (2021). Condition monitoring and vibration analysis of asynchronous generator of the wind turbine at high uncertain windy regions in India. *Materials Today: Proceedings*, *46*, 3639–3643. https://doi.org/10.1016/j.matpr.2021.01.656

Suppes, G. J., and Storvick, T. S. (2016). The new electric vehicle society. *Sustainable Power Technologies and Infrastructure*, 161–190. https://doi.org/10.1016/b978-0-12-803909-0.00005-1

Surendra, T. V., Espenti, C. S., and Arunachalam, S. V. (2020). Nanostructured materials for photocatalytic energy conversion. In A. Pandikumar and P. Rameshkumar (Eds.), *Nanostructured, Functional, and Flexible Materials for Energy Conversion and Storage Systems* (pp. 325–343). Elsevier. https://doi.org/10.1016/B978-0-12-819552-9.00010-5

Sustainability, E. (2020). Energy sources. In *Energy Sustainability* (pp. 19–58). Elsevier Inc. https://doi.org/10.1016/B978-0-12-819556-7.00002-4

Szlufcik, J., Sivoththaman, S., Nijs, J. F., Mertens, R. P., and Van Overstraeten, R. (2018). Low-cost industrial technologies for crystalline silicon solar cells. In S. A. Kalogirou (Ed.), *McEvoy's Handbook of Photovoltaics (3rd Edition)* (pp. 129–158). Academic Press. https://doi.org/10.1016/B978-0-12-809921-6.00006-9

Tamošiūnas, A., and Jeguirim, M. (2019). Char gasification. In M. Jeguirim and L. Limousy (Eds.), *Char and Carbon Materials Derived from Biomass* (pp. 187–228). Elsevier. https://doi.org/10.1016/B978-0-12-814893-8.00006-7

Taplin, R. (2012). People, policy and politics in future climates. In A. Henderson-Sellers and K. McGuffie (Eds.), *The Future of the World's Climate (2nd Edition)* (pp. 29–46). Elsevier. https://doi.org/10.1016/B978-0-12-386917-3.00002-6

Taubenblat, P. W., Smith, W. E., and Evans, C. E. (1972). Production of P/M parts from copper powder. *Precision Metal.*

Tawfik, A., and Salem, A. (2012). The effect of organic loading rate on bio-hydrogen production from pre-treated rice straw waste via mesophilic up-flow anaerobic reactor. *Bioresource Technology*, *107*, 186–190. https://doi.org/10.1016/j.biortech.2011.11.086

Teddy. (2010). *TERI Energy data directory and year book.*

Teng, Z., Hua, J., Wang, C., and Lu, X. (2014). Design and optimization principles of biogas reactors in large scale applications. In F. Shi (Ed.), *Reactor and Process Design in Sustainable Energy Technology* (pp. 99–134). Elsevier. https://doi.org/10.1016/B978-0-444-59566-9.00004-1

Tennessee Valley Authority. (2000). *Hydroelectric dam.* https://commons.wikimedia.org/wiki/File:Hydroelectric_dam.svg

Thapar, S., and Sharma, S. (2020). Factors impacting wind and solar power sectors in India: A survey-based analysis. *Sustainable Production and Consumption*, *21*, 204–215. https://doi.org/10.1016/j.spc.2020.01.001

The World Energy Conference. (1985). In the World Energy Conference (Ed.), *Energy Balances and Energy Accounting* (pp. 14–51). Pergamon. https://doi.org/10.1016/B978-0-08-033439-4.50012-5

Theristis, M., Venizelou, V., Makrides, G., and Georghiou, G. E. (2018). Energy yield in photovoltaic systems. In S. A. Kalogirou (Ed.), *McEvoy's Handbook of Photovoltaics (3rd Edition)* (pp. 671–713). Academic Press. https://doi.org/10.1016/B978-0-12-809921-6.00017-3

Thollander, P., Karlsson, M., Rohdin, P., Wollin, J., and Rosenqvist, J. (2020). Energy management using Lean. In P. Thollander, M. Karlsson, P. Rohdin, J. Wollin, and J. Rosenqvist (Eds.), *Introduction to Industrial Energy Efficiency* (pp. 259–287). Academic Press. https://doi.org/10.1016/B978-0-12-817247-6.00014-6

Tinnesand, H., and Sethuraman, L. (2019). *Distributed Wind Resource Assessment Framework: Functional Requirements and Metrics for Performance and Reliability Modeling*. https://www.nrel.gov/docs/fy19osti/72523.pdf

Tomar, S., and Sharma, N. (2021). A systematic review of agricultural policies in terms of drivers, enablers, and bottlenecks: Comparison of three Indian states and a model bio-energy village located in different agro climatic regions. *Groundwater for Sustainable Development*, *15*, 100683. https://doi.org/10.1016/j.gsd.2021.100683

Tompkins, E. L., and Amundsen, H. (2008). Perceptions of the effectiveness of the United Nations Framework Convention on Climate Change in advancing national action on climate change. *Environmental Science and Policy, 11*(1), 1–13. https://doi.org/10.1016/j.envsci.2007.06.004

Toth, A., and Bobok, E. (2017). What is geothermal energy? In A. Toth and E. Bobok (Eds.), *Flow and Heat Transfer in Geothermal Systems* (pp. 1–19). Elsevier. https://doi.org/10.1016/B978-0-12-800277-3.00001-3

Towler, B. F. (2014). Geothermal energy. In B. F. Towler (Ed.), *The Future of Energy* (pp. 237–256). Academic Press. https://doi.org/10.1016/B978-0-12-801027-3.00011-7

Travelarz. (n.d.). *Hyundai H350 Fuel Cell.* https://creativecommons.org/licenses/by-sa/3.0/pl/deed.en

Trop, P., Agrez, M., Urbancl, D., and Goricanec, D. (2016). Co-gasification of torrefied wood biomass and sewage sludge. In Z. Kravanja and M. Bogataj (Eds.), *26th European Symposium on Computer Aided Process Engineering* (Vol. 38, pp. 2229–2234). Elsevier. https://doi.org/10.1016/B978-0-444-63428-3.50376-3

Tsvetkova, O., and Ouarda, T. B. M. J. (2021). A review of sensitivity analysis practices in wind resource assessment. *Energy Conversion and Management, 238*, 114112. https://doi.org/10.1016/j.enconman.2021.114112

Tu, Q. (2019). Life cycle assessment of biodiesel. *World Biodiesel Policies and Production, February*, 175–200. https://doi.org/10.1201/9780429282881-7

Tummala, A., Velamati, R. K., Sinha, D. K., Indraja, V., and Krishna, V. H. (2016). A review on small scale wind turbines. *Renewable and Sustainable Energy Reviews, 56*, 1351–1371. https://doi.org/10.1016/j.rser.2015.12.027

Uçkun Kiran, E., Stamatelatou, K., Antonopoulou, G., and Lyberatos, G. (2016). Production of biogas via anaerobic digestion. In R. Luque, C. S. K. Lin, K. Wilson, and J. Clark (Eds.), *Handbook of Biofuels Production (2nd Edition)* (pp. 259–301). Woodhead Publishing. https://doi.org/10.1016/B978-0-08-100455-5.00010-2

Uma, V. S., and Dineshbabu, G. (2020). Biobased fats and oils from microalgae. In V. K. Gupta, H. Treichel, R. C. Kuhad, and S. Rodriguez-Cout (Eds.), *Recent Developments in Bioenergy Research* (pp. 273–298). Elsevier. https://doi.org/10.1016/B978-0-12-819597-0.00014-3

United Nations Framework Convention on Climate Change. (2015). Paris Agreement. Retrieved January 2, 2022, from https://unfccc.int/files/essential_background/convention/application/pdf/english_paris_agreement.pdf

United Nations Framework Convention on Climate Change. (2002). Third Meeting of the Executive Board. Retrieved from https://cdm.unfccc.int/EB/003/index.html

U.S. Army Corps of Engineers, (2004). Hydraulic turbine and electrical generator, Available at https://commons.wikimedia.org/w/index.php?curid=18581340, in Public Domain via Wikimedia Commons.

U.S. Department of Energy, Geothermal Research: Enhanced Geothermal Systems. Available at: https://www.nrel.gov/geothermal/assets/images/egs-source-mb.jpg [Access date: 06.12.2022]

U.S. Environmental Protection Agency, Geothermal Heating and Cooling Technologies. Available at: https://www.epa.gov/rhc/geothermal-heating-and-coolingtechnologies [Access date: 06.12.2022]

U.S. Geological Survey, (1999). Cutaway views showing the internal structure of the Earth: Volcano Hazards, Available at: https://www.usgs.gov/media/images/cutaway-views-showing-internal-structure-earth-left [Access date: 06.12.2022]

User:Dani 7C3. (2004). *Rance tidal power plant*. https://commons.wikimedia.org/wiki/File:Rance_tidal_power_plant.JPG

Usmani, Z., Kumar, V., Varjani, S., Gupta, P., Rani, R., and Chandra, A. (2020). Municipal solid waste to clean energy system: a contribution toward sustainable development. In S. Varjani, A. Pandey, E. Gnansounou, S. K. Khanal, and S. Raveendran (Eds.), *Current Developments in Biotechnology and Bioengineering* (pp. 217–231). Elsevier. https://doi.org/10.1016/B978-0-444-64321-6.00011-2

Vakkilainen, E. K. (2017a). *Steam Generation from Biomass* (pp. 18–210). Butterworth-Heinemann.

van Grieken, M., and Dower, B. (2017). Wind turbines and landscape. In T. M. Letcher (Ed.), *Wind Energy Engineering* (pp. 493–515). Academic Press. https://doi.org/10.1016/B978-0-12-809451-8.00023-0

Varanda, M. G., Pinto, G., and Martins, F. (2011). Life cycle analysis of biodiesel production. *Fuel Processing Technology*, *92*(5), 1087–1094. https://doi.org/10.1016/j.fuproc.2011.01.003

Vázquez, N., and Vázquez, J. (2018). Photovoltaic system conversion. In M. H. Rashid (Ed.), *Power Electronics Handbook (4th Edition)* (pp. 767–781). Butterworth-Heinemann. https://doi.org/10.1016/B978-0-12-811407-0.00026-X

Vera, D., Jurado, F., Torreglosa, J. P., and Ortega, M. (2018). Biomass gasification for power generation applications: A modeling, economic, and experimental study. In I. Yahyaoui (Ed.), *Advances in Renewable Energies and Power Technologies* (pp. 87–121). Elsevier. https://doi.org/10.1016/B978-0-12-813185-5.00003-6

Verardi, A., Lopresto, C. G., Blasi, A., Chakraborty, S., and Calabrò, V. (2020). Bioconversion of lignocellulosic biomass to bioethanol and biobutanol. In A. Yousuf, D. Pirozzi, and F. Sannino (Eds.), *Lignocellulosic Biomass to Liquid Biofuels* (pp. 67–125). Academic Press. https://doi.org/10.1016/B978-0-12-815936-1.00003-4

Verma, S. R., Mittal, J. P., and Singh, S. (1994). *Energy management and conservation in agricultural production and food processing*. USG Printers and distributors, Ludhiana.

Vijay, V., Subbarao, P. M. V, and Chandra, R. (2021). An evaluation on energy self–sufficiency model of a rural cluster through utilization of biomass residue resources: A case study in India. *Energy and Climate Change, 2*, 100036. https://doi.org/10.1016/j.egycc.2021.100036

Vijayalakshmi, S., Anand, M., and Ranjitha, J. (2020). Microalgae-based biofuel production using low-cost nanobiocatalysts. In A. Yousuf (Ed.), *Microalgae Cultivation for Biofuels Production* (pp. 251–263). Academic Press. https://doi.org/10.1016/B978-0-12-817536-1.00016-3

Viswanathan, B. (2017). Petroleum. *Energy Sources*, 29–57. https://doi.org/10.1016/B978-0-444-56353-8.00002-2

Viva, F. A. (2020). Platinum-based cathode catalyst systems for direct methanol fuel cells. In K. Dutta (Ed.), *Direct Methanol Fuel Cell Technology* (pp. 257–287). Elsevier. https://doi.org/10.1016/B978-0-12-819158-3.00009-4

Vrabie, V., Scarpete, D., and Zbarcea, O. (2016). Vegetable oils as alternative fuel for new generation of diesel engines. A review. *International Scientific Journal "Trans Motauto World," 1*(3), 18–22. https://stumejournals.com/journals/tm/2016/3/18.full.pdf

Wang, C., Zhang, S., Wu, S., Sun, M., and Lyu, J. (2020). Multi-purpose production with valorization of wood vinegar and briquette fuels from wood sawdust by hydrothermal process. *Fuel, 282*, 118775. https://doi.org/10.1016/j.fuel.2020.118775

Wang, P., and Lü, X. (2021). General introduction to biofuels and bioethanol. In X. Lü (Ed.), *Advances in 2nd Generation of Bioethanol Production* (pp. 1–7). Woodhead Publishing. https://doi.org/10.1016/B978-0-12-818862-0.00006-6

Wang, Xu. (2016). Ocean wave energy conversion analysis. In Xu Wang (Ed.), *Frequency Analysis of Vibration Energy Harvesting Systems* (pp. 249–269). Academic Press. https://doi.org/10.1016/B978-0-12-802321-1.00011-X

Wang, Xuefei, Zeng, X., Li, J., Yang, X., and Wang, H. (2018). A review on recent advancements of substructures for offshore wind turbines. *Energy Conversion and Management, 158*, 103–119. https://doi.org/10.1016/j.enconman.2017.12.061

Wang, Z. (2019). The solar resource and meteorological parameters. In Z. Wang (Ed.), *Design of Solar Thermal Power Plants* (pp. 47–115). Academic Press. https://doi.org/10.1016/B978-0-12-815613-1.00002-X

Ward, V. C. A. (2020). Biofuels. In A. G. Rodrigues (Ed.), *New and Future Developments in Microbial Biotechnology and Bioengineering* (pp. 153–162). Elsevier. https://doi.org/10.1016/B978-0-444-64301-8.00007-X

Weiss, J., and Tsuchida, T. B. (2017). Integration into national grids. In T. M. Letcher (Ed.), *Wind Energy Engineering* (pp. 419–436). Academic Press. https://doi.org/10.1016/B978-0-12-809451-8.00020-5

Whittlesey, R. (2017). Vertical axis wind turbines: Farm and turbine design. In T. M.

Letcher (Ed.), *Wind Energy Engineering* (pp. 185–202). Academic Press. https://doi.org/10.1016/B978-0-12-809451-8.00010-2

Williams, C. L., Dahiya, A., and Porter, P. (2020). Introduction to bioenergy and waste to energy. In A. Dahiya (Ed.), *Bioenergy (2nd Edition)* (pp. 5–44). Academic Press. https://doi.org/10.1016/B978-0-12-815497-7.00001-4

Williams, M. C. (2011). Fuel cells. In D. Shekhawat, J. J. Spivey, and D. A. Berry (Eds.), *Fuel Cells: Technologies for Fuel Processing* (pp. 11–27). Elsevier. https://doi.org/10.1016/B978-0-444-53563-4.10002-1

Wu, Y., Xia, T., Wang, Y., Zhang, H., Feng, X., Song, X., and Shibasaki, R. (2022). A synchronization methodology for 3D offshore wind farm layout optimization with multi-type wind turbines and obstacle-avoiding cable network. *Renewable Energy*, *185*, 302–320. https://doi.org/10.1016/j.renene.2021.12.057

Xing, J., Luo, K., Wang, H., Gao, Z., and Fan, J. (2019). A comprehensive study on estimating higher heating value of biomass from proximate and ultimate analysis with machine learning approaches. *Energy*, *188*, 116077. https://doi.org/10.1016/j.energy.2019.116077

Yadala, S., and Cremaschi, S. (2016). A dynamic optimization model for designing open-channel raceway ponds for batch production of algal biomass. *Processes*, *4*(2). https://doi.org/10.3390/pr4020010

Yahia, E. M., Carrillo-López, A., Barrera, G. M., Suzán-Azpiri, H., and Bolaños, M. Q. (2019). Photosynthesis. In E. M. Yahia (Ed.), *Postharvest Physiology and Biochemistry of Fruits and Vegetables* (pp. 47–72). Woodhead Publishing. https://doi.org/10.1016/B978-0-12-813278-4.00003-8

Yakushido, K., Kobayashi, Y., and Kato, H. (2014). Thermochemical transformation of biomass. In S. Tojo and T. Hirasawa (Eds.), *Research Approaches to Sustainable Biomass Systems* (pp. 269–278). Academic Press. https://doi.org/10.1016/B978-0-12-404609-2.00012-X

Yamori, W. (2016). Photosynthesis and respiration. In T. Kozai, G. Niu, and M. Takagaki (Eds.), *Plant Factory* (pp. 141–150). Academic Press. https://doi.org/10.1016/B978-0-12-801775-3.00009-3

Yamori, W. (2020). Photosynthesis and respiration. In T. Kozai, G. Niu, and M. Takagaki (Eds.), *Plant Factory (2nd Edition)* (pp. 197–206). Academic Press. https://doi.org/10.1016/B978-0-12-816691-8.00012-1

Yang, H., and Chen, H. (2015). Biomass gasification for synthetic liquid fuel production. In R. Luque and J. G. Speight (Eds.), *Gasification for Synthetic Fuel Production* (pp. 241–275). Woodhead Publishing. https://doi.org/10.1016/B978-0-85709-802-3.00011-4

Yang, Zeyu, Wang, Z., Hollebone, B. P., Yang, C., and Brown, C. E. (2016). Forensic fingerprinting of biodiesel and its blends with petroleum oil. In *Standard Handbook Oil Spill Environmental Forensics: Fingerprinting and Source Identification, (2nd Edition*, Issue 2). Elsevier Inc. https://doi.org/10.1016/B978-0-12-809659-8.00012-7

Yang, Zhixue, Ren, Z., Li, Z., Xu, Y., Li, H., Li, W., and Hu, X. (2022). A comprehensive analysis method for levelized cost of energy in tidal current power generation farms. *Renewable Energy*, *182*, 982–991. https://doi.org/10.1016/j.renene.2021.11.026

Yen, H. W., Hu, I. C., Chen, C. Y., Nagarajan, D., and Chang, J. S. (2019). Design of photobioreactors for algal cultivation. In *Biofuels from Algae (2nd Edition).* Elsevier B.V. https://doi.org/10.1016/b978-0-444-64192-2.00010-x

Yi, Y. (2021). Tiny bugs play big role: Microorganisms' contribution to biofuel production. In X. Lü (Ed.), *Advances in 2nd Generation of Bioethanol Production* (pp. 113–136). Woodhead Publishing. https://doi.org/10.1016/B978-0-12-818862-0.00007-8

Yin, C. Y. (2011). Prediction of higher heating values of biomass from proximate and ultimate analyses. *Fuel*, *90*(3), 1128–1132. https://doi.org/10.1016/j.fuel.2010.11.031

Yoneda, Y. (n.d.). *Fry Power: How to Convert Your Car to Run on Vegetable Oil.* 2009. Retrieved October 7, 2021, from https://inhabitat.com/converting-your-vehicle-to-run-on-veggie-oil/

You, Z., Badalyan, A., Yang, Y., and Bedrikovetsky, P. (2018). Formation damage challenges in geothermal reservoirs. In B. Yuan and D. A. Wood (Eds.), *Formation Damage During Improved Oil Recovery* (pp. 447–497). Gulf Professional Publishing. https://doi.org/10.1016/B978-0-12-813782-6.00012-9

Yousuf, A. (2020). Fundamentals of microalgae cultivation. In A. Yousuf (Ed.), *Microalgae Cultivation for Biofuels Production* (pp. 1–9). Academic Press. https://doi.org/10.1016/B978-0-12-817536-1.00001-1

Yuebo, X., Kabo-bah, A. T., Kabo-bah, K. J., and Domfeh, M. K. (2018). Hydropower development—review of the successes and failures in the world. In A. Kabo-Bah and C. J. Diji (Eds.), *Sustainable Hydropower in West Africa* (pp. 53–61). Academic Press. https://doi.org/10.1016/B978-0-12-813016-2.00004-6

Zakaria, B. S., Lin, L., Chung, T., and Dhar, B. R. (2020). *An overview of complementary microbial electrochemical technologies for advancing anaerobic digestion* (Y. Li and S. K. Khanal (Eds.); Vol. 5, pp. 129–167). Elsevier. https://doi.org/10.1016/bs.aibe.2020.04.004

Zarrouk, S. J., and McLean, K. (2019). *Geothermal Well Test Analysis* (pp. 13–61). Academic Press.

Zekry, A., Shaker, A., and Salem, M. (2018). Solar cells and arrays: Principles, analysis, and design. In I. Yahyaoui (Ed.), *Advances in Renewable Energies and Power Technologies* (pp. 3–56). Elsevier. https://doi.org/10.1016/B978-0-12-812959-3.00001-0

Zener, C., Fetkovich, J. (1975). Foam solar sea power plant. Science. 189(4199):294-295. Available at: https://www.science.org/doi/10.1126/science.189.4199.294

Zhang, H., Yang, J., Ren, X., Wu, Q., Zhou, D., and Elahi, E. (2020). How to accommodate curtailed wind power: A comparative analysis between the US, Germany, India and China. *Energy Strategy Reviews*, *32*, 100538. https://doi.org/10.1016/j.esr.2020.100538

Zhang, Jiajun, and Zhang, X. (2019). The thermochemical conversion of biomass into biofuels. In D. Verma, E. Fortunati, S. Jain, and X. Zhang (Eds.), *Biomass, Biopolymer-Based Materials, and Bioenergy* (pp. 327–368). Woodhead Publishing. https://doi.org/10.1016/B978-0-08-102426-3.00015-1

Zhang, Jie, and Hodge, B.-M. (2014). Forecastability as a design criterion in wind resource assessment. In M. R. Eden, J. D. Siirola, and G. P. Towler (Eds.), *Proceedings of the 8th International Conference on Foundations of Computer-Aided Process Design* (Vol. 34, pp. 663–668). Elsevier. https://doi.org/10.1016/B978-0-444-63433-7.50095-X

Zhang, Q., and Zhang, Z. (2018). Biological hydrogen production from renewable resources by photofermentation (Y. Li and X. Ge (Eds.); Vol. 3, pp. 137–160). Elsevier. https://doi.org/10.1016/bs.aibe.2018.03.001

Zhang, Y., Cui, Y., Chen, P., Liu, S., Zhou, N., Ding, K., Fan, L., Peng, P., Min, M., Cheng, Y., Wang, Y., Wan, Y., Liu, Y., Li, B., and Ruan, R. (2019). Gasification technologies and their energy potentials. In M. J. Taherzadeh, K. Bolton, J. Wong, and A. Pandey (Eds.), *Sustainable Resource Recovery and Zero Waste Approaches* (pp. 193–206). Elsevier. https://doi.org/10.1016/B978-0-444-64200-4.00014-1

Zhang, Z., Zhang, Y., Zhou, Y., Ahmad, R., Pemberton-Pigott, C., Annegarn, H., and Dong, R. (2017). Systematic and conceptual errors in standards and protocols for thermal performance of biomass stoves. *Renewable and Sustainable Energy Reviews*, *72*, 1343–1354. https://doi.org/10.1016/j.rser.2016.10.037

Zhao, D., Han, N., Goh, E., Cater, J., and Reinecke, A. (2019a). *Wind Turbines and Aerodynamics Energy Harvesters* (pp. 1–20, 431–461). Academic Press.

Zhao, F., Yi, Y., and Lü, X. (2021). Essential process and key barriers for converting plant biomass into biofuels. In X. Lü (Ed.), *Advances in 2nd Generation of Bioethanol Production* (pp. 53–70). Woodhead Publishing. https://doi.org/10.1016/B978-0-12-818862-0.00009-1

Zhao, M., Ji, J., Ding, H., and Chen, L.-Q. (2020). The effects of bending moments on the dynamics of a wind turbine planetary gearbox. In H. R. Karimi (Ed.), *Vibration Control and Actuation of Large-Scale Systems* (pp. 321–359). Academic Press. https://doi.org/10.1016/B978-0-12-821194-6.00012-3

Zhao, Q., and Zhao, Y. (2016). An optimal scheduling of the heliostat field in a solar tower power plant. In Z. Kravanja and M. Bogataj (Eds.), *26th European Symposium on Computer Aided Process Engineering* (Vol. 38, pp. 1021–1026). Elsevier. https://doi.org/10.1016/B978-0-444-63428-3.50175-2

Zhen, X. (2018). Methanol as an internal combustion on engine fuel. In A. Basile and F. Dalena (Eds.), *Methanol* (pp. 313–337). Elsevier. https://doi.org/10.1016/B978-0-444-63903-5.00011-X

Zhou, H., Long, Y., Meng, A., Li, Q., and Zhang, Y. (2015). Classification of municipal solid waste components for thermal conversion in waste-to-energy research. *Fuel*, *145*, 151–157. https://doi.org/10.1016/j.fuel.2014.12.015

Zhou, M., and Lü, X. (2021). Strategies on simultaneous fermentation of pentose and hexose to bioethanol. In X. Lü (Ed.), *Advances in 2nd Generation of Bioethanol Production* (pp. 161–211). Woodhead Publishing. https://doi.org/10.1016/B978-0-12-818862-0.00010-8

Zhou, W., Lu, Q., Han, P., and Li, J. (2020). Microalgae cultivation and photobioreactor design. In A. Yousuf (Ed.), *Microalgae Cultivation for Biofuels Production* (pp. 31–50). Academic Press. https://doi.org/10.1016/B978-0-12-817536-1.00003-5

Zhuo, Y., Xie, Z., and Shen, Y. (2021). Model study of carbonisation of low rank coal briquettes: Effect of briquettes shape. *Powder Technology*, *385*, 120–130. https://doi.org/10.1016/j.powtec.2021.02.071

Zubkova, V., and Strojwas, A. (2020). Influence of briquetting on thermal behavior of bituminous coal oxidized during storage. *Journal of Analytical and Applied Pyrolysis*, *152*, 104969. https://doi.org/10.1016/j.jaap.2020.104969

Zuccaro, G., Yousuf, A., Pollio, A., and Steyer, J. P. (2020). Microalgae cultivation systems. In A. Yousuf (Ed.), *Microalgae Cultivation for Biofuels Production* (pp. 11–29). Academic Press. https://doi.org/10.1016/B978-0-12-817536-1.00002-3

Zulqarnain, Ayoub, M., Yusoff, M. H. M., Nazir, M. H., Zahid, I., Ameen, M., Sher, F., Floresyona, D., and Budi Nursanto, E. (2021). A comprehensive review on oil extraction and biodiesel production technologies. *Sustainability (Switzerland)*, *13*(2), 1–28. https://doi.org/10.3390/su13020788

APPENDIX

Popular units of Energy

1 pico-joule (pJ)	$= 10^{-12}$ J
1 nano-joule (nJ)	$= 10^{-9}$ J
1 micro-joule (μJ)	$= 10^{-6}$ J
1 milli-joule (mJ)	$= 10^{-3}$ J
1 kilo-joule (kJ)	$= 10^{3}$ J
1 mega-joule (MJ)	$= 10^{6}$ J
1 giga-joule (GJ)	$= 10^{9}$ J
1 tera-joule (TJ)	$= 10^{12}$ J
1 peta-joule (PJ)	$= 10^{15}$ J
1 exa-joule (EJ)	$= 10^{18}$ J